AF412520

THE NEAR EARTH ASTEROID RENDEZVOUS MISSION

Cover figure: Artist's conception of the NEAR spacecraft in orbit about Eros, conceived and executed by Pat Rawlings of Science Applications International Corporation. Copyright, NASA.

THE NEAR EARTH ASTEROID RENDEZVOUS MISSION

Edited by

C. T. RUSSELL

Institute of Geophysics and Planetary Physics,
University of California,
Los Angeles, U.S.A.

Reprinted from Space Science Reviews, Vol. 82, Nos. 1–2, 1997

KLUWER ACADEMIC PUBLISHERS

DORDRECHT / BOSTON / LONDON

A C.I.P. Catalogue record for this book is available from the Library of Congress.

ISBN 0-7923-4957-1

Published by Kluwer Academic Publishers,
P.O. Box 17, 3300 AA Dordrecht, The Netherlands

Sold and distributed in North, Central and Latin America
by Kluwer Academic Publishers,
101 Philip Drive, Norwell, MA 02061, U.S.A.

In all other countries, sold and distributed
by Kluwer Academic Publishers,
P.O. Box 322, 3300 AH Dordrecht, The Netherlands

Printed on acid-free paper

All rights reserved
©1997 Kluwer Academic Publishers
No part of the material protected by this copyright notice may be reproduced or
utilized in any form or by any means, electronic or mechanical,
including photocopying, recording or by any information storage and
retrieval system, without written permission from the copyright owner.

Printed in the Netherlands

Table of Contents

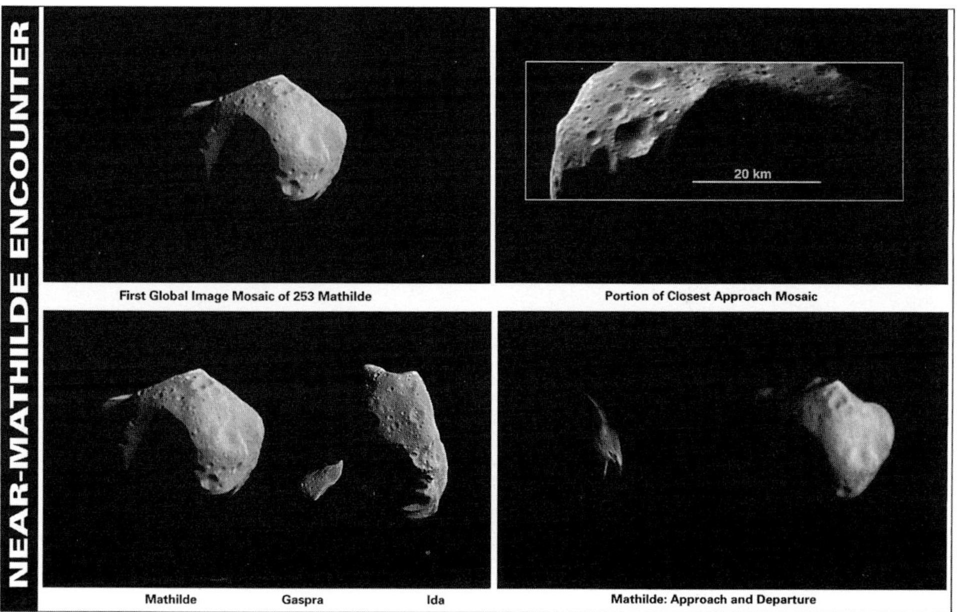

Legends to this figure:

Views of asteroid 253 Mathilde obtained by the NEAR spacecraft on June 27, 1997. Upper left panel shows a mosaic of four images taken at 2400 kilometers. Sunlight is from upper right. Details as small as 380 meters across can be discerned. Lower left panel compares, on same scale, Mathilde with asteroids Gaspra (middle) and Ida (right) earlier imaged by the Galileo spacecraft. The visible part of Mathilde measures 59 by 47 kilometers. Upper right panel shows Mathilde at closest approach (1200 kilometers). Illumination is from upper left. Lower right panel shows two additional views of Mathilde at phase angles of 136° (left) and 43° (right).
Copyright, NASA.

Plate 2. Multi-spectral images covering 16 to 18 September 1995. Upper left panel shows a tropical storm. Upper right panel shows Meteosat visible imagery (0.6–0.7 μm) and the tropical storm at quiescence. The visible part of Meteosat imagery shows illumination from the left. Lower right panel shows the chlorophyll stock in the ocean. Also see Figures 13.10, 13.11 and 14.1 in the text.

Copyright ESA.

FOREWORD

Even before the present Administrator of NASA, Daniel Goldin, made the phrase 'better, faster, cheaper' the slogan of at least the Office of Space Science, that same office under the Associate Administrator of Lennard Fisk and its Division of Solar System Exploration under the direction of Wes Huntress had begun a series of planetary spacecraft whose developmental cost, phase C/D in the parlance of the trade, was to be held to under $150M. In order to get the program underway rapidly they chose two missions without the open solicitation now the hallmark of the program. One of these two missions, JPL' s Mars Pathfinder, was to be a technology demonstration mission with little immediate science return that would enable later high priority science missions to Mars. Many of the science investigations that were included had significant foreign contributions to keep NASA's cost of the mission within the Discovery budget. The second of these missions and the first to be launched was the Near Earth Asteroid Rendezvous mission, or NEAR, awarded to Johns Hopkins University's Applied Physics Laboratory. This mission was quite different than Mars Pathfinder, being taken from the list of high priority objectives of the science community and emphasizing the science return and not the technology development of the mission. This mission was also to prove to be well under the $150M phase C/D cap.

NEAR was launched on February 17, 1996 passing by asteroid 253 Mathilde on June 27, 1997, and is scheduled to reach and orbit 433 Eros in January 1999? Eros was discovered by G. Witt in 1898 and was the first known asteroid to cross both the orbits of Mars and the Earth. Thus it is most appropriate that it be the first asteroid to be orbited and that this encounter take place in 1999, shortly following the 100th anniversary of its discovery. The asteroid, Eros, was named after the son of Aphrodite, whose mortal wife, Psyche, was forbidden to gaze upon him by the light of day. Even though scientists have no such restriction to pique their curiosity and even though Eros, the asteroid, may not prove to be as pleasing to the eye as Eros, the god, they are still intensely eager to examine it and have installed six scientific instruments on NEAR with which to 'view' the asteroid: a multispectral imager, a near-infrared spectrometer an X-ray spectrometer, a gamma-ray spectrometer, a laser range finder and a fluxgate magnetometer.

In this volume we describe the various elements of the mission. We begin with an overview of the mission by the project scientist A. F. Cheng. This overview is followed by a series of articles on the instruments by those most familiar with the development of these 'facility' instruments that were well into construction prior to the selection of the science teams associated with each. The first instrument paper by S. E. Hawkins et al. describes the multispectral imager. It is followed by a description of the near-infrared spectrometer authored by J. W. Hawkins and colleagues. The X-ray and the gamma-ray spectrometers are described in the same

Space Science Reviews **82**: 1–2, 1997.

paper by J. O. Goldsten et al. The laser altimeter is described by T. D. Cole et al. and the magnetometer by D. A. Lohr and colleagues. The final paper by K. Heeres et al. describes the NEAR Science Data Center.

The successful construction, launch, and initial operation have been due to the heroic efforts of many individuals. Two notable individuals in addition to the authors of the articles included herein are the project manager, T. B. Coughlin, and the mission director, R. W. Farquhar. Since shortly before launch the science team has gradually begun to take a greater role in the science planning for the mission. The NEAR science team consists of J. Veverka, J. F. Bell III, C. R. Chapman, M. C. Malin, L. A. McFadden, M. S. Robinson, P. C. Thomas, J. I. Trombka, W. V. Boynton, J. Bruckner, S. W. Squyres, M. H. Acuna, M. T. Zuber, D. K. Yeomans, J.-P. Barriot, A. S. Konopliv, A. F. Cheng and C. T. Russell.

The articles in this volume have all been refereed by two experts on the subject matter of each article. Generally one of these persons was familiar with the NEAR mission and the other was learning about the mission for the first time. To these referees we owe a debt of gratitude for their diligence in helping to make these articles both readable and informative. The authors too deserve our thanks for their efforts undertaken at a very busy time for the mission. Finally I would like to express my thanks to Anne McGlynn who assisted me with the editorial duties for this volume.

<table>
<tr><td>Institute of Geophysics and Planetary Physics,</td><td>C. T. RUSSELL</td></tr>
<tr><td>Department of Earth and Space Sciences,</td><td></td></tr>
<tr><td>University of California,</td><td></td></tr>
<tr><td>Los Angeles, U.S.A.</td><td></td></tr>
</table>

June 1997

NEAR EARTH ASTEROID RENDEZVOUS: MISSION OVERVIEW

ANDREW F. CHENG

The Johns Hopkins University Applied Physics Laboratory, Laurel, MD 20723, U.S.A.

(Received January, 1997)

Abstract. The Near Earth Asteroid Rendezvous (NEAR) mission launched successfully on February 17, 1996 aboard a Delta II-7925. NEAR will be the first mission to orbit an asteroid and will make the first comprehensive scientific measurements of an asteroid's surface composition, geology, physical properties, and internal structure. It will orbit the unusually large near-Earth asteroid 433 Eros for about one year, at a minimum altitude of about 15 km from the surface. NEAR will also make the first reconnaissance of a C-type asteroid during its flyby of the unusual main belt asteroid 253 Mathilde. The NEAR instrument payload is: a multispectral imager (MSI), a near infrared spectrometer (NIS), an X-ray/gamma ray spectrometer (XRS/GRS), a magnetometer (MAG), and a laser rangefinder (NLR), while a radio science investigation (RS) uses the coherent X-band transponder. NEAR will improve our understanding of planetary formation processes in the early solar system and clarify the relationships between asteroids and meteorites. The Mathilde flyby will occur on June 27, 1997, and the Eros rendezvous will take place during February 1999 through February 2000.

1. Introduction

Numerous small bodies – asteroids, comets, and meteoroids – pervade the inner solar system. They have collided with each other and with the terrestrial planets throughout geologic time. Those that come within 1.3 AU of the Sun are called Near-Earth Objects. Their importance for solar system exploration is based upon the following. First, some of them are primitive objects that may be remnants from the epoch of planet formation in the early solar system, so they may preserve evidence of the nature of the materials from which planets were built; some of these small bodies, on the other hand, are fragments of evolved objects. Second, the impacts of comets and asteroids shaped the surfaces of airless bodies like the Moon and Mercury, and they influenced the evolution of the atmospheres of the inner planets as well as the biosphere of Earth. The impact of a Near-Earth Object some 65 million years ago caused the extinction of the dinosaurs (Alvarez et al., 1980; Rampino and Haggerty, 1994; Smit, 1994).

Asteroids, comets, and meteoroids are primitive bodies that underwent most of their evolution in the first few hundred million years of the solar system. There are no widely accepted quantitative criteria for distinguishing among them. A significant fraction of the more than 7000 objects now denoted as asteroids may be extinct or dormant comets (Weissman et al. 1989). Most asteroids are found in the main belt between the orbits of Mars and Jupiter, but some asteroids have perihelia within 1.3 AU of the Sun and are known as Near-Earth Asteroids (NEAs). The NEAs are a dynamically young population, whose orbits evolve on hundred million year time scales because of collisions and gravitational interactions with

Space Science Reviews **82:** 3–29, 1997.
© 1997 *Kluwer Academic Publishers. Printed in Belgium.*

planets. Some NEAs are already in Earth-crossing orbits, and those that are not have an appreciable probability for evolving into one. About 400 NEAs are known to date. The NEA population appears to be a broad sample of the main belt population (McFadden et al., 1988). Most or all asteroid types found in the main belt are also represented among the NEAs. The present day orbit of an NEA does not necessarily indicate where it was formed.

Our current knowledge of the nature of asteroids comes from three sources: Earth-based remote sensing, data from the Galileo spacecraft flybys of the two main belt asteroids 951 Gaspra and 243 Ida, and laboratory analyses of meteorites, whch are believed to be collisional fragments of asteroids. Gaspra, Ida, and the NEAR target 433 Eros are all S-type (stony) asteroids, the most common asteroid type in the inner part of the asteroid belt. Meteorites may be a biased and incomplete sample of the materials actually found in asteroids, and it has proved difficult to establish firm links between meteorite types and asteroid types (Gaffey et al., 1993a). The uncommon meteorite type eucrite (a basaltic achondrite) has been linked by visible and near-infrared reflectance measurements to the relatively rare asteroid type V (McCord et al., 1970; Binzel and Xu, 1993). However there is still controversy over whether and how the most common meteorite types (the ordinary chondrites) may be linked to S asteroids (Bell et al., 1989; Gaffey et al., 1993b; Binzel et al., 1996). Such a link is expected because the S asteroids are the most common type in the most dynamically probable meteorite source regions.

The NEAR mission will spend approximately a year in orbit around Eros. It will make the first comprehensive, spatially resolved measurements of the geology, mineralogy, and elemental composition of an S asteroid. The S asteroids are known to contain the silicate minerals olivine and pyroxene plus an admixture of iron-nickel metal, but key questions remain as to the fundamental nature and evolution of S asteroids. The S asteroids are known to be a diverse class of objects (Gaffey et al., 1993b). Some of them appear to be fragments of differentiated bodies that underwent substantial melting and differentiation. Others may consist of primitive materials like ordinary chondrites that never underwent melting and that may preserve characteristics of the solid material from which the inner planets accreted.

NEAR will orbit Eros at low altitude, about one body radius above the surface, for more than six months. This long-duration, low altitude orbit provides the opportunity for the NEAR X-ray and gamma-ray spectrometers to measure the abundances of key elements at Eros with a maximum spatial resolution of about 2 km. These NEAR data, especially when combined with data from the *Galileo* flybys of Gaspra and Ida, will greatly advance our understanding of S asteroids and their possible relationships to other small bodies of the solar system. The *Galileo* instrument complement did not include any capability to make measurements of elemental composition, unlike NEAR. The *Galileo* flybys did provide the first high resolution images of asteroids, and they revealed complex surfaces covered by craters, fractures, grooves, and subtle color variations (Belton et al., 1992, 1994). Moreover, *Galileo* discovered a satellite of the asteroid Ida, which is a member

Table Ia
[433] Eros

Diameters of triaxial ellipsoid model	$40 \times 14 \times 14$ km
Heliocentric orbit period	1.76 year
Inclination of heliocentric orbit	10.8°
Perihelion, aphelion	1.133 AU, 1.784 AU
Rotation period	5.27011 hour
Rotation pole	ecliptic lon 16°, lat 11°
Spectral type	S: pyroxene, olivine, Fe-Ni metal
Geometric albedo	0.16

Table Ib
Main belt asteroid [253] Mathilde

[253] Mathilde	Encounter parameters
Type C	Flyby date June 27, 1997
Estimated diameter 61 km	Closest approach 1200 km
H magnitude 10.3	Approach phase angle 140°
Rotation period 17.4 days	Flyby speed 9.9 km s^{-1}
Perihelion 1.94 AU	Earth distance 2.20 AU
Aphelion 3.35 AU	Sun distance 1.99 AU
Inclination 6.71°	

of the Koronis asteroid family (Eros is not an asteroid family member). The near-infrared spectrum of Gaspra indicates a high olivine abundance such that Gaspra is inferred to be a fragment of a differentiated body. Ida and Eros, on the other hand, display infrared spectra that may be consistent with a silicate mineralogy like that in ordinary chondrites (Chapman, 1995; Murchie and Pieters, 1996). NEAR will also perform color imaging and near-infrared spectral mapping of Eros, at spatial resolutions more than an order of magnitude greater than *Galileo* achieved at Gaspra and Ida.

The NEAR flyby of 253 Mathilde provides a unique opportunity to obtain the first spatially resolved color images of a C asteroid, a completely different class of object from the S-asteroids Gaspra, Ida, and Eros. The C asteroids are dark and spectrally neutral in the visible band, unlike the relatively bright and spectrally red S asteroids. The C asteroids are the most common type in the mid-to-outer asteroid belt. They are believed to be primitive, carbonaceous and volatile rich bodies, but their nature and composition are fundamentally mysterious: how are they related to comets? to the dark satellites of the outer solar system?

2. NEAR Asteroid Targets: 433 Eros and 253 Mathilde

The NEAR rendezvous target, 433 Eros, is the second largest NEA and is intermediate in size between the *Galileo* flyby targets Gaspra and Ida. Eros is one of only three NEAs with maximum diameter above 10 km, and it is the only one of these whose heliocentric orbit is accessible enough to permit a rendezvous mission using the Delta 2 launch vehicle. The mean diameter of Eros, at about 20 km, is an order of magnitude larger than that of typical known NEAs.

Eros was discovered in 1898. Its vital statistics are given in Table I. Eros was the subject of a world-wide ground-based observing campaign in 1975 when it passed wthin 0.15 AU of Earth. Visible, infrared, and radar observations determined the approximate size, shape, rotation rate, and pole position of Eros and showed that a regolith was present on its surface (Zellner, 1976; Ostro et al., 1990; Yeomans, 1995). Eros is presently in a Mars-crossing but not Earth-crossing orbit. However, numerical simulations suggest that Eros may evolve into an Earth-crosser within 2 million years (Michel et al., 1996).

More recent spectroscopic analyses have found a hemispherical heterogeneity in the near infrared spectra of Eros (Murchie and Pieters, 1996). One side of Eros has a spectrum consistent with higher pyroxene content and a radar signature consistent with a facet-like surface; the other side displays higher olivine content and a convex-shaped surface. The detection of variations with rotation phase in disk-averaged data suggests that substantial geologic and compositional complexity may be found at higher spatial resolution by NEAR. Such diversity may arise in a number of different ways. One possibility is that Eros is a fragment of a large, differentiated parent body that did not survive intact to the present; this body may have had volcanic activity, for example, and Eros might preserve material from the surface and/or the interior of such a body. Alternatively, present-day Eros may have been assembled from fragments of many different parent bodies, and detailed study of variations across the surface of Eros may reveal evidence of such diverse origins.

The NEAR trajectory is targeted to pass very close to the main belt asteroid 253 Mathilde on June 27, 1997. Mathilde is a large Type C asteroid (of mean diameter $\sim$60 km) that has been recently discovered to have an anomalously large rotation period of 17.4 days (Mottola et al., 1995). This rotation is more than an order of magnitude slower than that of typical asteroids. While NEAR will obtain the first resolved images of a C-type asteroid and will make a mass determination during the Mathilde flyby, NEAR will not be able to measure the rotation state. Mathilde will be observable by NEAR's imager for up to a week; because of the 9.9 km s^{-1} flyby speed, the spacecraft moves out of range before Mathilde completes a single rotation. Earth-based radar observations are planned for the summer of 1997 to characterize more fully Mathilde's shape and rotation state.

Table IIa
Science objectives

– Characterize an asteroid's physical and geological properties and infer its elemental and mineralogical composition
– Clarify the relationships between asteroids, comets, and meteorites, and
– Advance the understanding of processes and conditions during the formation and early evolution of the planets

Table IIb
Measurement objectives

– Bulk properties – size, shape, mass, density, gravity field, and spin state
– Surface properties – elemental and mineralogical composition, geology, morphology, and texture
– Internal properties - search for heterogeneity and magnetic field

3. Science Objectives

The overall objectives of the NEAR mission are to rendezvous with a near-Earth asteroid, to achieve orbit around it, and to conduct a first systematic scientific exploration of it. NEAR will study the nature and evolution of S asteroids, improve understanding of processes and conditions relevant to formation of planets in the early solar system, and clarify relationships between asteroids and meteorites. The scientific objectives and measurement objectives for NEAR are listed in Table II.

There are six scientific instruments on NEAR, all of which are facility instruments:
- a multispectral imager (MSI) which is an eight-color CCD visible imager,
- a 64-channel near infrared spectrograph (NIS) covering 0.8 to 2.6 μ,
- an X-ray spectrometer (XRS) covering 1 to 10 keV,
- a gamma ray spectrometer (GRS) covering 0.3 to 10 MeV,
- a three-axis fluxgate magnetometer (MAG), and
- a Nd:YAG laser rangefinder (NLR).

There is also a radio science investigation (RS) using the spacecraft X-band telecommunications system. Table IIIa gives examples of science questions that will be addressed by NEAR, and Table IIIb summarizes how these questions are addressed by the instrument payload. A complete description of the science objectives, measurements, and observing strategies for each of the NEAR instruments is presented in a special issue of the Journal of Geophysical Research (Cheng et al., 1997; Veverka et al., 1997; Trombka et al., 1997; Acuna et al., 1997; Zuber et al., 1997; Yeomans et al., 1997).

Table IIIa

Examples of science questions for NEAR

- What are the characteristic morphology and texture of the surface and how do they compare with those on larger bodies?
- What is the elemental and mineralogical composition of the asteroid?
- Is there evidence of compositional or structural heterogeneity?
- Is it a solid fragment of a larger parent body, or a rubble pile?
- Were precursor body(ies) primitive or differentiated?
- Is there evidence for past or present cometary activity?
- Is the asteroid related to a meteorite type or types?
- Is there an intrinsic magnetic field? What is it like?
- Are there any asteroid satellites, and how might they compare with Eros?

Table IIIb

Closure of science questions and measurements

	MSI/NIS	XRS/GRS	MAG	NLR	RS
Surface morphology and texture	*			*	
Composition	*	*	*	*	*
Heterogeneity	*	*	*	*	*
Solid fragment or rubble pile	*	*	*	*	*
Differentiation	*	*	*	*	*
Cometary activity	*			*	
Meteorite link	*	*			
Magnetic field			*		
Asteroid satellites	*				

* = investigation addresses question

4. Mission Design Overview

The 1996 NEAR launch opportunity to 433 Eros took advantage of an orbital alignment that occurs only once every seven years (Farquhar et al., 1995). NEAR launched on February 17, 1996, the second day of its 16-day launch window, aboard a Delta-II 7925 rocket. NEAR was initially injected into a two-year, Earth-return trajectory that takes NEAR to its maximum distance from the Sun of 2.2 AU (see Figure 1). The 253 Mathilde flyby, on June 27, 1997, occurs during this 23-month orbital loop around the Sun. About a week after the Mathilde flyby, NEAR will perform a deep space Δv maneuver of 273 meters per second. On January 23, 1998, NEAR passes close to the Earth again for a gravity assist that raises the heliocentric orbit energy and changes the orbital inclination to match that of Eros.

During the Earth flyby NEAR will pass within 545 km of Earth's surface. This sets up the optimal geometry needed to make a slow approach to Eros in early January 1999. The first of four rendezvous burns is scheduled for December 20,

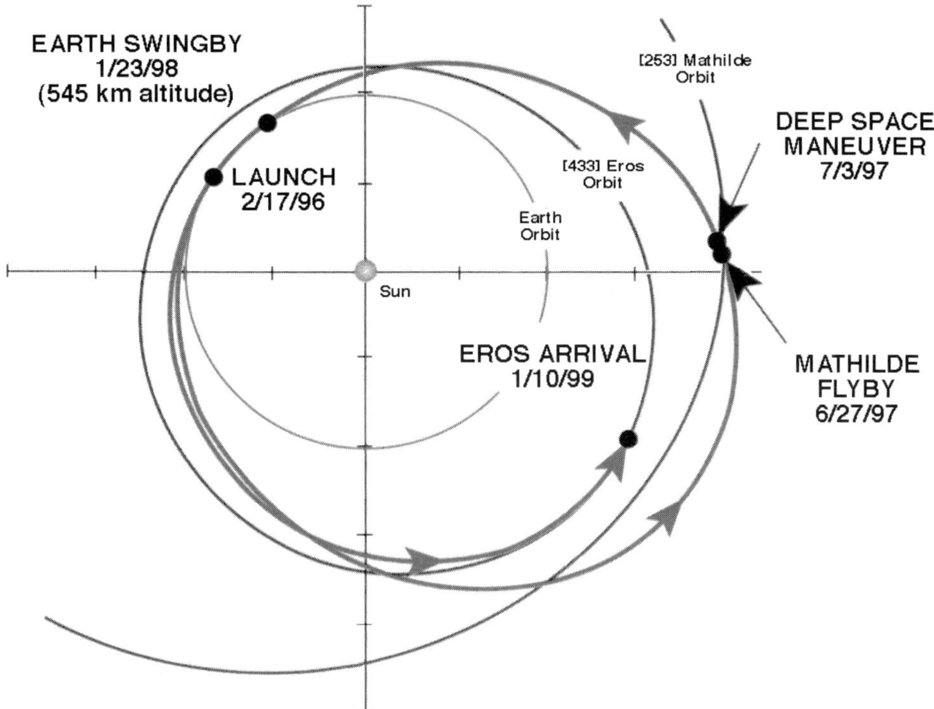

Figure 1. NEAR mission trajectory, ecliptic plane projection. The first day of the launch window was February 16, 1996, but the spacecraft actually launched on February 17, 1996 into a two-year, Earth return trajectory with Mathilde flyby on June 27, 1997. After an Earth gravity assist in January 1998, the spacecraft enters rendezvous with Eros in January 1999.

1998, reducing the spacecraft velocity relative to Eros to a final value of only 5 m s^{-1} for a slow flyby. On January 10, 1999 NEAR will fly past Eros at a closest approach distance of 500 km. Subsequently, NEAR will be injected into a series of transfer orbits and intermediate altitude circular mapping orbits, so that the spacecraft is gradually lowered to a circular mapping orbit of 35 km radius. A total of more than 6 months will be spent in the 35 km mapping orbits. End of mission will be on February 6, 2000.

5. Spacecraft Description

The NEAR spacecraft is a solar-powered, three-axis stabilized spacecraft (Santo et al., 1995). The launch mass, including propellant, was 805 kg, and the dry mass was 468 kg. The spacecraft is simple and highly redundant. NEAR uses X-band telemetry to the NASA Deep Space Network, with the data rates at Eros selectable in the range 2.9 to 8.8 kbps using a 34 m HEF antenna. With a 70 m antenna, the data rates from Eros range from 17.6 to 26.5 kbps. Two solid state recorders are

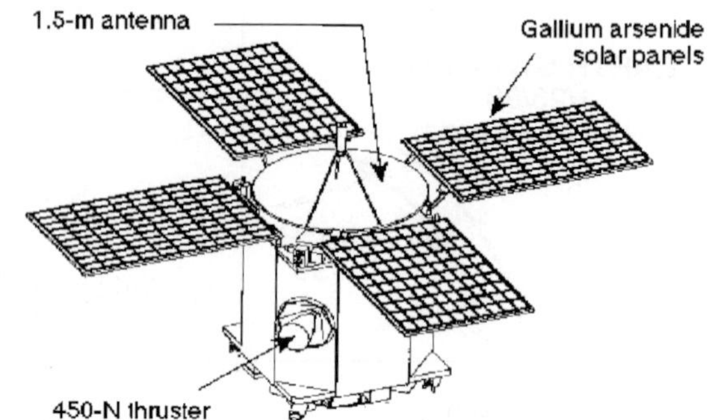

Figure 2. NEAR flight configuration and top-level requirements. The spacecraft is shown after deployment of the fixed solar panels. The solar panels and the fixed high gain antenna are mounted on the forward deck. The magnetometer is in the high gain antenna feed. The spacecraft significantly exceeds the required onboard data storage and downlink capability.

accommodated with a combined memory capacity of 1.8 Gb. The spacecraft attitude is determined using a star camera, a fully redundant inertial measurement unit, and redundant digital Sun sensors. The propulsion system is dual mode (hydrazine and bipropellant) and includes one 450 N bipropellant thruster for large maneuvers, four 21 N thrusters, and seven 3.5 N hydrazine thrusters for fine velocity control and momentum dumping. The spacecraft attitude can be controlled by a redundant set of four reaction wheels or by the thruster complement to within 1.7 mr. The line-of-sight pointing stability is within 50 μr over 1 s, and post-processing attitude knowledge is within 50 μr.

The NEAR spacecraft in the deployed flight configuration is shown in Figures 2 and 3. It is a simple design. The spacecraft structure is composed of forward and aft aluminum honeycomb decks connected with eight aluminum honeycomb side panels. Mounted on the outside of the forward deck are a fixed, 1.5 m diameter, X-band high gain antenna (HGA) and four fixed gallium arsenide (GaAs) solar panels. Also mounted on the outside of the forward deck is the X-ray solar monitor system. The HGA axis is defined as the spacecraft z-axis, which must be pointed to within 1° of Earth to use the high gain antenna for downlink. When the z-axis is pointed at the Sun, the solar panels are fully illuminated, and the Sun is in the center of the solar monitor field-of-view. The magnetometer is mounted on top of the high gain antenna feed, which is the location at which it is exposed to the minimum level of spacecraft-generated magnetic fields. No booms could be accommodated on the spacecraft. The spacecraft electronics are mounted on the inside of the forward deck and the inside of the aft deck. The remaining instruments MSI, NIS, NLR, XRS, and GRS are all mounted on the outside of the aft deck. They are on fixed mounts and are co-aligned to view a common boresight direction perpendicular to the spacecraft z-axis. The Near Infrared Spectrograph has a scan mirror that allows it to look more than 90° away from the common boresight.

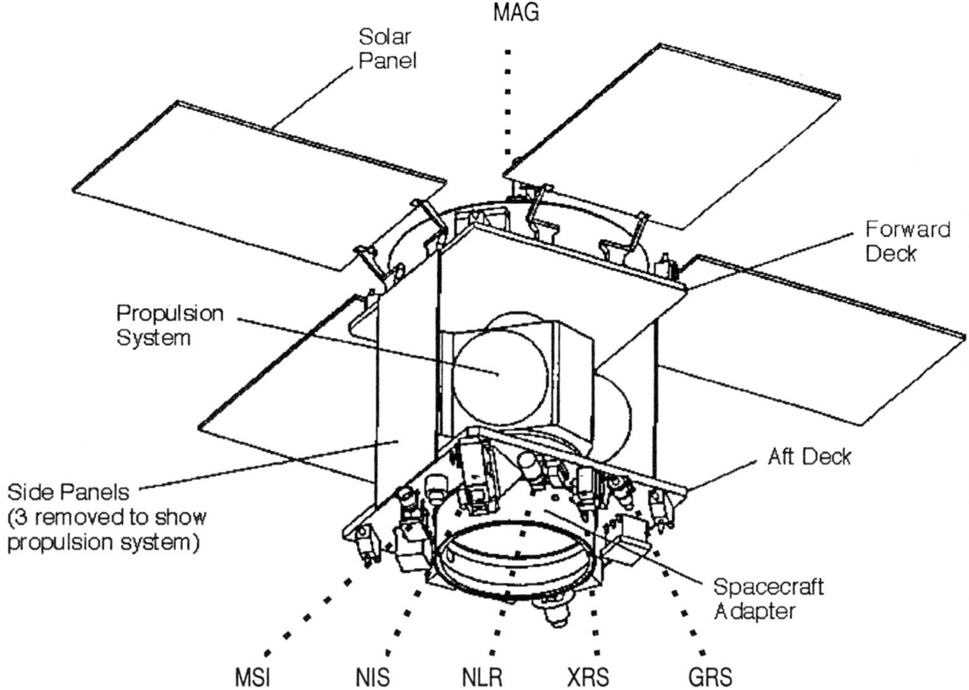

Figure 3. Spacecraft structure cutaway drawing and instruments. This view shows the aft deck of the spacecraft with instruments mounted in front of the spacecraft adapter. From left to right, they are: MSI, NIS in the corner facing the viewer, NLR, XRS, and GRS. Thruster pods occupy the other two visible corners of the aft deck, where additional electronics boxes and the aft low gain antenna are also mounted.

Table IV shows the mass and power summary of the spacecraft instruments and subsystems. The mass and power numbers are measured values on flight hardware. The power system uses four GaAs solar panels, each about 1.22×1.83 m, developing a total power of about 2000 W at 1 AU from the Sun. The system also includes a 9 A-hr, 22 cell super-nickel-cadmium battery. The battery is required to sustain the spacecraft during the eclipse phase shortly after launch. Table V summarizes the spacecraft guidance and control system.

The NEAR system block diagram is shown in Figure 4. The system is designed to be highly fault tolerant. Redundant subsystems include: the telecommunications system except for the fixed HGA and the fixed medium gain antenna; the solid state recorders; the command and telemetry processors; the 1553 data buses; the Attitude Interface Unit and the flight computers for guidance and control; and power subsystem electronics. The NEAR instruments are non-redundant. They use and are controlled by a total of four instrument Data Processing Units (DPUs), each driven by an RTX 2010 FORTH microprocessor. One DPU is shared by XRS and GRS, and another is shared by NIS and MAG. MSI has a dedicated DPU as does the NLR.

Table IV

NEAR mass and power summary

Component	Mass (kg)	Power (W)
Instruments		
Multi-spectral imager (MSI)	7.8	13.9
Near imaging spectrograph (NIS)	14.2	20.0
X-ray/gamma-ray spectrograph (XGRS)	27.3	31.3
Magnetometer	1.6	1.5
Laser rangefinder	5.1	26.8
Propulsion		
Propulsion structure	33.1	
Propulsion system	85.1	
Propellant and pressurant	319.7	
Power		
Solar panels	46.1	
Battery	12.2	4.3*
Power system electronics	6.1	2.5*
Telecommunication		
High gain antenna	6.5	
Medium/low gain antennas	0.7	
Solid state amplifiers (2)	4.1	38.7*
Transponders (2)	8.2	18.1*
Command detector units (2)	0.7	
Telemetry conditioner units (2)	1.7	3.8*
RF switches coaxial cables	3.0	
Guidance and control		
Reaction wheels (4)	12.9	20.0*
Star tracker	2.7	9.9*
Inertial measurement unit	5.3	21.4*
Digital sun sensors (5)	1.9	0.3*
Attitude interface unit	6.4	10.8*
Flight computers	4.7	8.0*
Command and data handling		
Command and telemetry processors (2)	9.8	18.2*
Solid state recorders (2)	3.0	6.4*
Power switching unit	5.9	0.7*
Mechanical		
Spacecraft primary structure	78.0	
Spacecraft secondary structure	18.1	
Despin mass and balance mass	6.1	
Thermal		
Thermal blankets, heaters, thermostats	11.0	
Propulsion survival heaters		75.8*
Spacecraft and instrument survival heaters		71.0*
Instrument operations heaters		40.2
Harness		
Harness and terminal boards	38.8	4.5*
Totals	787.8	314.4*

* indicates configuration at minimum power point.

Table V

Guidance and control system components

Item	Supplier	Characteristics
Inertial measurement unit	Delco	Gyros (4): 30 mm hemispherical resonator gyros Rate bias < 0.01 deg hr^{-1}, over 16 hr ARW < 0.001 deg hr$^{-1/2}$ Accelerometers (4): Sunstrant QA-2000 <100 μg RMS noise
Star tracker	Ball	FOV: $20 \times 20°$ Sensitivity: $+0.1$ to $+4.5$ M No. of stars tracked: 5 Output rate: 5 Hz
Reaction wheels	Ithaco	Brushless DC motor Momentum: 4 Nms (@ 5100 RPM) Torque: 0.025 Nm
Sun sensors	Adcole	Quantization: $0.5°$ Accuracy: $0.25°$
Attitude interface unit	JHU/APL	Clock: 6 MHz Memory (16 bit words): RAM: 64K EEPROM: 64K PROM: 2K Processor: RTX 2010
Flight computer	Honeywell	Clock: 9 MHz Memory (16 bit words): RAM: 512K EEPROM: 256K PROM: 16K Processor: MIL-STD-1750A

Figure 5 shows the X-band telecommunications system block diagram. Uplink is at 7.1820 GHz and downlink is at 8.4381 GHz. The normal uplink data rate is 125 bps, while emergency mode uplink is at 7.8 bps. Eight discrete downlink rates are supported: 9.9 bps (emergency mode), 39.4 bps, 1.1 kbps, 2.9 kbps, 4.4 kbps, 8.8 kbps, 17.6 kbps, and 26.5 kbps. The downlink will use either of two convolution codes: rate $\frac{1}{2}$ and $k = 7$, or rate $\frac{1}{6}$ and $k = 15$. The downlink data are also 8-bit Reed-Solomon encoded. The specified bit error rates are 10^{-6} both for uplink and for downlink. The coherent X-band transponder will support Radio Science measurements of the spacecraft radial velocity component relative to the Earth to within 0.1 mm s^{-1}.

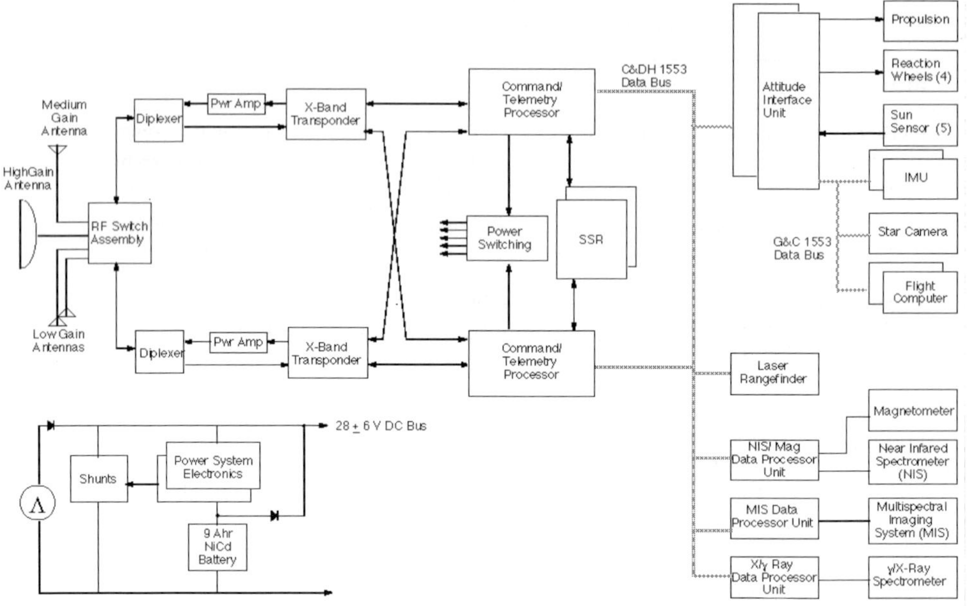

Figure 4. NEAR System Block Diagram. The system design is highly fault tolerant.

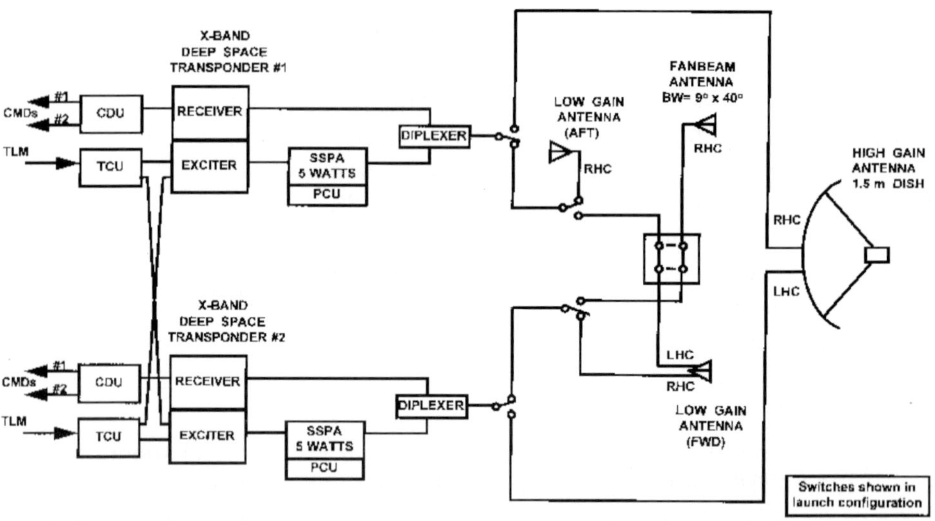

Figure 5. Telecommunication system block diagram. The telecommunications system is fully redundant and uses a 1.5 m high gain antenna, a medium gain fanbeam antenna, and two low gain antennas for omnidirectional coverage.

The predicted downlink data rates at Eros using the DSN 34 m HEF antennas are shown as a function of time during the mission in Figure 6. Actual link margins as determined from flight experience to date agree with predictions to within 1 db.

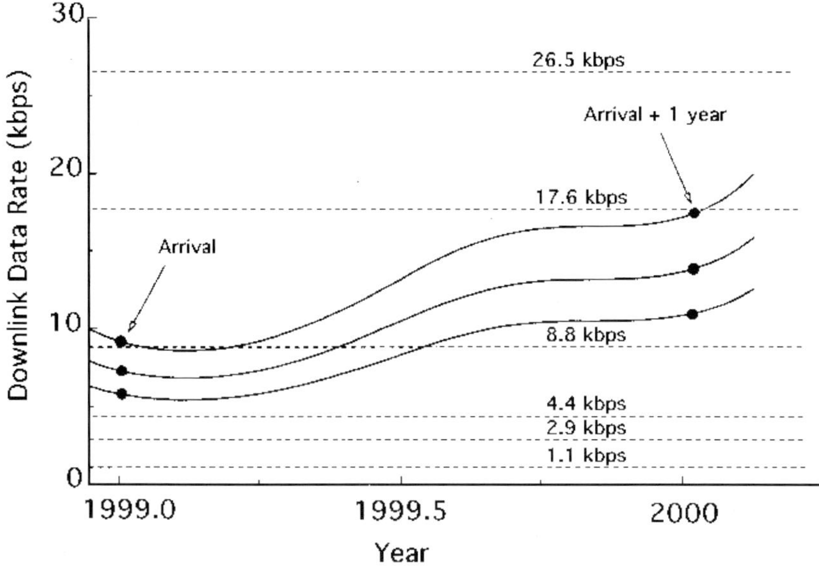

Figure 6. Predicted downlink data rates at Eros as a function of time during the asteroid rendezvous phase, assuming a DSN 34 m HEF antenna. Three curves are shown, for link margins of 1 db (highest curve), 2 db (middle curve), and 3 db (lowest curve). With a DSN 70 m antenna, data rates would be about four times greater.

Key properties of the mission design permit the use of a fixed spacecraft geometry for NEAR. Throughout the course of the rendezvous with Eros, the angle between the Sun and the Earth, as seen from the spacecraft, remains less than about 30°. In addition, the mission aphelion is reached during cruise. Hence, if the solar panels are sufficently large to sustain the spacecraft at aphelion, there will be sufficient power margin at Eros that the spacecraft can pull its solar panels 30° off full illumination in order to point the HGA at Earth. Moreover, the large propulsive burns are such that the required impulse is almost orthogonal to the Sun direction. Hence the large bipropellant thruster can be mounted perpendicular to the spacecraft z-axis.

Finally, during the rendezvous with Eros, the orbit plane will be maintained such that the orbit normal points approximately at the Sun. To take data while in orbit around Eros, the spacecraft will point its z-axis at or close to the Sun, so the solar panels are almost fully illuminated, and it will roll around its z-axis to point the common instrument boresight direction at the asteroid. In order to use the HGA for data transmission to Earth, the spacecraft z-axis must be pointed to Earth. There is ample power margin to support the required solar panel offset from the Sun for downlink, but the instruments cannot, in general, be pointed at Eros while HGA is pointed at Earth. It is therefore planned that the spacecraft will spend about 70% of the time during the rendezvous orbit acquiring data with the z-axis pointed at the

Sun and storing these data in the solid state memory. It will spend the remaining time transmitting data to the Earth.

In order to obtain global, three-color imaging coverage of Eros with ample margin from 40 km altitude or less during the Eros rendezvous, the spacecraft is required to be able to acquire and return to Earth at least 40 uncompressed images per day. To downlink this data volume within a single 10 hr, DSN 34 m HEF pass requires a minimum data rate of 3 kbps. As noted above, the NEAR spacecraft far surpasses these requirements. With the onboard image compression, the spacecraft can store more than 1000 images at once in its solid state memory. The spacecraft can also downlink 1000 images within 10 hr at its maximum data rate of 26.5 kbps. It is anticipated that during the rendezvous, NEAR will routinely have continuous 34m HEF DSN coverage plus one DSN 70 m pass per day. NEAR will then be able to return seven-color global maps of Eros from altitudes of 30 km or less, at spatial resolutions of 3×5 m pixel^{-1} or better.

6. Instruments and Science Goals

Only a brief summary of instrument characteristics will be given here for completeness. The accompanying articles in this volume give full descriptions of each science investigation and summarize the design and performance characteristics of each instrument. Each of the investigation teams has performed observations for calibration and test during the cruise to Eros; for details see accompanying papers (this issue).

(1) *Multispectral Imager* (MSI). The main goals of MSI are to determine the shape of Eros and to map the mineralogy and morphology of features on its surface at high spatial resolution. MSI is a 537×244 pixel charge-coupled device (CCD) camera with five-element, radiation-hard refractive optics. It covers the spectral range from 0.4 to 1.1 μm, and it has an eight position filter wheel with filters chosen to optimize the sensitivity of MSI to minerals expected or known to occur on Eros. The camera has a field-of-view of $2.9° \times 2.25°$ and a pixel resolution of 95×161 μrad. MSI has a maximum framing rate of 1 per second with images digitized to 12 bits. With its maximum exposure time of 1 s, MSI can detect a star of magnitude 10.5 through its broadband 700 nm filter. It has a dedicated DPU with an eight frame image buffer that is able to perform both lossless and lossy onboard image compression. The mass and power values given in Table IV include the MSI DPU.

(2) *Near-Infrared Spectrograph* (NIS). NIS will measure the spectrum of sunlight reflected from Eros in the near-infrared range from 0.8 to 2.6 μm in 64 channels, so as to determine the distribution and abundance of surface minerals like olivine and pyroxene. It is a grating spectrometer that disperses the light from the slit field-of-view across a pair of passively cooled, one-dimensional array detectors. The first is a germanium array covering the wavelengths 0.8 μm to 1.5 μm,

and the other is an indium-gallium-arsenide array covering 1.3 μm to 2.6 μm. The NIS slit field-of-view is 0.38° × 0.76° in its narrow position and 0.76° × 0.76° in its wide position. The slit can be closed for dark current measurement. NIS has a scan mirror that enables it to view the common boresight direction (within the MSI field-of-view) or directions more than 90° away. Spectral images can be built up by a combination of scan mirror and spacecraft motions. In addition, NIS has a gold calibration target that can be placed so as to scatter sunlight into the instrument and provide a quantitative, inflight spectral calibration. The NIS mass and peak power values in Table IV include its mounting bracket and a DPU that is shared with MAG.

(3) *X-Ray Spectrometer* (XRS). XRS is an X-ray resonance fluorescence spectrometer that detects the characteristic X-ray line emissions excited by solar X-rays from major elements in the asteroid surface. XRS covers X-rays in the energy range from 1 to 10 keV using three gas proportional counters. The balanced, differential filter technique is used to separate the closely spaced Mg, Al, and Si lines lying below 2 keV. The gas proportional counters directly resolve higher energy line emissions from Ca and Fe. A mechanical collimator gives XRS a 5° field-of-view, with which XRS will map the chemical composition at spatial resolutions as much as 2 km in the low orbits. XRS also includes a separate solar monitor system to measure continuously the incident spectrum of solar X-rays, using both a gas proportional counter and a high spectral resolution silicon X-ray detector. Inflight calibration capability is provided for XRS using a calibration rod with Fe55 sources that can be rotated into or out of the detector field-of-view.

(4) *Gamma Ray Spectrometer* (GRS). GRS detects characteristic gamma rays in the 0.3 to 10 MeV range emitted from specific elements in the surface. Some of these emissions are excited by cosmic rays and some arise from natural radioactivity in the asteroid. GRS uses a body-mounted, passively cooled, NaI scintillator detector with a bismuth germanate (BGO) anti-coincidence shield that defines a 45° field-of-view. Abundances of several important elements, such as K, Si, and Fe, will be measured in four quadrants of the asteroid. The NEAR GRS does not require a mounting boom because its BGO shield discriminates not only against solar particles and cosmic rays but also against gamma rays from the spacecraft. The NEAR GRS operates in flight at a temperature of 20 °C. Table IV shows mass and power values for XRS and GRS combined, including their shared DPU, mounting brackets, harness, and power supplies.

(5) *Magnetometer* (MAG). MAG is a three-axis, fluxgate magnetometer which uses ring core sensors made of highly magnetically permeable material. MAG will search for and map an intrinsic magnetic field of Eros. Magnetic fields at Eros as strong as those found on the Moon would easily be measured by NEAR. MAG is mounted in the HGA feed, where inflight calibrations to date have established that the spacecraft residual magnetic field is 170 nT. The residual field is sufficiently stable and well-characterized that the solar wind magnetic field near 1 AU can be detected. MAG is expected to reach a limiting sensitivity of about 1 nT. The

recent *Galileo* flybys of the S-asteroids Gaspra and Ida revealed evidence that both of these bodies are magnetic, but this evidence was ambiguous (Kivelson et al., 1993). Discovery of an intrinsic magnetic field at Eros would be the first definitive detection of magnetism at an asteroid and would have important implications for its thermal and geologic history. The mass and power values for MAG in Table IV include the sensor, harness, and electronics only.

(6) *NEAR Laser Rangefinder* (NLR). NLR is a laser altimeter that measures the distance from the spacecraft to the asteroid surface by sending out a short burst of laser light and then recording the time required for the signal to return from the asteroid. NLR uses a chromium-neodymium-yttrium-aluminum-garnet (Cr:Nd:YAG) solid state laser and a compact reflecting telescope. It sends a small portion of each emitted laser pulse through an optical fiber of known length and into the receiver, providing a continuous in-flight calibration of the timing circuit. The ranging data will be used to construct a global shape model and a global topographic map of Eros with horizontal resolution of about 300 m. NLR will also measure detailed topographic profiles of surface features on Eros with a best spatial resolution of about 5 m. The detailed topographic profiles will enhance and complement the study of surface morphology from imaging. The mass and power values in Table IV include power supplies and a dedicated DPU.

(7) *Radio Science* (RS). The coherent X-band transponder on NEAR will be used to conduct a radio science investigation by measuring the Doppler shift due to the spacecraft radial velocity component relative to the Earth. Accurate measurements of the Doppler shift as the spacecraft orbits the asteroid will allow the asteroid's gravity field to be mapped. In collaboration with the MSI/NIS and NLR teams, the gravity determinations will be combined with global shape and rotation state data to constrain the internal density structure of Eros, possibly revealing heterogeneity.

7. Overview of Spacecraft Operations

The NEAR spacecraft and mission have been designed to optimize the science return while minimizing the complexity of both spacecraft and mission operations. An important attribute of the selected mission geometry is that the Sun-vehicle-Earth (SVE) angle is always less than 40°, except during the first month after launch and the first two months after the Earth flyby. This reduces spacecraft complexity by enabling the use of a fixed high gain antenna and fixed solar panels. During the brief periods when the SVE angle is above 40°, the spacecraft can be operated with the solar panels pointed at the Sun, and it can communicate with Earth at low data rates using the medium gain and low gain antennas.

The spacecraft has fixed, body-mounted instruments. All mission science objectives can be met without an instrument scan platform. Instruments are pointed by slewing the spacecraft. Only the Near-Infrared Spectrograph has built-in capability to change its pointing by using a scan mirror. This capability is needed to obtain low phase angle coverage for spectral mapping.

Another consequence of the Sun-spacecraft-Earth geometry is that the autonomous Earth acquisition algorithm that is used in the event of an attitude anomaly can be greatly simplified. In a traditional planetary spacecraft the Earth acquisition algorithm, which may be triggered by an on-board fault or a watchdog time-out, needs two attitude reference vectors and Universal Time. The two vectors are used to determine the spacecraft's pointing attitude, and Universal Time is used to determine the position of Earth in the sky. For the NEAR mission the spacecraft can autonomously find Earth by simply finding the Sun. The procedure is to switch the communication system to the medium gain antenna, point the axis of the high gain antenna at the Sun, and rotate the spacecraft about the Sun-spacecraft line. The medium gain antenna has a gain pattern that covers an area $10°$ wide, extending from the HGA axis to $40°$ off the axis. Therefore, whenever the Earth is within 40 deg of the Sun, rotating the medium gain antenna about the Sun-spacecraft line in a 'radar sweep' fashion will guarantee Earth acquisition.

8. Overview of Eros Operations

The fixed geometry of the NEAR spacecraft will determine the nature of orbital operations around Eros. NEAR will be maintained in an orbit plane at Eros that is close to perpendicular to the line of sight to the Sun. The solar panels then provide more than adequate power for science operations. When data are to be downlinked to Earth, the spacecraft will be slewed if necessary to point the high-gain antenna at Earth. The instruments face $90°$ from the direction of the high-gain antenna, so they can point at Eros as the spacecraft rolls in its orbit. All or any combination of the instruments can operate simultaneously, taking data and storing data on the solid state recorders. The spacecraft can also take data and downlink data simultaneously, although the instruments cannot always be pointed at the asteroid during the downlink periods.

As Eros moves in its orbit over the course of the rendezvous, its rotation axis is approximately fixed in inertial space. However, the spacecraft orbit plane is maintained approximately normal to the solar line-of-sight, so the obliquity to the Eros rotation axis will change continually. Likewise, the angle between the orbit plane normal and the direction to Earth, which approximately equals the SVE angle, changes over the course of the rendezvous. The changing mission geometry drives the design of Eros rendezvous operations.

NEAR will first encounter Eros near its aphelion at 1.75 AU, and will subsequently spend about a year (two-thirds of an Eros year) in orbit around the asteroid. When NEAR first arrives at Eros, its southern rotational hemisphere is illuminated by the Sun, and much of the northern hemisphere never rotates into daylight. Only during the latter part of the year-long rendezvous will the northern hemisphere of Eros become illuminated. If the rendezvous were to terminate prematurely, much of the surface of Eros would never be observed under sunlight. The imager, infrared spectrograph, and X-ray spectrometer are all able to observe only the sunlit portions of Eros, although the gamma ray spectrometer, magnetometer and laser rangefinder are independent of sunlight. In order to make the full set of measurements over the entire surface, and in particular to be able to image all of Eros at highest resolution, NEAR must wait until the season changes as Eros moves in its orbit around the Sun. By about eight months after the rendezvous begins, all of Eros has become illuminated by the Sun.

Eros is far from spherical, and it has a large effective J_2 gravitational quadrupole moment because of its large axis ratios, tending to drive precession of NEAR's orbit plane around Eros's rotation axis. In order to maintain the NEAR orbit plane in the desired orientation, frequent small maneuvers will be needed. Both the frequency of required propulsive maneuvers, and the magnitude of the velocity changes, increase rapidly as the orbit is lowered, because of the strong radial dependence of the quadrupole interaction. The Δv cost to maintain the orbit varies roughly as the inverse fourth power of the orbit radius, for near-circular orbits. In addition, the cost to maintain the orbit, both in terms of maneuver frequency and propellant usage, is minimized during the specific periods when the desired orbit orientation, normal to the Sun direction, approximately coincides with the rotational equator of Eros. There are two such periods during the rendezvous orbit phase. These specific periods are chosen as the times when NEAR will be at its minimum orbit altitude in the 35 km, nominally circular, mapping orbits.

The total propellant budget for Eros rendezvous operations (after capture orbit insertion on January 12, 1999) corresponds to a Δv of 80 m s^{-1}. In addition, the frequency of propulsive events will be limited to one per week or fewer, because of the small size of the Mission Operations team at rendezvous (33 people at APL). Detailed simulations of the 35 km mapping orbit show that the propellant cost and the frequency of required orbital maintenance maneuvers remain within these limits for two specific periods, totaling more than 120 days, during the rendezvous year. The periods spent in 35 km orbit will divided into disjoint time segments, the first extending roughly from April 10 through June 18, 1999, and the second from about November 10 through end of mission on February 6, 2000. The intervening months will be spent in higher orbits.

During the 35 km-orbit time segments, NEAR will be in nearly equatorial orbits around Eros, so remote sensing coverage of the polar regions will be poor. However, during the interval between the two 35 km-orbit time segments, NEAR will be in high inclination orbits relative to Eros's equator. During this interval, NEAR will obtain its best coverage of the polar regions. It is currently planned that NEAR will spend from August 28, 1999 to October 6, 1999, in a 40 km radius, nominally circular polar orbit.

Finally, the irregular shape of Eros requires that NEAR remain in retrograde orbit relative to the asteroid spin. Prograde orbits tend to be unstable in the sense that the spacecraft would typically be ejected from orbit or caused to impact the surface. An orbital plane flip maneuver circa August 23, 1999, is required to maintain a retrograde orbit. This maneuver involves transfer of the spacecraft to high altitude and performance of a small burn near apoapsis to reverse the orbital angular momentum.

The detailed design of the plane flip maneuver is not yet final. The plane flip maneuver provides an opportunity to perform a second slow flyby of Eros with different portions of the surface illuminated than for the first flyby, but the time to perform the second flyby must be taken from the time that would otherwise be spent in low altitude orbit. The time required for the plane flip will depend on the fuel allocated for the maneuver. The final design of the maneuver will be undertaken after the start of the rendezvous with input from the Science Team.

9. Mission Phases and Science Operations

To simplify science operations, the mission will be divided into distinct phases (see Table VII). During each phase, particular aspects of the science will be emphasized. The highest priority science will vary by mission phase and during the course of the year in rendezvous orbit around Eros. The dates and orbit parameters shown in Table VII are preliminary and may be subject to future modification, because of present uncertainties in the values of Eros mass and J_2 as well as spacecraft performance.

The first detection of Eros by the imager is expected some 200 days prior to the unperturbed flyby closest approach date (i.e., the date that flyby would occur in the absence of rendezvous burns). Following this first detection, imaging of Eros will be undertaken for optical navigation and for initial shape and rotation studies.

The rendezvous burn sequence is targeted to put NEAR into an initial slow flyby trajectory past Eros. This initial flyby is at a nominal speed of 5 m s^{-1} with a miss distance of 500 km, taking the spacecraft through zero solar phase angle (Sun-asteroid-spacecraft angle). The purpose of this initial flyby is to enable critical initial science observations. Since the nominal rendezvous orbit plane will be near the Eros terminator, most of the rendezvous observations will be made at large phase angles that are favorable for imaging but unfavorable for IR spectral mapping. The initial flyby will allow more than 30 hr of observation at low phase angles that are not accessible within the nominal rendezvous geometry. Hence it will provide an important opportunity to obtain global IR spectral maps under optimal lighting conditions.

The approach to Eros and the initial flyby will also be used to perform an optical search for satellites of Eros. It is expected that any satellite with the albedo of Eros and within 200 km of Eros would be detected if the diameter is at least 12 m (Veverka et al., 1996). The satellite Dactyl of 243 Ida was discovered by *Galileo* (Belton et al., 1994). If a satellite like Dactyl is present at Eros, it will not prevent

Table VI

Mission phases

Cruise phase
    From interplanetary injection until first detection of Eros by MSI
Far encounter phase
    From first detection until start of rendezvous burn sequence
Approach to Eros and initial flyby
    From start of rendezvous burn sequence until injection to initial high altitude
    parking orbit around Eros
Asteroid orbit phase
    From Eros orbit injection to end of mission

NEAR from entering low altitude orbit and achieving its primary science goals. Needless to say, if an Eros satellite is discovered, the mission plans will be modified to include appropriate studies of the satellite.

NEAR will remain in a bound orbit around Eros for more than ten months. The spacecraft will spend at least 120 days in 35×35 km orbit around Eros, during which time the highest priority science will be measurement of elemental composition, although every instrument will be in operation. Much of the remaining time in orbit around Eros will be spent at semi-major axes of 50 km or less. Again, all instruments will be operating during these periods, but imaging and spectral mapping will have increased priority. The various mission phases are summarized in Table VI, and Table VII gives a brief summary of science operations anticipated in the various mission phases. A more complete discussion of science operations is given in the accompanying articles.

10. Science Operations Planning

Since the mass of Eros is presently unknown, and available shape and rotation pole estimates are subject to significant uncertainties, it is not possible to plan a detailed 'tour' of Eros in the sense that a tour can be constructed of a planetary system. Mission simulations performed to date have established that the real-time navigation accuracy and the predictability of the spacecraft ephemeris will be adequate for safe operations in nominal rendezvous orbit at a few body radii, but the detailed mission operations and science sequences cannot be developed until shortly (a few weeks) prior to actual execution.

Mission operations and science sequence development will be carried out with active interaction between the NEAR Science Team and the Operations Team (Landshof and Cheng, 1995). This interaction must be ongoing throughout the rendezvous, owing to the rapid pace of operations at the asteroid. Science sequences for NEAR will be developed on a nominal weekly cycle.

Table VII

Science operations summary

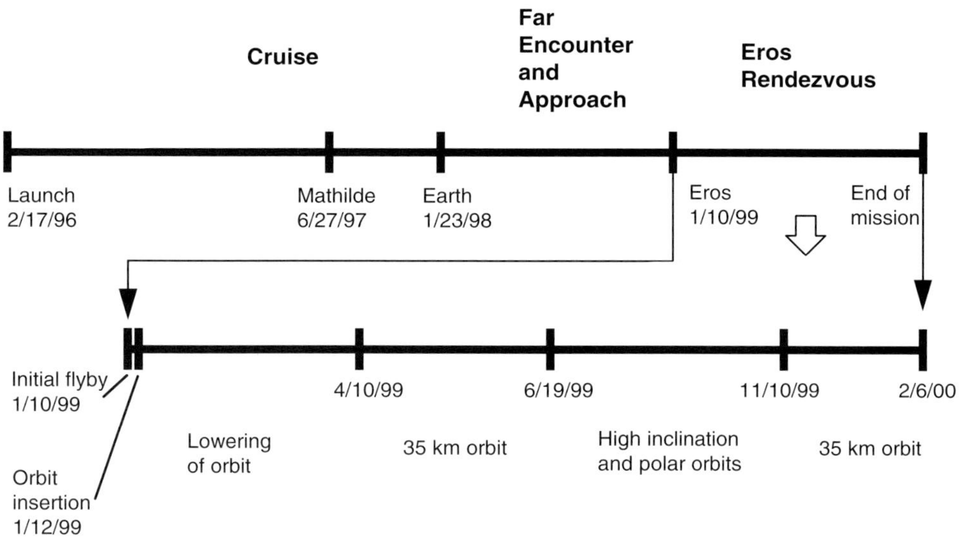

Overall priorities for science operations are set on a monthly basis by the NEAR Project Scientist and Science Team Leaders (one for each of the five investigations). This group will meet to review science, spacecraft and mission status, and to allocate spacecraft resources among the science investigations. Data rate allocations for each instrument will be adjusted over the course of the mission in response to changing science priorities and the decreasing Earth-spacecraft range. Lists of desired observations and targets, including desired lighting conditions and coordinated experiments with multiple instruments, will also be maintained. The same group will convene 'weekly' sequencing meetings, in order to review results of previous operations, approve the next sequence to be uplinked, and develop future sequences. The nominal weekly cycle of sequence generation and review is summarized in Table VIII. Figure 7 depicts the functional flow of science operations planning.

Table VIII

Weekly sequence generation

Review results of science operations from previous uplink ($n - 1$)
Approve next sequence (n) to be uplinked
Revise planned sequence ($n + 1$) for following uplink
Preliminary event script ($n + 1$) already developed by Operations Team
Develop first cut at sequence ($n + 2$) for submission to Operations Team

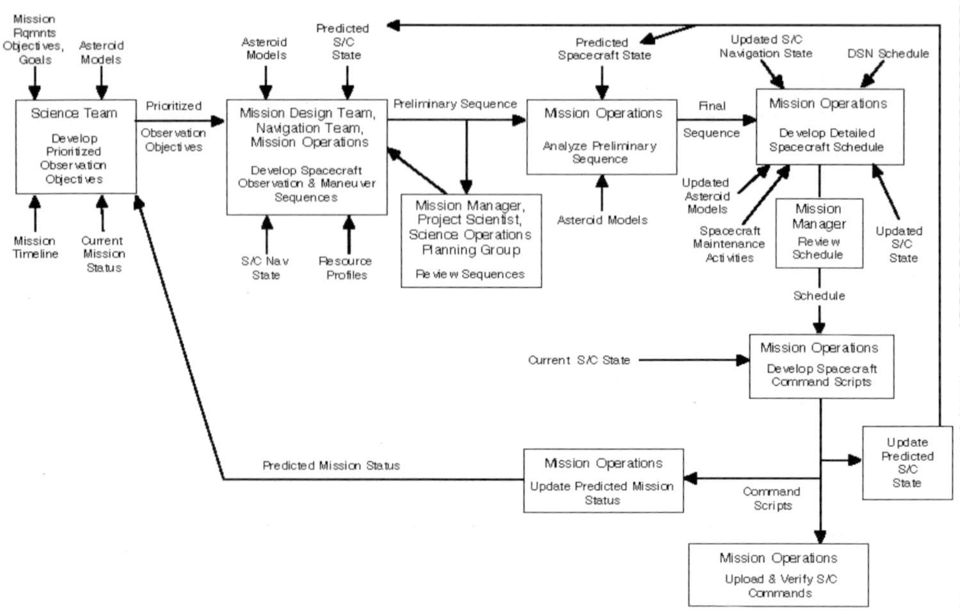

Figure 7. Science Operations Planning Process, showing the interaction between the Science Team and the Mission Operations Team.

11. Overview of Data Flow

All data from the NEAR mission will be downlinked to the NASA Deep Space Network and then forwarded to the Mission Operations Center (MOC) at the Applied Physics Laboratory (APL). Doppler and ranging data from the spacecraft will be analyzed primarily by the NEAR navigation team at the Jet Propulsion Laboratory (JPL) and processed to determine the spacecraft ephemeris as well as to perform Radio Science investigations. The entire spacecraft telemetry stream, including spacecraft and instrument housekeeping data and all science data, will be forwarded to the APL MOC together with the radiometric Doppler and/or range data as well as results of navigation solutions. Navigation data will be forwarded to MOC in the form of SPICE kernels (SPICE is an information system developed by the Navigation Ancillary Information Facility at JPL, and it consists of data

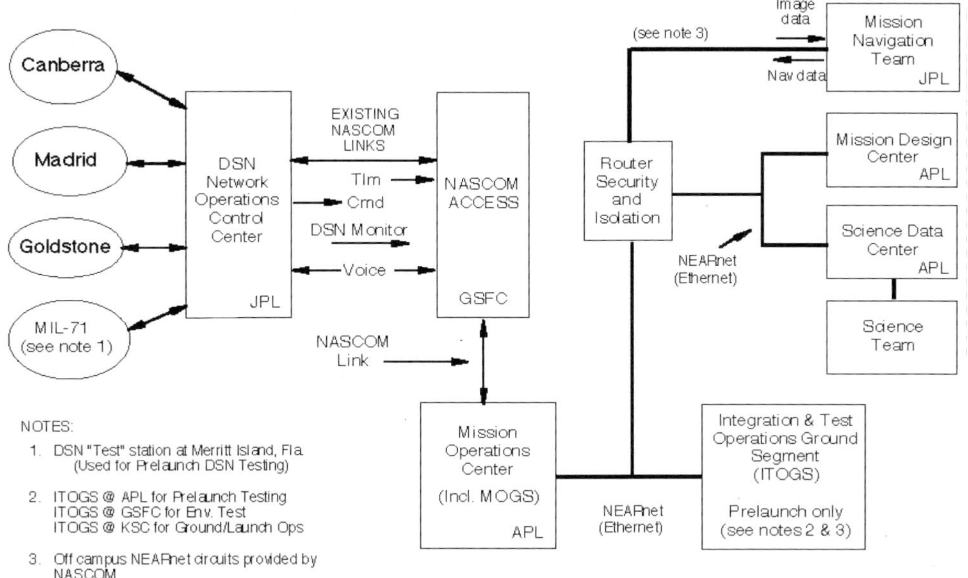

Figure 8. NEAR Mission Data Flow, showing information flows among the Deep Space Network, the Mission Operations Team at APL, the navigation team at JPL, the Science Data Center at APL, and the Science Team.

files and software for managing navigation-related data including spacecraft and planetary ephemerides, spacecraft pointing, timekeeping, gravity data, etc.). The NEAR mission data flow is depicted in Figure 8.

From the APL MOC the spacecraft telemetry stream is passed to the Science Data Center (SDC). The SDC is the Project facility responsible for low level processing of spacecraft telemetry, data distribution, and data archiving. As such, the SDC is the Project facility that supports the activities of the Science Team in data analysis and mission planning. The SDC will create and maintain a Science Archive, which will be the central Project repository for science data products such as images, asteroid models, asteroid maps, etc. that will be stored in standard formats agreed upon with the Science Team. The SDC will provide for easy access to mission data sets by members of the Science Team and by others, and it will collect observing requests and science priorities from the Science Team. It will maintain a telemetry archive, a record of instrument and spacecraft commands as executed, and records of science sequences as requested and as executed. It will provide ancillary data (spacecraft and planetary ephemerides, spacecraft and planetary attitudes, shape and gravity files, and spacecraft clock files) in the form of SPICE kernels to the Science Team. Finally, SDC will create and maintain a database, called the Science Data Catalog, to facilitate access to science data based upon specified criteria such as observing conditions, instrument status or mode,

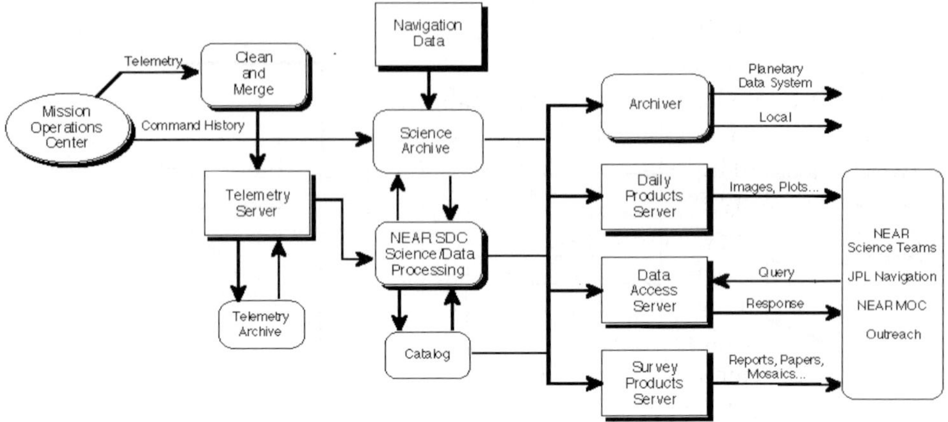

Figure 9. NEAR SDC Data Flow, showing low level telemetry processing, data archiving, and distribution of data to the Science Team, the navigation team, the Mission Operations team, and the public.

mission phase, target of observation, etc. Data flow within the SDC is shown in Figure 9.

The NEAR spacecraft will use packetized telemetry. The telemetry stream is divided into units called transfer frames, each 8832 bits long. Each transfer frame contains a frame synchronization section, two headers (a primary header mainly for identification and a secondary header that contains, among other engineering data, the time tag for creation of the transfer frame), and three data packets each 2864 bits long. The packets are the fundamental units that actually contain instrument or housekeeping data. Each packet also has a primary and a secondary header, each 48 bits long, plus a data segment of 2768 bits. For instrument packets, the secondary packet header contains the time at which the data were sampled.

The SDC will maintain the entire NEAR data set available on-line. The size of the total data set will depend on the actual realized DSN coverage, but is anticipated to be <100 Gb. SDC will support distributed data analysis by scientists at their home institutions using the Internet to access the Science Data Catalog and to transfer data files, ancillary files, software, etc. SDC will also provide dedicated facilities at APL for members of the Science Team and other scientists working on-site during the mission.

The data policy of the NEAR Project is such that all data from every instrument will be made available on-line to every member of the science team and associated scientists. The Science Team Leaders, one for each of the five investigations MSI/NIS, XGRS, NLR, MAG, and RS, are responsible for coordinating and approving plans for data analysis and publication by their team members and associates. Any use of data from a given instrument must be approved by the respective Team Leader, and any conflicts will be resolved by the Project Scientist.

The Experiment Data Records (EDRs), or raw science data records as maintained by SDC within the Science Archive, will contain uncalibrated data. These data will have been cleaned and merged, time-ordered and separated by instrument and instrument mode. This means that duplicate data will have been deleted, missing data will have been padded out, and the data will have been organized by days. The EDR data will be in scientifically useful form, e.g., as individual image frames or individual X-ray spectra, but will not be calibrated. Calibrations, corrections, and conversion to scientific units will be performed upon access to the science data records. The Science Team Leaders for each instrument are responsible for providing the SDC with up-to-date algorithms and/or software to accomplish the calibrations, corrections, and conversion to scientific units. Science Team members and associated scientists with access to the Science Archive will thereby be able to obtain fully corrected and calibrated data from any of the instruments.

The SDC is also responsible for providing access to NEAR data for the scientific community at large and for the public. Uncalibrated data, including images, will be released publically over the Internet as soon as validated. The Internet World Wide Web (WWW) will be used for data released by the Project, at the Uniform Resource Locator: http://sd-www.jhuapl.edu/NEAR/
The NEAR Project will also release data via print and other electronic media.

The Science Data Catalog will be developed and maintained by the SDC to enable easy access to EDRs, to support creation of data products, and to facilitate searching for applicable data. The Science Data Catalog will be a relational database and will contain pointers to the EDRs and to ancillary data files, rather than the actual data themselves. The Science Data Catalog will support on-line queries to find all data sets satisfying specified criteria, e.g., all data when the spacecraft is in specified locations; all data from specified instrument(s) in particular modes; all observations of a specified target on the asteroid; all data taken under specified illumination conditions; all data taken at specific times and/or during specified mission phases. The product of a query will be a list of data product identification numbers, which are the specific pointers to the EDRs and/or ancillary files requested.

Finally, the SDC will be the primary Project facility for archiving NEAR data with the Planetary Data System (PDS). All information in the Science Archive, including all EDRs, ancillary data files, instrument calibration algorithms and software, plus higher level data products developed by Science Team members and forwarded to the SDC for distribution, will be archived in PDS. Additional data products may be archived directly in PDS by members of the Science Team.

12. Summary

The NEAR mission will substantially increase our knowledge of primitive bodies in the solar system by providing a long, up-close look at the S-type asteroid 433

Eros and the first resolved images of the C-type asteroid 253 Mathilde. It will be an auspicious beginning for the NASA Discovery Program.

Acknowledgements

We thank the many members of the NEAR team at APL, NASA, universities, and industry for their hard work and dedication. This work was supported by NASA.

References

Acuna, M. et al.: 1997, 'The NEAR Magnetic Field Investigation: Science Objectives at Asteroid 433 Eros and Experimental Approach', *J. Geophys. Res.* **102**, 23751–23759.

Alvarez, L., Alvarez, W., Asaro, F., and Michel, H. (1980), 'Extraterrestrial Cause for the Cretaceous-Tertiary Extinction', *Science* **208**, 1095–1108.

Bell, J., Davis, D., Hartmann, W., and Gaffey, M.: 1989, in R. Binzel, T. Gehrels, and M. Matthews (eds.), 'Asteroids: The Big Picture', *Asteroids II*, University of Arizona Press, Tucson, pp. 921–945.

Belton, M. et al.: 1994, 'First Images of Asteroid 243 Ida', *Science* **265**, 1543–1547.

Belton, M. et al.: 1992, '*Galileo* Encounter with 951 Gaspra: First Pictures of an Asteroid', *Science* **257**, 1647–1652.

Binzel, R. and Xu, S.: 1993, 'Chips Off of Asteroid 4 Vesta: Evidence for the Parent Body of Basaltic Achondrite Meteorites', *Science* **260**, 186–191.

Binzel, R. P., Burbine, T., and Bus, S.: 1996, 'Ground-Based Reconnaissance of 253 Mathilde: Visible Wavelength Spectrum and Meteorite Comparison', *Icarus* **119**, 447–449.

Binzel, R. P., Bus, S., Burbine, T., and Sunshine, J.: 1996, 'Spectral Properties of Near-Earth Asteroids: Evidence for Sources of Ordinary Chondrite Meteorites', *Science* **273**, 946–948.

Chapman, C.: 1995, 'Near Earth Asteroid Rendezvous: Eros as the Key to the S-type Conundrum', *LSPC XXVI*, pp. 229–230.

Cheng, A. F., Veverka, J., Pilcher, C., and Farquhar, R.: 1994, in T. Gehrels (ed.), 'Missions to Near-Earth Objects', *Hazards due to Comets and Asteroids*, University of Arizona Press, Tucson, pp. 651–670.

Cheng, A. F., Santo, A., Heeres, K., Landshof, J., Farquhar, R., Gold, R., and Lee, S.: 1997, 'Near-Earth Asteroid Rendezvous: Mission Overview', *J. Geophys. Res.* **102**, 23695–23708.

Farquhar, R. W., Dunham, D., and McAdams, J.: 1995, 'NEAR Mission Overview and Trajectory Design', *J. Astron. Sci.* **43**, 353.

Gaffey, M., Burbine, T. H., and Binzel, R.: 1993a, 'Asteroid Spectroscopy: Progress and Perspectives', *Meteoritics* **28**, 161–187.

Gaffey, M., Bell, J., Brown, R., Burbine, T., Piatek, J., Reed, K., and Chaky, D.: 1993b, 'Mineralogic Variations within the S-Type Asteroid Class', *Icarus* **106**, 573–602.

Kivelson, M., Bargatze, L., Khurana, K., Southwood, D., Walker, R., and Coleman, P.: 1993, 'Magnetic Field Signatures Near *Galileo*'s Closest Approach to Gaspra', *Science* **261**, 331–334.

Landshof, J. A. and Cheng, A. F.: 1995, 'NEAR Mission and Science Operations at Eros', *J. Astron. Sci.* **43**, 477.

McCord, T., Adams, J. B., and Johnson, T. V.: 1970, 'Asteroid Vesta: Spectral Reflectivity and Compositional Implications', *Science* **168**, 1445–1447.

McFadden, L., Tholen, D., and Veeder, G.: 1989, in R. Binzel, T. Gehrels, and M. Matthews (eds), 'Physical Properties of Aten, Apollo, and Amor Asteroids', *Asteroids II*, University of Arizona Press, Tucson, pp. 442–467.

Michel, P., Farinella, P., and Froeschle, C.: 1996, 'The Orbital Evolution of the Asteroid Eros and Implications for Collision with the Earth', *Nature* **380**, 689–691.

Mottola, S. et al.: 1995, 'The Slow Rotation of 253 Mathilde', *Planetary Space Sci.* **43**, 1609–1613.

Murchie, S. L. and Pieters, C.: 1996, 'Spectral Properties and Rotational Spectral Heterogeneity of 433 Eros', *J. Geophys. Res.* **101**, 2201–2214.

Ostro, S., Rosema, K., and Jurgens, R.: 1990, 'The Shape of Eros', *Icarus* **84**, 334–351.

Rampino, M. R. and Haggerty, B. M.: 1994, in T. Gehrels (ed.), 'Extraterrestrial Impacts and Mass Extinctions of Life', *Hazards Due to Comets and Asteroids*, University of Arizona Press, Tucson, pp. 827–858.

Santo, A. G., Lee, G. C., and Gold, R. E.: 1995, 'NEAR Spacecraft and Instrumentation', *J. Astron. Sci.* **43**, 373.

Smit, J.: 1994, in T. Gehrels (ed.), 'Extinctions at the Cretaceous-Tertiary Boundary: the Link to the Chicxlub Impact', *Hazards due to Comets and Asteroids*, University of Arizona Press, Tucson, pp. 859–878.

Trombka, J. et al.: 1997, 'Compositional Mapping with the NEAR X-ray/Gamma-ray Spectrometer', *J. Geophys. Res.* **102**, 23729–23759.

Veverka, J. et al.: 1997, 'An Overview of the NEAR Multispectral Imager NEAR Infrared Spectrometer Investigation', *J. Geophys. Res.* **102**, 23775–23780.

Veverka, J., Langevin, Y., Farquhar, R., and Fulchignoni, M.: 1994, in T. Gehrels (ed.), 'Spacecraft Exploration of Asteroids: the 1988 Perspective', *Hazards Due to Comets and Asteroids*, University of Arizona Press, Tucson, pp. 970–996.

Weissman, P. R., A'Hearn, M. F., McFadden, L., and Rickman, H.: 1989, in R. Binzel, T. Gehrels, and M. Matthews (eds.), 'Evolution of Comets Into Asteroids', *Asteroids II*, University of Arizona Press, Tucson, pp. 880–920.

Yeomans, D. et al.: 1997, 'The NEAR Radio Science Investigation', *J. Geophys. Res.* **102**, 23775–23780.

Yeomans, D. K.: 1995, 'Asteroid 433 Eros: the Target Body of the NEAR Mission', *J. Astron. Sci.* **43**, 417.

Zellner, B.: 1976, 'Physical Properties of Asteroid 433 Eros', *Icarus* **28**, 149–153.

Zuber, M. et al.: 1997, 'The NEAR Laser Ranging Investigation', *J. Geophys. Res.* **102**, 23761–23773.

MULTI-SPECTRAL IMAGER ON THE NEAR EARTH ASTEROID RENDEZVOUS MISSION

S. EDWARD HAWKINS III, E. HUGO DARLINGTON, SCOTT L. MURCHIE,
KEITH PEACOCK, TERRY J. HARRIS, CHRISTOPHER B. HERSMAN,
MICHAEL J. ELKO, DANIEL T. PRENDERGAST, BENJAMIN W. BALLARD and
ROBERT E. GOLD

*The Johns Hopkins University, Applied Physics Laboratory, Johns Hopkins Road, Laurel,
MD 20723–6099, U.S.A.*

JOSEPH VEVERKA

Cornell University, Space Sciences Building, Ithaca, NY 14853, U.S.A.

MARK S. ROBINSON

US Geological Survey, 2234 North Gemini Drive, Flagstaff, AZ 86001, U.S.A.

(Received July, 1996)

Abstract. A multispectral imager has been developed for a rendezvous mission with the near-Earth asteroid, 433 Eros. The Multi-Spectral Imager (MSI) on the Near-Earth Asteroid Rendezvous (NEAR) spacecraft uses a five-element refractive optical telescope, has a field of view of $2.93 \times 2.25°$, a focal length of 167.35 mm, and has a spatial resolution of 16.1×9.5 m at a range of 100 km. The spectral sensitivity of the instrument spans visible to near infrared wavelengths, and was designed to provide insight into the nature and fundamental properties of asteroids and comets. Seven narrow band spectral filters were chosen to provide multicolor imaging and to make comparative studies with previous observations of S asteroids and measurements of the characteristic absorption in Fe minerals near 1 μm. An eighth filter with a much wider spectral passband will be used for optical navigation and for imaging faint objects, down to visual magnitude of $+10.5$. The camera has a fixed 1 Hz frame rate and the signal intensities are digitized to 12 bits. The detector, a Thomson-CSF TH7866A Charge-Coupled Device, permits electronic shuttering which effectively varies the dynamic range over an additional three orders of magnitude. Communication with the NEAR spacecraft occurs via a MIL-STD-1553 bus interface, and a high speed serial interface permits rapid transmission of images to the spacecraft solid state recorder. Onboard image processing consists of a multi-tiered data compression scheme. The instrument was extensively tested and calibrated prior to launch; some inflight calibrations have already been completed. This paper presents a detailed overview of the Multi-Spectral Imager and its objectives, design, construction, testing and calibration.

1. Introduction

The first planetary mission intended to encounter and orbit a small body blasted off from Cape Canaveral on 17 February 1996. The Near Earth Asteroid Rendezvous mission, or NEAR, was the first launch under NASA's Discovery Program of small planetary spacecraft to be developed in under 36 months, with a cost cap of $150 million which includes all development costs, construction of the spacecraft, and 30 days of post-launch operations. NEAR satisfied these requirements with substantial margins: the final cost was well below the cost cap, and the development time from the start of the program to launch was only 27 months.

Space Science Reviews **82:** 31–100, 1997.
© 1997 *Kluwer Academic Publishers. Printed in Belgium.*

The primary mission of NEAR – to provide insight into the nature, origin, and evolution of near-Earth asteroids – will be achieved during a rendezvous with the largest near-Earth asteroid, 433 Eros, discovered in 1898 by G. Witt. The 805 kg spacecraft was launched from a Delta II rocket into a 2 year ΔVEGA trajectory (Farquhar et al., 1995; Santo et al., 1995). This trajectory will provide an opportunity to flyby a large main-belt C asteroid (253 Mathilde) in June 1997. A subsequent Earth swingby will alter the spacecraft's velocity vector and place NEAR in the same plane as Eros. Encounter with the asteroid and the beginning of the primary orbital mission is scheduled to begin between February 1999 (Farquhar et al., 1995) and February 2000.

Once the primary mission begins, coordinated observations of Eros will be made by the complement of scientific instruments onboard NEAR which consists of a Multi-Spectral Imager (MSI), a Near-Infrared Spectrograph (NIS), X-Ray and Gamma-Ray Spectrometers (XGRS), a Magnetometer (MAG), and a NEAR Laser Rangefinder (NLR). MSI will measure the shape and rotation of Eros, the morphology of the surface at small scales, and the spectral properties and spectral heterogeneity of its surface, and will search for natural satellites. NIS (Warren et al., this volume) will measure the distribution of minerals in the surface at spatial scales $\sim$300 m using near-infrared reflectance spectroscopy. XGRS (Goldsten et al., this volume) will measure the elemental abundances in the upper millimeter of the surface of Eros through 1 to 10 keV X-ray fluorescence with a spatial resolution on the order of 2 km. At a much coarser spatial resolution the surface elemental composition will be measured down to about 100 mm for naturally occurring radioactive elements and cosmic-ray excited elements using 0.1 to 10 MeV gamma-ray spectroscopy. The combined observations of XGRS, NIS, and MSI should fully map the surface composition of Eros and the configuration of its compositional units.

En route to Eros, MSI will be used for optical navigation to the main-belt asteroid 253 Mathilde. During the 1200 km flyby of this object on 27 June 1997 (Harch et al., 1996), MSI will provide the first high resolution images of a type C asteroid. Optical navigation using MSI will also be important in placing NEAR into its preliminary orbit around Eros, because of the asteroid's irregular shape and the uncertainty in its mass and mass distribution. Details of the spacecraft's trajectory will not be known until an accurate mass is deduced. During the initial very high orbits around Eros, the shape and volume of the asteroid will be measured by MSI and its bulk density will be determined by combining this information and the mass deduced by the radio science experiment through high precision tracking of the spacecraft (Farquhar et al., 1995). Later, comparison of the asteroid's gravity field derived from tracking at lower orbits with predictions based on the shape from MSI and the NLR will constrain the internal density structure. Accurate shape and density will both be necessary for conducting later phases of the mission, when the orbital radius will be reduced to as low as 35 km.

Spectral mapping of Eros by MSI requires precise data covering wavelengths which convey key information needed to infer mineralogy and to compare Eros with other S asteroids. The 12-bit digitization of the MSI images supports spectrometer-quality data output, in contrast to the less precise 8-bit digitization typical of previous planetary imagers (e.g., *Galileo*). Both relative and absolute calibration of images through different filters are tracked from launch to Eros orbit, using a careful strategy of inflight observations of spectral standards including the Moon and bright stars. Tables I and II summarize these planned inflight calibrations.

The new low-cost, fast development initiative of the Discovery Program provided the opportunity for us to once again return to planetary science through *in situ* measurements made by spacecraft. The NEAR program schedule was such that the instruments had to be available for integration with the spacecraft 20 months after the program start. Consequently, most of the instruments had to be largely developed from existing designs rather than completely new ones. This necessit-ated some compromises between specific instrument parameters and the desire to change the heritage designs as little as possible.

This paper fully describes the Multi-Spectral Imager on NEAR, including its objectives and how the science requirements are met, detailed descriptions of the design, assembly, and calibration of the instrument both onground and inflight, all within the context of the Discovery Program constraints. A more detailed review of the combined science investigation using the Multi-Spectral Imager and Near-Infrared Spectrograph is found in Veverka et al. (1997).

2. Science Background and Requirements

2.1. KEY SCIENCE ISSUES

Most asteroids reside in the 'main belt' between Mars and Jupiter, and are thought to be collisional fragments of a few dozen larger precursor bodies. Near-Earth asteroids, with a perihelion distance of less than 1.3 AU, are thought to repre-sent mainbelt asteroids ejected from their original orbits by Jupiter's gravitational influence, as well as extinct Jupiter-family comets (McFadden et al., 1989).

The near-Earth asteroid 433 Eros is a member of the 'S' class of asteroids, one of several classes that has been defined based on astronomical measurements of albedo, spectral, and polarimetric properties (Chapman et al., 1975). The basic spectral attributes of the class are a moderate albedo, a relatively red spectral continuum, and broad absorptions near 1 μm and 2 μm attributed to ferrous-iron containing silicates. S asteroids dominate the inner main belt and constitute a large fraction of near-Earth asteroids. With dimensions of approximately $40 \times 14 \times 14$ km, Eros is the largest of the near-Earth asteroids.

Astronomical spectroscopy has been the major technique for identifying the mineral compositions of S asteroids. Transition metal ions in silicate minerals, in

Table I

MSI inflight calibration plan during cruise (major events noted)

Time	Target	Calibration objectives	Procedure
Post-launch 20 Feb. 1996	Moon I	Check radiometric response of imager Provide data for check of long-term stability Check operation of compression and automatic exposure control routines	Image moon through all filters with several exposure times; use various combinations of compression algorithms and automatic exposure control
Early cruise 1 May 1996	Bright star I (Canopus)	Star images near in time to moon at post-launch (for relative calibration of moon star) Begin tracking inflight stability of radiometric response, focus	Acquire multiple exposures through each filter, some with star image blurred over several pixels using spacecraft geometric scan
2 May 1996		**MSI lens cover open**	
Early cruise 2 May 1966	Bright star II (Canopus)	Compare with earlier Canopus images to determine effect of removing lens cover Track inflight stability of radiometric respose, focus	Acquire multiple exposures through each filter, some with star image blurred over several pixels using spacecraft geometric scan
	Star cluster I (Praesepe)	Check geometric distortion of imager	Acquire images with star cluster occupying different portions of field of view
	Star cluster II (Eta Carinae)	Check geometric distortion of imager	Acquire images with star cluster occupying different portions of field of view
Early-mid cruise	2 Pallas	Test ability to track and identify faint object	Image on successive days through broad-band filter
Early-mid cruise	Bright star III (Arcturus)	Track inflight sabilty of radiometric response, focus	Acquire multiple exposures through each filter. Software checkout
		Deep-space maneuver (DSM)	
Mid-cruise	Bright star IV (Canopus)	Track inflight radiometric stability, focus	Acquire multiple exposures through each filter, some with star image blurred over several pixels using spacecraft geometric scan
Earth swingby	Bright star V (Canopus)	Track inflight radiometric stability, focus Star images near in time to moon at swingby (for relative calibration of moon to star)	Acquire multiple exposures through each filter, some with star image blurred over several pixels using spacecraft geometric scan
Earth swingby	Moon II	Verify flat-field calibration Radiometric calibration Quantify magnitude of scattered and stray light	Image moon through all filters at several locations in-field, at different distances out of field
Eros approach	Bright star VI (Canopus)	Track inflight radiometric stability, focus Star images near in time to Eros (for calibration to moon)	Acquire multiple exposures through each filter, some with star image blurred over several pixels using spacecraft geometric scan

Table II
MSI inflight calibration plan at Eros (major events noted)

Time	Target	Calibration objectives	Procedure
Bipropellant burn to slow spacecraft for approach to Eros			
Eros approach	Eros	Coalignment of NIS and MSI	Take data with both instruments while slowly scanning across Eros in 2 orthogonal directions
High orbit at Eros	Eros	Verify flat-field calibration	Image Eros through all filters at several locations in-field, at different distances out of field
High orbit at Eros	Star cluster III (Pleiades)	Check geometric distortion of imager	Acquire images with star cluster occupying different portions of field of view
Every 3 months in Eros orbit	Bright star VII-X (Canopus, Vega)	Track inflight radiometric stability, focus Star images near in time to Eros (for calibration to moon)	Acquire multiple exposures through each filter, some with star image blurred over several pixels using spacecraft geometric scan
Low orbits at Eros	Eros, night side	MSI-NLR coalignment	Fire NLR in rapid burst mode while acquiring long-exposure images through 1050 nm filter
All orbital altitudes	Eros limb	Quantify magnitude of scattered and stray light	Image swath across Eros limb using all filters

general, have a characteristic absorption band in the 1000 nm wavelength region; the exact position depends upon the individual mineral, such that the wavelength of the absorption band center serves as a 'fingerprint' to that mineral. In the silicate mineral olivine, a strong composite Fe absorption is located at 1050 nm. In the mineral pyroxene, the absorption varies in position from 900–1050 nm. Unlike olivine, pyroxene also exhibits an additional Fe absorption feature at 1900–2300 nm. The wavelength positions of the two pyroxene absorptions vary together systematically as a function of the balance of Fe, Mg, and Ca cations in the pyroxene crystals, allowing detection of differences in pyroxene composition as well as mixing of olivine with pyroxene (Adams, 1974; Cloutis et al., 1986; Cloutis and Gaffey, 1991). Several studies using these techniques (e.g., Gaffey et al., 1989, 1993) have shown that the overall character of S asteroid absorption features is consistent with mixtures of olivine and pyroxene with Ni-Fe metal. However compositions of S asteroids are highly varied, and the mineralogic hcterogeneity within the class is indicative of a variety of rock types on the different S asteroids.

The metal-olivine-pyroxene mineral assemblage on S asteroids is present in meteorite types with highly divergent histories, and substantial debate has arisen regarding the connection between S asteroids and various classes of meteorites. One hypothesis holds that many S asteroids consist of the most common type of meteorites: primitive, undifferentiated ordinary chondrites (e.g., Wetherill and Chapman, 1988). However, spectrally these S asteroids exhibit a redder spectral continuum and weaker absorption bands than do ordinary chondrites measured in terrestrial laboratories. This difference in spectral properties has been attributed to optical alteration of the asteroids' surfaces by the space environment, or 'space weathering'. An alternative hypothesis is that the spectral contrast between S asteroids and ordinary chondrites occurs because the S asteroids consist predominantly of different rock types from ordinary chondrite (Bell et al., 1989; Gaffey et al., 1989, 1993). The meteorites most closely resembling the bulk of S asteroids spectrally are stony-iron and primitive achondritic meteorites, which have undergone varying degrees of igneous melting and separation of mineral phases. The implications of either model for the origin and evolution of S asteroids, their relationship to meteorites, and the evolution of the early solar system are profound: if S asteroids consist mainly of undifferentiated ordinary chondrite, they would be primitive bodies largely unchanged since accretion; if S asteroids are differentiated, they instead represent collisional fragments of parent bodies that experienced a much greater degree of geologic evolution early in solar system history. In either case, the diversity of S asteroid compositions suggests that there is not a single source for all asteroids of this class.

Prior to 1991, when *Galileo* flew by the S asteroid 951 Gaspra, the best information on surface processes on small, rocky bodies was from spacecraft encounters with Phobos and Deimos, the two tiny satellites of Mars. The dramatically different surface properties of these two moons, coupled with the long history of remote measurements of asteroids and analysis of Apollo lunar samples, have stimulated some of the major questions about asteroids in general including S-types (e.g., Veverka and Thomas, 1979).

Are S asteroids coherent rock fragments or collections of loosely bound rubble? Some asteroids undoubtedly are single, coherent 'chunks', but some may have been disrupted by impacts and re-accreted as loose rubble piles.

What processes have physically modified asteroidal surfaces? Both Phobos and Deimos show clear evidence for a fragmented surface layer, or regolith, including downslope movement of loose material and infilling of older craters. However, the degree and nature of regolith development on asteroids in general is uncertain (McKay et al., 1989). Phobos exhibits a well-developed system of topographic grooves (Veverka and Thomas, 1979), at least some of which are thought to originate by drainage of regolith into open fractures; groove formation may be an important surface process on asteroids.

How have the S asteroids' surface materials been affected by exposure to the space environment? On the Moon, reduction of ferrous iron in the regolith into free metal and formation of glassy agglutinates have had the effect on soils older than $\sim 10^8$ years of reducing albedo, subduing spectral absorption features due to minerals, and reddening the spectral continuum (Morris, 1977; Pieters et al., 1993a; Fischer and Pieters, 1994; Allen et al., 1994). This relevance of space weathering to asteroids has been vigorously debated (e.g., Wetherill and Chapman, 1988; Bell et al., 1989; Gaffey et al., 1989, 1993). Do asteroids have satellites, and if so how abundant are they and how did they form? Asteroidal occultations of stars have provided tantalizing evidence for even smaller bodies in orbit around some asteroids, but the physical mechanisms by which asteroidal satellites may form is unclear (Weidenschilling et al. 1989).

2.2. LESSONS LEARNED FROM GASPRA AND IDA

Flybys of the S asteroids 951 Gaspra (Belton et al., 1992; Veverka et al., 1994) and 243 Ida (Belton et al., 1994) by *Galileo* have provided revealing data on two specific asteroids in the same spectral class as Eros. The detailed data from these asteroids refines the questions that NEAR will address.

Gaspra, at $18 \times 11 \times 9$ km, is somewhat smaller than Eros. Its surface exhibits a number of grooves, interpreted as with the grooves on Phobos, to have formed by drainage of regolith into open fractures. Groove orientations parallel the facets of the asteroid's highly angular shape. Both the angular shape and the coherence of groove orientations with elements of that shape are interpreted to indicate that Gaspra is a coherent fragment of a parent body large enough for its component materials to have annealed or solidified sufficiently for fracture planes to have formed (Veverka et al., 1994). Gaspra's reflectance spectrum suggests an olivine/pyroxene ratio much greater than that in ordinary chondrites (Granahan et al., 1995), also suggesting that Gaspra originated as a fragment of a larger body in which minerals had been segregated by thermal evolution.

Ida's size ($60 \times 25 \times 19$ km) is larger than Eros and has a much more irregular shape than Gaspra and lacks Gaspra's well-defined angularity and facets. Ida is a member of the 'Koronis' family – asteroids which share similar orbital elements and may be fragments of a precursor body. Its natural satellite Dactyl (diameter, 1.5 km) is the first confirmed satellite of an asteroidal body. The spectroscopically estimated proportion of olivine/pyroxene ratio on Ida is consistent with that found in ordinary chondrites (Granahan et al., 1995); however, Dactyl appears to be enriched in calcium-containing pyroxene compared with Ida, which would imply some igneous processing (Veverka et al., 1996). Ida's density, derived from estimates in volume through imaging and mass from Dactyl's orbital characteristics, is 2.5 ± 0.4 g cm^{-3}, significantly lower than that of plausible meteorite analogs (≥ 3.5 g cm^{-3}) (Wasson, 1974), and suggests considerable porosity in the interior (Belton et al., 1994).

Gaspra and Ida are highly cratered. The number density of craters per unit area is about four times greater on Ida than on Gaspra; this may mean that Ida's surface is four times older than Gaspra's (Belton et al., 1994). On small bodies such as these, 'surface age' is the time since the last major disruptive global event, such as shattering of a pre-existing parent asteroid by a severe impact. On both Ida and Gaspra, the smallest, geologically youngest craters exhibit deeper Fe-mineral absorptions and less red spectral continuums, indicating fresher material less affected by space weathering. Such observations imply the existence of substantial regolith (Helfenstein et al., 1994; Sullivan et al., 1996). Smaller, older craters generally have a lower depth-to-diameter ratio than larger craters, consistent with partial infilling of the small craters by the inferred fragmental layer of regolith (Carr et al., 1994; Sullivan et al., 1996).

2.3. EROS

433 Eros, the target of NEAR, is one of the most well-characterized near-Earth asteroids, by virtue of its regular close approaches to Earth. The dimensions of Eros have been estimated as $36 \times 15 \times 13$ km (Zellner, 1976) or 39×16 km (Lebofsky and Rieke, 1979) using polarimetry and photometry, 35×16 km using radar (Ostro et al., 1990), and 40×14 km using speckle interferometry (Drummond et al., 1985). The shape is non-axisymmetric; radar data suggest that one of the elongate faces is relatively flat and the other highly convex (Ostro et al., 1990). Eros' albedo at 560 nm is estimated as 0.18 ± 0.03 (Jones and Morrison, 1974; Morrison, 1976), 0.17 ± 0.01 (Zellner and Gradie, 1976), 0.19 ± 0.01 (Zellner, 1976), and 0.125 ± 0.025 (Lebofsky and Rieke, 1979). Thermal and polarimetric properties are consistent with a surface of medium-grained regolith (i.e., predominantly sand-sized particles), with a component of either larger fragments or bare rock (Zellner and Gradie, 1976; Morrison, 1976; Lebofsky and Rieke, 1979).

Visible and near-infrared reflectance spectra of Eros place it squarely in the 'S' spectral class (Millis et al., 1976; Wisniewski, 1976; Miner and Young, 1976; Zellner et al., 1985). Furthermore, mineral abundances estimated from the spectra suggest olivine and pyroxene in relative amounts comparable to those found in the H or L subclasses of ordinary chondrites (Pieters et al., 1976; Murchie and Pieters, 1996). However, Eros' spectrum varies with rotational phase of the asteroid, implying some type of compositional heterogeneity. The flat face has a shorter wavelength 1000 nm absorption and stronger 2000 nm absorption than the convex face (Figure 1), suggesting an $\sim$20% greater abundance of pyroxene relative to olivine on the flat face than on the convex face. This degree of compositional heterogeneity, if real, implies compositional variations larger than those within individual petrologic types of ordinary chondrite, and perhaps a degree of differentiation into separate rock units (Murchie and Pieters, 1996).

Given the evidence for geologic heterogeneity of Eros, despite its overall similarity to ordinary chondrites, this asteroid is an excellent target for detailed study of the origin of S asteroids in general and their relationship to meteorites in particular. The correlation of spectral variations with the asteroid shape suggests that Eros' surface may contain geologic boundaries which, when imaged at high spatial resolution, will reveal important details of the asteroid's evolution.

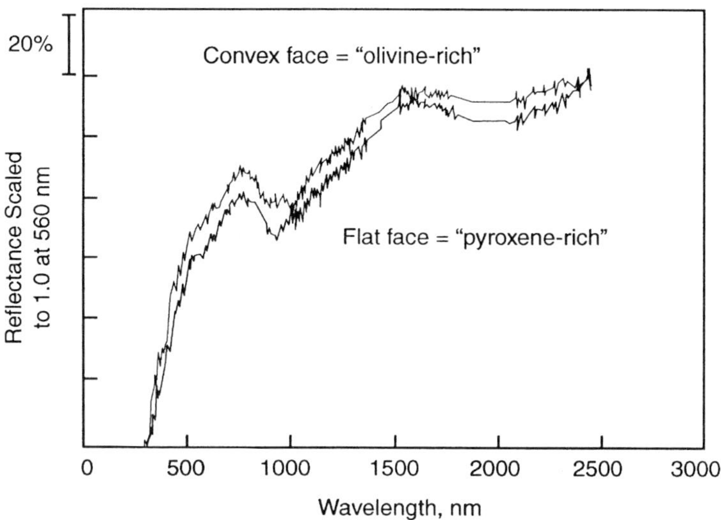

Figure 1. Composite visible-near-infrared spectra of opposite faces of Eros. This presentation highlights differences in strengths and positions of the absorption features due to olivine and pyroxene.

2.4. MEASUREMENT REQUIREMENTS

Four key measurement objectives of MSI at Eros address these science questions outlined above. These are: (a) to determine Eros' bulk properties (shape, rotation, mass), from which density will be derived; (b) to determine the morphology of the surface at small scales and to constrain the history of surface processes; (c) to map the spectral properties and heterogeneity of the surface so that compositional properties and variations determined by NIS and XGRS can be extrapolated to smaller spatial scales and correlated with the three-dimensional configuration of geologic units; and (d) to determine the existence, sizes, spectral and morphologic characteristics, and orbits of any natural satellites. These requirements are met through a combination of instrumentation (Sections 3 and 4), a comprehensive strategy of imager testing and calibration (Section 5), and a carefully planned sequence of imaging from different altitudes during Eros approach and orbit, which affords a factor of $\sim$100 variation in spatial resolution (Section 6).

2.4.1. *Rotation, Shape, and Density*

The rotational pole of Eros and the shape of the asteroid at the southern and equatorial latitudes will be determined by images taken from MSI during the approach phase and from high orbital altitudes early in the rendezvous portion of the mission. After July 1999, the illuminated northern latitudes will be imaged in detail to establish the precise shape of that region. Inflight geometric calibration of the imager assures distortion of <50 μrad, minimizing uncertainty in shape determination.

The density of Eros might lie in the range 2.1 to 6 g cm^{-3}, with the lower and upper ranges corresponding with the minimum density of Ida and the density of the higher density, non-porous meteorite analogs to S asteroids. Mass determination will be provided by the radio science experiment; volume determination is described above. An initial estimate of Eros' density may be expected early in the orbital mission and this will be refined with progressively more complete latitudinal knowledge of the asteroid's shape.

2.4.2. *Surface Morphology*

For purposes of determination of surface morphology, the MSI investigation may be divided into two phases: first, approach imaging from 850 to 3000 km range, and early orbital imaging acquired as altitude is lowered stepwise through 1000, 400, 200, and 100 km; and second, high resolution imaging acquired from orbital radii of 50 and 35 km, acquired during the main part of the orbital mission. During the first phase, Eros will be mapped completely through all spectral filters each time the spatial resolution improves by a factor of two, culminating in a global map at ~12 m spatial resolution. This resolution is several times higher than the best *Galileo* images of Ida, and suffices to characterize the morphology, distribution, and orientation of craters, grooves, and other geologic features. During the second phase, imaging will be through selected filters at spatial resolutions as high as 3–5 m, with highest priority given to coverage of regions which exhibit geologically important features in low-resolution images. This second phase of imaging will address focused questions and will complement earlier imaging by resolving small spatial scale features (e.g., blocks, layering in crater walls, groove interiors). MSI images will reveal the thickness of the regolith, the history of impact cratering, the character and location of fractures in the asteroid's body, and processes which have redistributed the regolith such as downslope movement.

2.4.3. *Spectral Mapping*

The wavelengths of the MSI filters were chosen to satisfy several objectives. First, reconstruction of 'visible color' is supported by 450, 550, and 760 nm filters, which can be used to construct color composites not unlike those that would be perceived by the human eye. Second, the filters support comparison with color properties of other S asteroids. Five of the filters (450, 550, 700, 950, and 1050 nm) approximately match those used in the eight-color asteroid survey of Zellner et al. (1985), one of the most comprehensive collections of asteroidal color data. The 550, 760, 900, and 1000 nm filters approximately match bandpasses of four of the six filters used by the *Galileo* Solid State Imager (SSI) to acquire high spatial resolution color images of S asteroids Gaspra and Ida.

Most importantly, the filters convey mineralogic information. Figure 2 compares the measured response of the MSI filters with the reflectance spectra of olivine and the Fe, Mg-rich variety of pyroxene (orthopyroxene), which both occur on S asteroids. The 900, 950, 1000, and 1050 nm filters cover the region of important

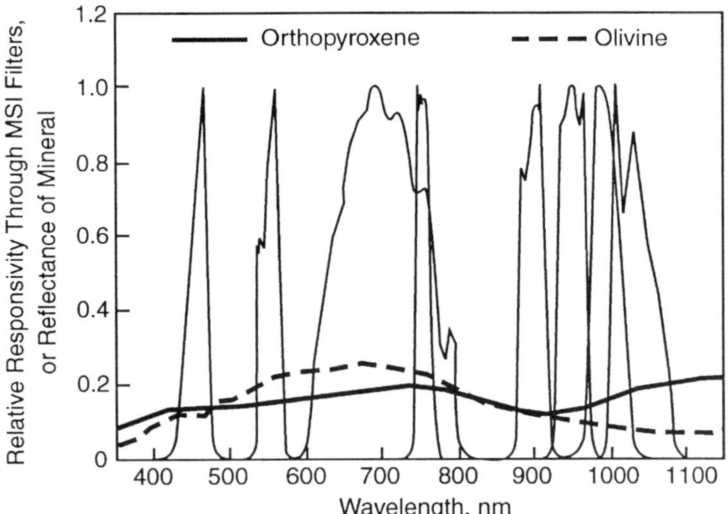

Figure 2. Measured response of MSI filters showing effective bandpasses, compared with laboratory reflectance spectra of the major Fe-containing silicates known to occur on S asteroids.

absorptions in these minerals, and can be used to estimate relative abundances. This capability allows more detailed spectroscopic measurements of asteroid mineralogy from the NIS to be extrapolated to the 70 times higher spatial resolution of MSI. The degree of space weathering of Eros' surface will be evident from spatial variations in the spectral continuum and in depth of the 1000 nm Fe-mineral absorption. Both the spectral continuum and the depth of the 1 μm absorption can be assessed in summary fashion using the three filters at 450, 760, and 950 nm.

2.4.4. *Satellite Search*

The detailed strategy for conducting a search for natural satellites of Eros is explained by Veverka et al. (1997). In summary, several mosaics covering Eros' sphere of influence will be acquired several days before closest approach, prior to orbit insertion, using the broadband filter. Assuming Eros-like photometric properties of a natural satellite, this search provides 3σ confidence in detecting a satellite approximately 15 m in diameter.

3. Instrument Overview

The overall philosophy of the NEAR instrument designs was to develop the most reliable hardware and software possible and satisfy the time and cost constraints of the Discovery Program (Farquhar et al., 1995). Whenever possible, proven flight designs were used as a foundation for the new designs (Hawkins, 1996; Hersman et al., 1996), provided that they met the science needs, even if they were not ideally

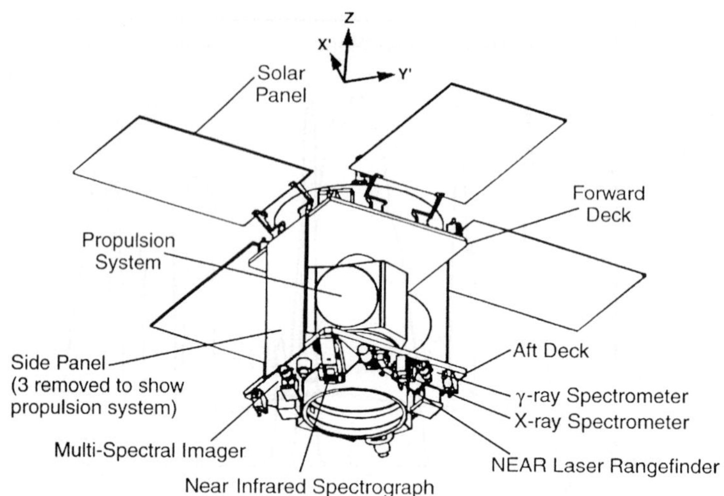

Figure 3. View the Near Earth Asteroid Rendezvous spacecraft. The aft deck contains the instrument palette, and is nearly completed in shadow. The spacecraft Z-axis is directed through the high gain antenna, nominally pointed toward the Earth. The instruments are oriented so that their boresights are nearly coaligned with the spacecraft's X'-axis.

suited for them. A case in point is the choice of a CCD with rectangular pixels; this CCD provides spatial resolution, responsivity, and dynamic range all well suited to MSI's investigation, but the images require post-processing for geometric rectification

Two major subassemblies make up the MSI instrument: a camera and a Data Processing Unit, or DPU. These two assemblies are physically separated by about 100 mm and are located on the aft deck of the NEAR spacecraft, with the camera's optical axis parallel to the X'-axis of the spacecraft (Figure 3). A refractive optical telescope, a filter wheel, and a detector with its associated electronics are all part of the camera. The DPU provides a digital interface to the spacecraft and supplies power and the master timing to the camera. Figure 4 shows a functional block diagram of MSI.

3.1. CAMERA ASSEMBLY

The camera specifications are summarized in Table III. It uses a five-element refractive optical design with a mass of 3.7 kg and power consumption of about 1.5 W. Seven narrow passband optical filters permit spectral imaging, and one broadband filter is used for optical navigation and faint object imaging. A frame transfer silicon Charge-Coupled Device (CCD) converts the optical signal into an electrical one which is then digitized to 12 bits in the Focal Plane Detector (FPD). These 12-bit pixels are transferred in parallel to the DPU. MSI provides a field of view of $2.93 \times 2.25°$ and a spatial resolution of 16.1×9.5 m at a range of 100 km.

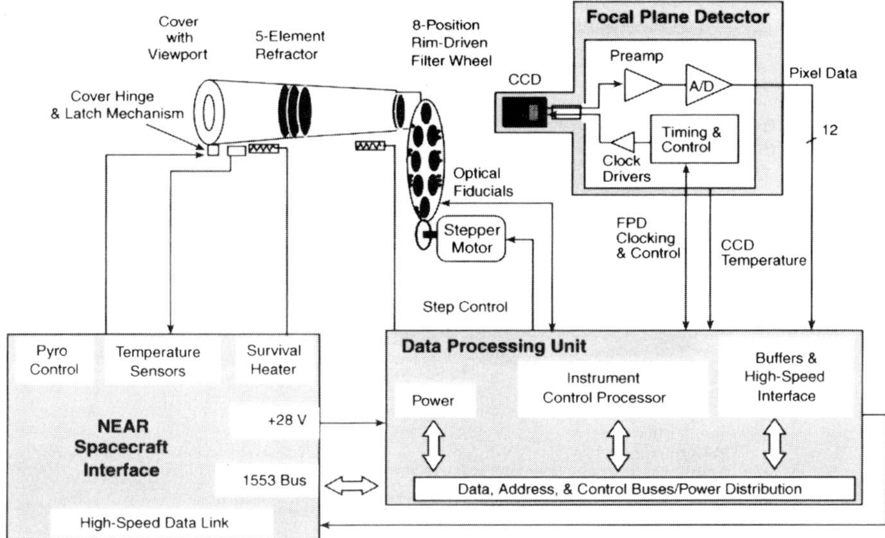

Figure 4. Functional block diagram of MSI.

Table III

Multi-spectral imager camera characteristics

Mass:	Camera	3.7 kg
	DPU	4.0 kg
Power:	Camera	1.43 W
	DPU	5.49 W
FOV		$2.93° \times 2.25°$
Spectral range		400–1100 nm
Refractive optics		5 elements
Focal length		167.35 mm
Clear aperture (no cover)		18.6 cm^2
Clear aperture (with cover)		4.35 cm^2
Frame size		537×244
Frame rate		1 Hz
Frame size (no compression)		1.6 Mbits
Quantization		12 bits
Exposure control		1 ms to 999 ms
Filter wheel		8 position
Broadband ('clear')		700 nm
Green		550 nm
Blue		450 nm
Red		760 nm
IR1		950 nm
IR2		900 nm
IR3		1000 nm
IR4		1050 nm

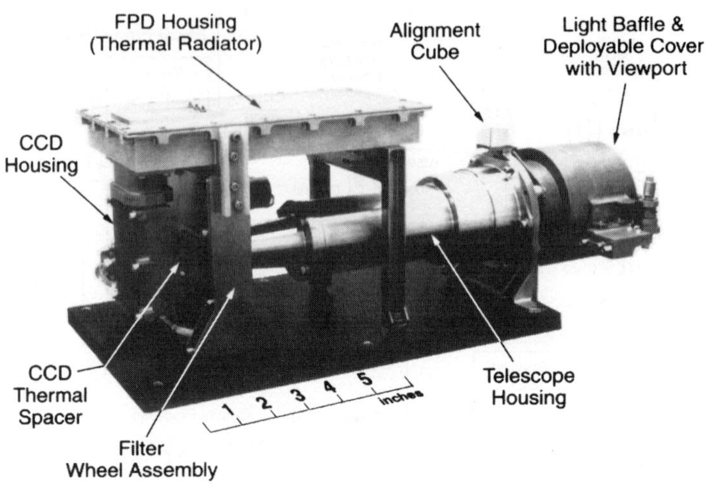

Figure 5. Photograph of MSI's camera assembly.

The camera meets the science requirements of a rendezvous mission, within the constraints of the NASA Discovery Initiative. Because of these programmatic restrictions, coupled with the limited number of opportunities to rendezvous with a near-Earth asteroid of sufficient size to satisfy all of the science objectives (Farquhar et al., 1995), one of the primary drivers to the instrument design was schedule. In order to facilitate all of the necessary optical characterizations required by an imaging instrument, MSI's camera assembly is thermally isolated from the spacecraft, and the various components of the camera were fabricated from materials specifically selected to keep the camera in focus over a wide range of operating temperatures ($-50\,°$C to $+20\,°$C). Compensation heaters maintain the optical telescope near room temperature ($\simeq 15\,°$C), and the lid of the FPD electronics serves as the thermal radiator to passively cool the CCD to its operating temperature (nominally $-30\,°$C). By cooling the CCD, the amount of dark current background signal is reduced (see Section 5.2.1.2). Figure 5, a photograph of the MSI camera, identifies major components.

3.2. DATA PROCESSING UNIT ASSEMBLY

The MSI Data Processing Unit (DPU) provides the necessary digital interface to the camera's FPD, and houses all of MSI's power converters, filter wheel control electronics, and the master clocking required by the camera. A microprocessor within the DPU receives commands from one of the two redundant spacecraft Command Telemetry Processors (CTPs). The CTP distributes commands and collects data from all instruments, and formats the data for recording and downlinking. Communication with the spacecraft takes place via a MIL-STD-1553 bus or alternatively through a high speed serial link (2 Mbits s^{-1}). Eight image buffers within the DPU allow temporary storage of the image data, permitting a variety of data compression algorithms to be applied to the data. Implementation of a multi-tiered compression scheme allows any algorithm to be applied to the data individually, or in combination. Both lossy and lossless compression algorithms are available. Images acquired and digitized by the camera are transferred to the DPU in parallel through a short harness.

A 1 Hz timing signal from the spacecraft synchronizes commands from the DPU to the camera. Integration times may be commanded from 1 to 999 ms effectively varying the sensitivity of the instrument by nearly three orders of magnitude. Each full image is made up of 244×537 pixels and contains a header of all the parameters associated with the image, including the time the image was taken, CCD temperature, exposure time, filter, data compression information, etc. The data are packetized within the DPU, and in addition to the image data, each packet also includes information to locate its data within the overall image. This design minimizes data loss in the event of a single lost or corrupted packet.

4. Instrument Design

4.1. OPTICAL COMPONENTS

Based on the science requirements, the optical design was specified to have a field of view of $2.93 \times 2.25°$ with a resolution limited by the CCD. In an effort to minimize the overall mass of the instrument, a compact design was most desirable. Although a reflective system would accommodate the broad spectral coverage of the imager better, the relatively wide field of view and high resolution of MSI would require a reflecting system that was bulky, with a complex three-mirror design. Instead, a refractive system was selected as the starting point of the optical design. The recent design of the wide-field visible imager from the Ultraviolet and Visible Imaging Spectrographs and Imagers (UVISI) instrument on the Mid-Course Space Experiment satellite (Carbary et al., 1994; Mill et al., 1994) shared a sufficient number of design parameters, so its lens design was the starting point of the MSI optical design.

The degrees of freedom in a refracting design are the radii of curvature of the lens surfaces, the lens thicknesses and relative spacings, and the index of refraction characteristics of the available glasses. As with any optical design process, materials are selected and the variables are adjusted to meet the principal requirements of focal length, field of view, collecting area, and image quality over the entire spectral band.

The five-element lens design implemented in MSI offers degrees of freedom sufficient to correct geometrical aberrations and to temper the chromatic aberration to levels providing adequate image quality. Disadvantages to the design are that it involves many lens surfaces, causing transmission sacrifices and potentially

Table IV

Spectral passbands and optical thicknesses of MSI's filters

Filter number	Spectral coverage (nm)	Optical thickness (mm)
0	700 ± 100	6.4201
1	550 ± 15	7.0952
2	450 ± 25	6.1264
3	760 ± 10	5.9387
4	950 ± 20	4.0582
5	900 ± 20	4.5969
6	1000 ± 25	3.4895
7	1050 ± 40	2.8962

causing some stray light concern. However, the optical system is compact, axially symmetric, centered, and all of the lens surfaces are spherical – all features that contribute to structural, mechanical, manufacturing, alignment, and adjustment benefits.

One of the most important features of the optical design is the location of the spectral filters behind the lens elements. Although this arrangement is usually not optimal for optical designs, by doing so, we satisfied all of the design requirements and minimized the size and mass of the filters and wheel. Filter placement in a converging beam typically results in some additional spherical aberration and somewhat degraded spectral performance of the filters. Residual chromatic aberration and wavelength dependent focus were corrected in MSI by specifying the thickness of each particular filter. The spectral passbands and optical thicknesses for each filter are given in Table IV.

The field of view is mapped onto the flat CCD solid state detector which has a pixel array of 244×537 elements. The physical pixel dimensions are $27 \times 16\ \mu$m (i.e., $161 \times 95\ \mu$rad per pixel) with a photosensitive region of $27 \times 12\ \mu$m organized such that every other pixel column has an $8\ \mu$m inactive region to accommodate the anti-blooming feature of the CCD (see Section 4.3, below).

The $F/3.44$ telescope design, coupled with the small pixel dimensions, resulted in an image depth of focus of $\pm 20\ \mu$m. This extremely small value proved to be one of the most challenging aspects of the mechanical design and dictated the choice of materials of the telescope, filter wheel, and thermal spacer which are depicted in Figure 5.

4.2. OPTICAL DESIGN DETAILS

Computer-aided optimization and exact ray tracing were used to clarify the details of the optical design; the complete lens prescription is given in Table V. For the

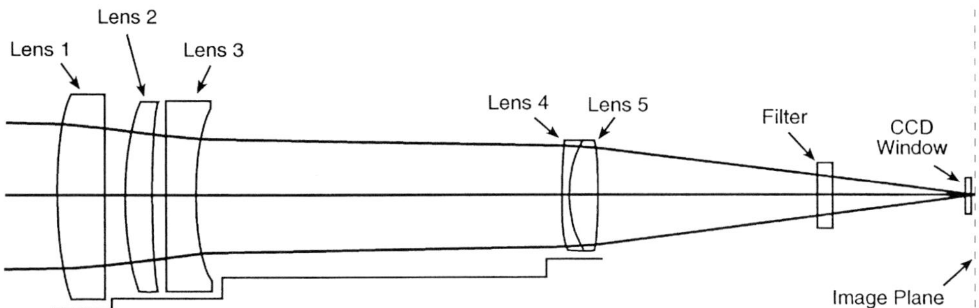

Figure 6. Optical layout of MSI. The five-element refractor uses radiation tolerant lenses, and an 8-position filter wheel permits multispectral imaging. All components, including the detector use anti-reflective coatings.

chosen focal length and image format, an optimization process was performed. The lens surface curvatures, thicknesses, and the relative spacings of the lenses were varied to achieve the best image quality over the broad spectral range. Exact ray tracing of marginal rays at the limits of the field of view determined the clear apertures of the lens elements.

Figure 6 shows the optical layout of MSI's telescope. Each element was ground from radiation hardened glass. Normal optical glasses exposed to radiation may become discolored and cloudy, altering the transmission characteristics of the glasses (Pellicori, et al., 1979; D. Duncan and T. Cotter, private communication). Radiation damage can largely be eliminated in some optical glasses, like the ones used in the lenses on MSI, by stabilizing the glass with cerium doping. Transmission measurements made on irradiated high index glasses show markedly reduced transmission after 10 krads of gamma radiation (D. Duncan and T. Cotter, private communication). A drawback of the cerium stabilized glasses (especially the heavy flint glasses) used in this design is that they do not transmit well in the shorter end of the visible wavelength region, most cut on around 400–450 nm. This sets the short wavelength limit for the imager, and the CCD response also falls off rapidly in this same region.

The NEAR imager incorporates radiation resistant glass for each of the lens elements. The optical filters, however, contain substrate and filter glasses that are not specifically radiation resistant and therefore could suffer long-term transmission degradation. The manufacturer, Schott Glaswerke, reports measurements on ionically and colloidally colored filter glasses (Schott, 1979) similar to those flown on MSI. The measurements indicate a few percent transmission reduction for total dose radiation levels somewhat above the NEAR exposure. Radiation effects are not expected to be severe on the NEAR mission; the total dose level is predicted to be less than 3 krads. (The NEAR mission takes place early in the rise of the Solar Cycle 23.)

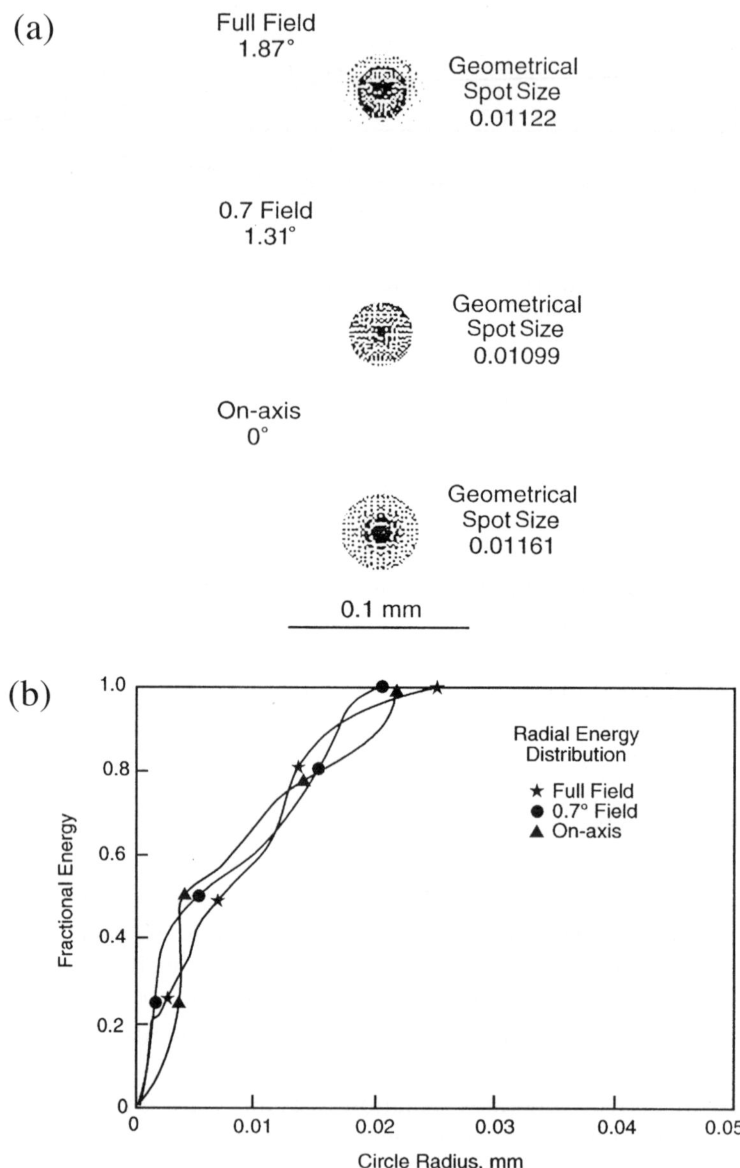

Figure 7. (a) Spot diagram for three field positions computed from exact ray tracing through the broadband filter using wavelengths 623, 700, and 777 nm. The size of the spot corresponds to the r.m.s. value of the ray trace in units of millimeters. (b) Radial energy profile of the spots shown in (a). Note that because of the large passband of this filter, it represents the worst case focus, with approximately 45% of the spot energy falling within a single pixel; this is in good agreement with the measured response (cf. Table IX).

Table V
Multi-Spectral Imager lens prescription

Element No. 1	
Front radius of curvature	127.269 mm convex
Rear radius of curvature	12710 mm concave
Lens thickness	14.473 mm
Glass type (index, V-number)	Schott Bak1 G12 (1.572, 57.8)
Spacing to next element	6.285 mm
Clear aperture radius	32.0 mm
Element No. 2	
Front radius of curvature	95.677 mm convex
Rear radius of curvature	215.514 mm concave
Lens thickness	8.565 mm
Glass type (index, V-number)	Schott Bak1 G12 (1.572, 57.8)
Spacing to next element	3.951 mm
Clear aperture radius	29.3 mm
Element No. 3	
Front radius of curvature	3548.1 mm convex
Rear radius of curvature	67.307 mm concave
Lens thickness	9.496 mm
Glass type (index, V-number)	Schott LF5 G15 (1.583, 40.9)
Spacing to next element	111.432 mm
Clear aperture	26.3 mm
Element No. 4	
Front radius of curvature	109.553 mm convex
Rear radius of curvature	38.019 mm concave
Lens thickness	1.751 mm
Glass type (index, V-number)	Schott LF5 G15 (1.583, 40.9)
Spacing to next element	0.636 mm
Clear aperture radius	17.0 mm
Element No. 5	
Front radius of curvature	38.283 mm convex
Rear radius of curvature	−138.643 mm convex
Lens thickness	8.451 mm
Glass type (index, V-number)	Schott Bk7 G18 (1.520, 63.6)
Spacing to filter	67.0 mm
Clear aperture radius	17.0 mm

Figure 7(a) shows an array of spot diagrams for the broadband filter, calculated by the exact ray tracing for three field locations. The RMS spot sizes shown are given in millimeters. The broadband filter was used because its resolution response is worse than all of the other narrower passband spectral filters. Figure 7(b) plots the radial energy distribution for the three spots of Figure 7(a). This plot shows

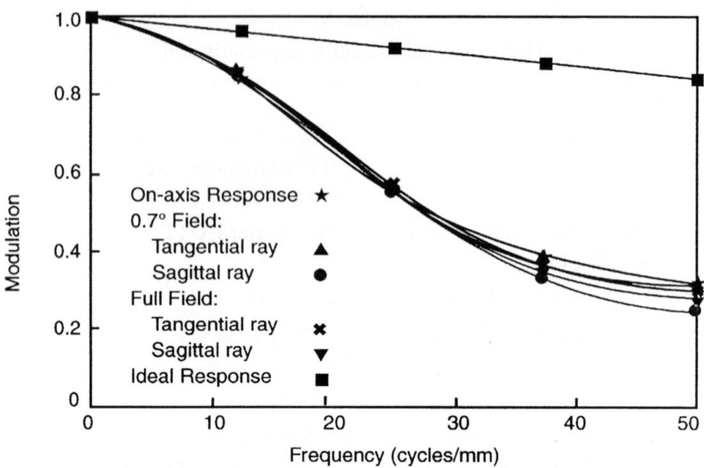

Figure 8. Modulation Transfer Function of the broadband filter.

that for the narrow pixel dimension in the broadest spectral filter, approximately 45% of the energy falls on a single pixel. This theoretical result is in good agreement with the experimentally determined resolution response of MSI, discussed in Section 5.2.1.3.

Figure 8 shows the modulation transfer function (MTF) for the broadband filter. The MTF is a measure of the response of an optical system to a sine wave intensity pattern. As the spatial frequency increases, the contrast – or modulation – will continue to decrease until the spatial frequencies can no longer be resolved. The straight line on the MTF shown in Figure 8 represents a diffraction limited optical system.

4.3. FOCAL PLANE DETECTOR ELECTRONICS

The MSI Focal Plane Detector, or FPD, is the name given to the assembly of the analog and digital electronics required to operate the electro-optic sensor of MSI: a Thomson-CSF CCD chip. The photosensitive elements of this chip consist of an area array of metal-oxide semiconductor (MOS) capacitors which accumulate photoelectrons. The ensuing pixel charge packets are manipulated by altering the potentials on a series of 'gate' electrodes. Four phases, or control voltages, are used to control each charge packet in the Thomson chip. This CCD contains an image region and a storage region, as shown if Figure 9. The storage region is identical to the image region except that an aluminum mask prevents any illumination from falling on the chip. When a 'picture' of charge packets has been accumulated in the image region, it is transferred rapidly to the adjacent storage region also referred to as the memory zone. From this region the charge packets are transferred one line at a time to a serial analog register. The line is then read out of the chip one packet at a time through a single amplifier which converts a charge packet (pixel)

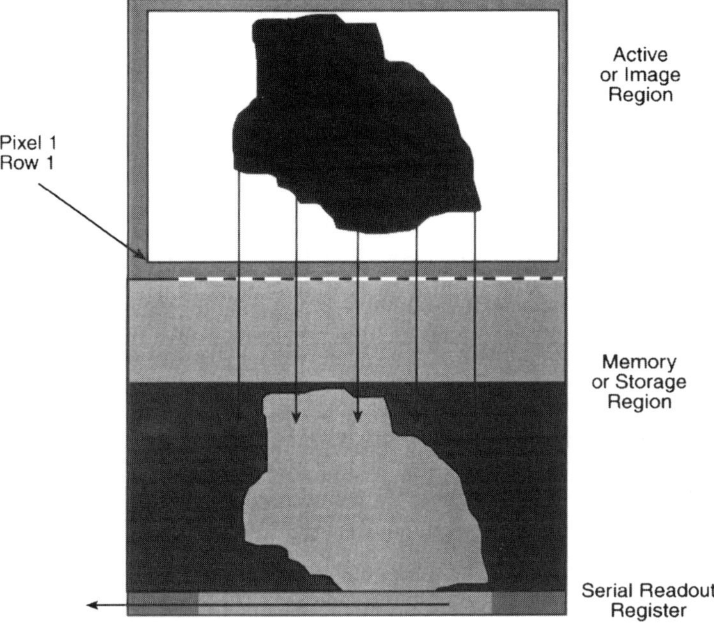

Figure 9. Schematic representation of a frame transfer CCD showing the three regions: active zone, memory zone, and serial readout register. Photons incident on the Active Region create photoelectrons, and these charge packets move as shown by varying bias voltages on the chip.

into an output voltage level. The Thomson chip uses two phase clocks for this line readout. All these functions – image acquisition, analog storage, data serialization, and charge to voltage conversion – are implemented in the CCD.

The particular Thomson CCD used, TH7866A, was originally designed to conform to the NTSC (National Television Standards Committee) format. It has a useful photosensitive array of 244 lines, each of 550 pixels (MSI uses all 244 lines but only the central 537 pixels of the array). In addition to the useful pixels, there are extra pixels and lines around the image area which must be clocked to maintain image registration.

Exposure time can be adjusted by using an extra gate to shunt all image area charge to an extra electrode until it is desired to begin accumulation. This same gate also allows excess charge to be drained from a well with a high signal on it, rather than spilling into neighboring pixels and causing an image defect called 'blooming'.

The voltage signal from the CCD, containing serial pixel data in analog form, is first referenced to a precise potential before the pixel value appears at the output. This is necessary because the CCD charge to voltage output converter has a large and variable offset. The reference potential also acts as a baseline to measure the output signal using a process called correlated double sampling which removes noise resulting from the on-chip reset of the charge to voltage amplifier. The

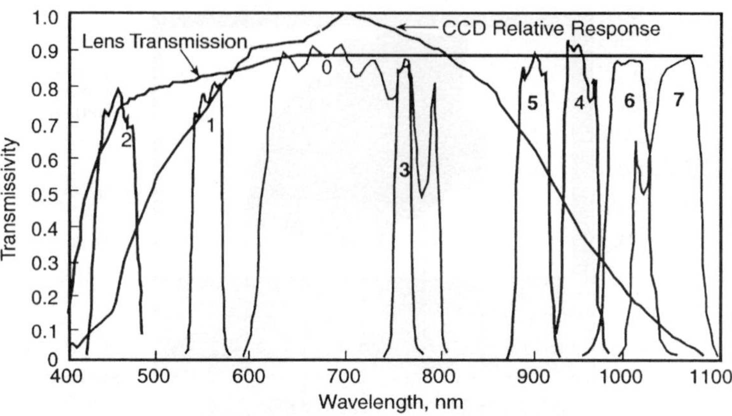

Figure 10. Manufacturer's measured data of transmission efficiency of MSI optics. The response of the CCD is superposed on top of this showing the peak sensitivity to be around 700 nm.

difference between the baseline value and the pixel output voltage is then sampled and converted to a 12-bit digital word. The conversion process has a speed of 156,250 samples per second, and this speed sets the minimum time needed to read out and digitize a complete image ($\sim$1 s). The analog storage portion of the CCD retains the pixel data until the converter can digitize it.

The CCD requires four phase image zone clocks, four phase storage (memory) zone clocks, two phase line clocks, and clocks to control the shutter (electronic exposure), bloom control and output amplifier functions. All of these clocks are generated as standard logic levels with precise timing control since relative phases are very important. These logic levels are then amplified to the voltage levels required by the chip and conditioned so that they can drive the capacitive clock lines on the CCD.

The complete system has a nominal frame rate of one per second, producing 244 lines of 550 pixels each, with an exposure controllable from 1 ms to nearly 1 s. There are no other user operable controls on the CCD and camera electronics. The gain of the system is such that $\sim$60 electrons corresponds to one least significant bit of the analog to digital converter. The full 12 bits correspond to about 240,000 electrons. The conversion of photons to electrons (quantum efficiency) peaks at 25%, but is strongly wavelength dependent. The response of the CCD, as a function of wavelength, is shown in Figure 10; also shown are the manufacturers' transmission data for the MSI lenses and spectral filters. The anti-blooming function can degrade the linearity of intensity response and must be set to balance good blooming protection and good linearity. It is adjusted during construction so that the response is reasonably linear up to full output. This level is fixed by the components selected in the design and cannot be altered.

4.3.1. *CCD Clocking*

Most of the clock drivers require voltage low levels close to 0 and high levels close to 10 V. The loads are capacitive and there are significant capacitances between different clock phases. The CCD has gate protection diodes which are reverse biased in normal operation. When these diodes conduct from overshoot or capacitive cross-coupling, they disrupt normal CCD operation. To avoid this diode conduction, particularly when other phases switch states, requires a careful and more complicated design of the phase drivers than would otherwise be the case.

The design uses Micrel drivers designed to switch the capacitive gates of power MOS Field-Effect Transistors (MOSFETs); these devices have excellent characteristics for fast clocking. One of the phases in the image zones has an intermediate potential during image accumulation to reduce the size of the barrier zone between pixels to a minimum. This intermediate level and small variations in the resting low clock level are provided by discrete components.

The clock which controls the blooming and shutter action has a high level when the 'shutter' is closed, an intermediate level during image accumulation, and a low level during readout. This tri-state clock is generated from the Micrel drivers and discrete components. A potential higher than any of the normal CCD biases must be provided to drain away electrons when the image zone is desensitized, and this potential is provided by a small charge pump circuit.

To control the rise and fall times of the gate voltages, series resistors are included in the drive lines. This is particularly important for the two phase line clocks where there is no phase overlap. It is also quite critical that the clock lines be kept short and uniform. This complicates the mechanical and thermal layout of the electronics where a considerable temperature difference between the CCD and the rest of the electronics may occur. Since the CCD must be aligned with respect to the optical system, it is physically separated from the rest of the electronics. The CCD is also thermally isolated in order to maintain the CCD's operating temperature of $-30\,^{\circ}$C. As described below, this conflict between short well-defined lines and mechanical and thermal isolation was resolved by using a rigid-flex board design.

4.3.2. *Analog Electronics*

The output from the CCD has a maximum amplitude of approximately 0.5 V peak to peak with an offset of about 9 V. This output is AC coupled to an amplifier with a voltage gain of 10. Thus, all subsequent drift occurs after voltage amplification. The effects of AC coupling and drift are removed in the DC restoration switch immediately following the gain stage. The DC restoration stores a value of the offset voltage on a capacitor which is in the signal path, and this value is re-sampled with a CMOS (complementary MOS) switch between each pixel. The DC restored signal is buffered and then fed to a hold capacitor via a CMOS switch.

Another buffer precedes the analog to digital converter which is an Analog Devices successive approximation device using BiCMOS technology. All analog offsets and temperature effects after the DC restoration capacitor can produce errors in the digital output. These effects are minimized by providing gain before this capacitor. The initial amplification stage and the buffers use National Semiconductor and Analog Devices FET input operational amplifiers. The initial gain stage uses a decompensated version to maintain system bandwidth. These amplifiers combine reasonable speed with low bias current and offset voltages.

Since the converter will not digitize negative voltages, provision is made in the electronics to provide a positive offset voltage in the analog chain. The final offset is made large enough to ensure that even with component aging and drift, a negative voltage should never be presented to the converter. No provision is made to shift the analog signal negative since the only consequence of failing to do this is a slight reduction in maximum signal. The magnitude of amplifier offset and temperature coefficients are easy to calculate, but the effects of switch capacitance and aging are more difficult. In the original design on which the NEAR electronics were based, it was possible to hold the clock state stable while the converter digitized the held pixel voltage. In the NEAR camera this was not possible and there was some feedthrough to the hold capacitor. As an example of the residual effect of this feedthrough, a signal which should be correctly digitized to 2047 DN (data number), instead gets recorded as 2048 DN. (The digital output from the analog to digital converter is referred to as a data number or DN.) A proportion of the digitized codes of 2045 and 2046 may also get incorrectly reported as 2048. This causes a slight reduction of data accuracy for the few values affected. There was not time in the compressed NEAR schedule to redesign the circuit, but the performance was improved by slightly modifying the timing of the clocking sequences.

The noise level of the system is determined by CCD read noise, analog electronics noise, and analog to digital converter noise. These contributions were not measured individually, but the total noise is about 2 DN or about 120 electrons for small signals. For large signals or high temperatures, shot noise in the signal or dark current, respectively, can increase this level. A signal equivalent to a full scale digital output has a signal dependent shot noise of 8 DN, and this term varies as the square-root of the signal level. The dark current shot noise contribution is much less than 1 DN at all operating temperatures. The accumulation of dark current in the image and storage zones is not compensated by the electronics. At high temperatures it is difficult to remove this contribution from the picture, especially because it is not spatially uniform. The resulting temperature-dependent fixed-pattern noise limits operation to low temperatures as much as rising shot noise and restricted dynamic range do. The dynamic range is reduced by the dark current subtracting from the full well size. To enable dark current correction at moderately high temperatures, a temperature sensor is mounted next to the CCD.

Another consequence of the particular CCD chosen for this design is different response from alternate columns. After removal of the fixed-pattern temperature dependent effects in an image, a fairly uniform variation exists between odd and even columns. This fixed frequency variation is easily correctable and is due to small errors in mask alignment during CCD manufacture. These alignment errors

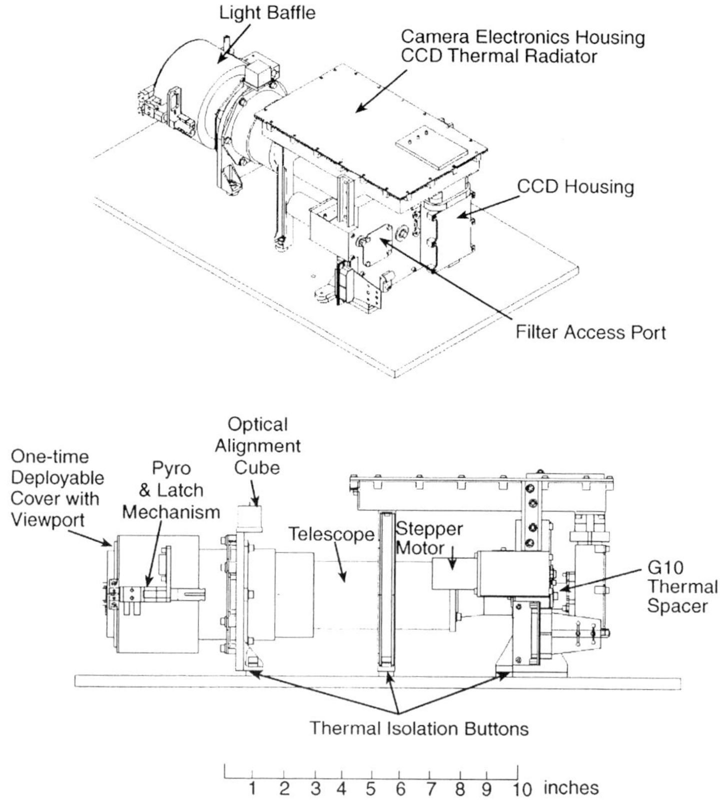

Figure 11. Mechanical outline drawing of the MSI camera assembly.

cause an odd/even difference because the anti-blooming drain occurs only between alternate pixels.

4.4. MECHANICAL DETAILS

4.4.1. *Telescope*

Figure 11 shows a mechanical outline drawing of the camera. The telescope is thermally isolated from the deck and the FPD. Operational heaters are thermostatically controlled to maintain the temperature of the telescope to be nominally at about +15 °C. In order to minimize the effect of thermal gradients and changes in the linear dimensions of the housing over temperature, the telescope was machined from titanium; this material provides the low coefficients of thermal expansion and thermal conductivity necessary to facilitate testing of the instrument at room temperature and ambient pressure while providing optimal performance at its operating temperatures.

The interior surface of the titanium telescope is painted black using Aeroglaze Z306. The lenses are held fast with blackened, threaded retaining rings bonded in place. Assembly of the telescope took place in a class 100 clean room, and the optical components were cleaned until a Level 1000 or better surface cleanliness was achieved. (A Level 1000 surface is qualitatively described as one in which a few particles are visible with the unaided eye per square foot (929 cm^2).) This goal of keeping surface particulate contamination to a minimum was set in order to minimize scattered light. Direct sunlight or spacecraft glint into the optical system aperture is not a major concern for MSI because of the imager's location on the spacecraft. The sunward pointing solar panels and instrument deck provide most of the glint protection (cf. Figure 3). In addition, to minimize residual stray light, MSI includes a light baffle made of magnesium with an outer finish of Iridite 15 and inner surfaces painted black using Aeroglaze Z306. This assembly is thermally isolated from the rest of the telescope assembly and covered with thermal blankets.

A protective cover with a reduced aperture viewport provided the capability to operate the camera during ground-based tests, outside of the clean room environment, as well as during the first 73 days of the mission. Through this viewport, the full $2.93 \times 2.25°$ field of view was available, however the collecting area was reduced from 18.6 cm^2 to 4.35 cm^2. On 1 May 1996, the cover was permanently opened inflight by firing a wire-cutter type pyrotechnic device. Once the beryllium-copper wire latching the cover closed was cut, two stiff springs forced the cover open. After swinging open approximately 135°, the cover struck a mechanical stop mounted to the instrument deck.

4.4.2. *Filter Wheel Assembly*

The filter wheel assembly houses a rim driven wheel containing eight spectral filters, and four optical switches which provide information about the wheel position. The titanium filter wheel housing is the primary support of the camera, and mounts between the telescope and the CCD housing. A 1.8° hybrid stepper motor controls the motion of the wheel through a Delrin pinion gear which drives the filter wheel with a gear ratio of 4:1. The wheel may be turned to an adjacent filter position within 1 frame period (1 s).

The wheel itself is a high quality commercial aluminum gear machined to hold the eight color filters. An alignment ring rigidly mounts to the 100 mm diameter wheel, and a binary code correlates each filter position to the ring. Three of the four optical switches mounted in the filter wheel housing determine the approximate location of the filter position (slots in the alignment ring are three motor steps wide). The fourth optical switch serves as a fiducial – a single step wide hole for each filter.

The telescope and filter wheel housing are thermally isolated from the spacecraft deck by Ultem 1000 thermal spacers under each mounting foot. An epoxy glass thermal spacer supports the CCD housing and maintains the large thermal gradient ($\Delta T \approx 50$ °C) between the CCD and filter wheel housings. Because of the very small depth of focus, the material selected for this spacer was required to have a minimal coefficient of thermal expansion but have sufficient strength to support the CCD housing. The epoxy glass thermal spacer satisfied these requirements and thus permitted assembly and testing of the imager at room temperature, since the instrument remained in focus with the CCD warm ($+20$ °C) as well as at its operating temperature (-30 °C).

4.4.3. *Focal Plane Detector Assembly*

The FPD may be divided into two parts: an aluminum camera electronics housing, and a magnesium CCD housing. Because of thermal and mechanical stress, the CCD housing had to be as small and light as possible. The two housings are coupled very loosely mechanically through an electromagnetic interference (EMI) gasket. Four high precision standoffs fix the location of the CCD heatsink in the CCD housing. An assembly fixture was used to align the CCD on its heatsink prior to the bonding procedure. A cotherm gasket between the heatsink and the CCD electronics board served as the mechanical strain relief. The silicon chip is protected by a fused silica cover glass, which is coated with an anti-reflective coating which minimizes the scattering of light.

In order to simplify the internal harness made up of the large number of wires required between the camera electronics board and the CCD board, a rigid-flex design was selected over a more traditional harness. The camera electronics board uses a conventional multilayer printed circuit design. Figure 12 shows a top view of the camera electronics board.

MSI's detector is passively cooled with the lid of the FPD housing serving as the thermal radiator; a copper thermal strap couples the CCD heatsink to the radiator. Silver-teflon tape on the radiator increases the emissivity of the surface, enhancing its cooling capability. Figure 13 shows the CCD mounted on its heatsink and the rigid-flex board design.

4.5. DATA PROCESSING UNIT

The DPU performs a number of tasks in the MSI instrument. It accepts all imager commands from the spacecraft and executes them by controlling the camera and processing its raw pixel data. The images may be compressed using a variety of algorithms. After tagging each image with a unique header, the data are packetized and transferred to the Command and Telemetry Processor (CTP). In addition to these functions, the DPU provides power to the camera and controls the filter wheel.

4.5.1. *DPU Interfaces*

The interfaces necessary to perform the required DPU functions are shown in Figure 14 and are grouped into three interface categories: imager, spacecraft, and test. The imager interfaces include all connections between the DPU and the camera assembly. All control and timing signals needed to operate the camera pass through the imager control interface. A variable width pulse control signal, is fed into the FPD electronics and establishes the exposure or integration time of the image.

Figure 12. Camera electronics board as mounted in the Focal Plane Detector of MSI. The connector located on the right side of the board is for the rigid-flex CCD board.

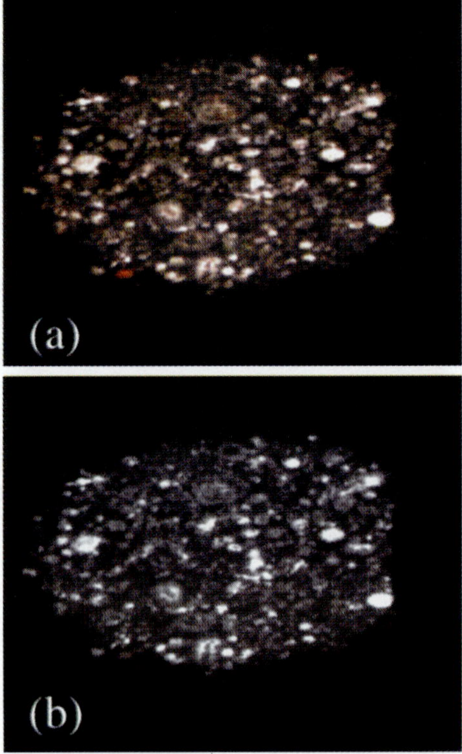

Figure 36. (a) False color image of sample of the Allende carbonaceous chondrite meteorite, constructed from individual images taken through the 450-nm, 760-nm, and 1000-nm filters. Pyroxene crystals appear as red in this representation, olivine as green, and Ca-Al silicate grains as white. (b) Monochrome image of sample of the Allende carbonaceous chondrite meteorite.

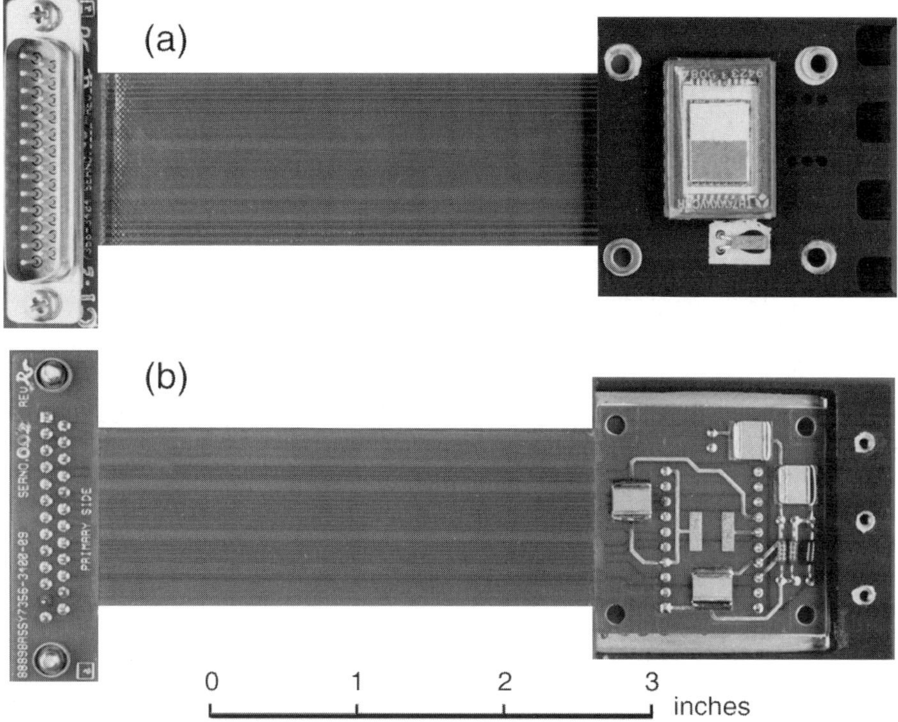

Figure 13. (a) The Thomson-CSF TH7866A CCD mounted to its heat sink. Next to the CCD is an AD590 temperature sensor. (b) Reverse view of showing structure of the rigid-flex board design.

The data returned from the imager are gathered via the imager data interface, which consists of 12 bit parallel data and control signals. A single analog signal is provided from the imager to determine the CCD detector temperature. The filter wheel interface controls the motion of the two-phase stepper motor. As described above, position information is determined from the fiducial sensors in the filter wheel assembly. The power interface provides +15 and −12 V to the FPD.

The spacecraft interfaces contain the necessary signals for communicating with the spacecraft either through the transformer-coupled 1553 bus or via the high speed serial link. Commands and telemetry are sent over the 1553 bus to and from the DPU, which operates as a remote terminal. The high speed link is unique to MSI and is used for rapid movement of data – up to one full uncompressed image in 1 s – to one of the two solid state recorders.

Test interfaces include a processor and imager test ports. The processor test port was used during ground testing for downloading and troubleshooting software. The imager test port provided the capability for real-time display of images during preflight testing. Neither of these interfaces is used in flight.

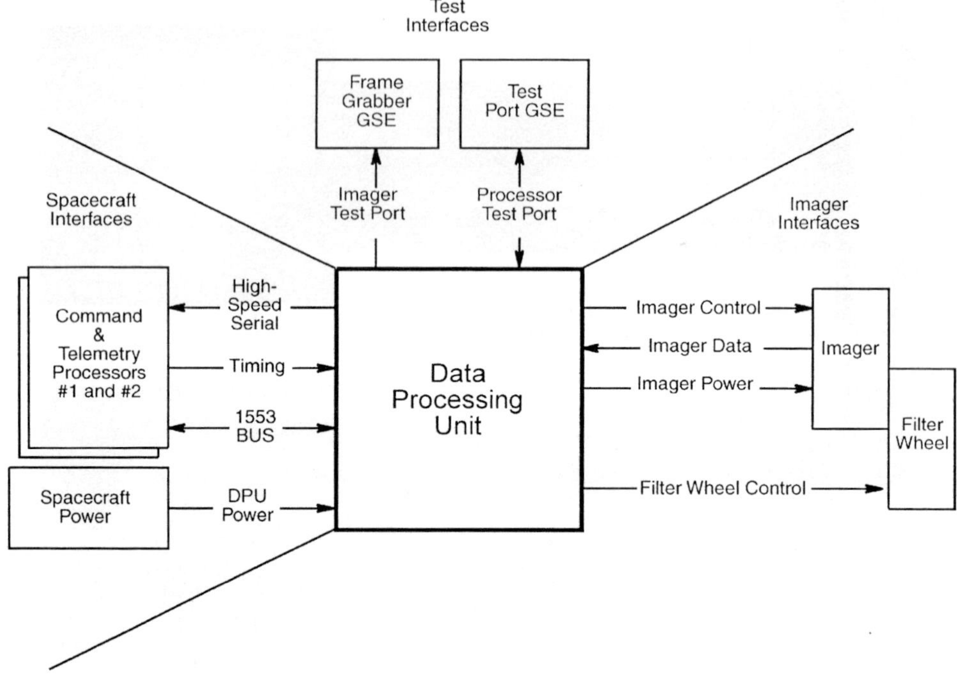

Figure 14. Data Processing Unit interface diagram.

4.5.2. *DPU Design*

The DPU was designed as a modular system to support many instruments on the NEAR spacecraft. By separating functions onto different printed circuit boards, the DPU can be assembled to meet a variety of system requirements. As shown in Figure 15, the MSI DPU was developed by adding imager specific boards to the common DPU core. The functions of each board are described below.

The main processor is a Harris RTX2010RH running at 6 MHz. As shown in Figure 16, the processor board is equipped with 4K words of programmable read-only memory (PROM) for booting the central processing unit. Operational software is stored in 32K words of electrically erasable PROM (EEPROM). After booting the processor, the operational code is copied to the 96K words of random access memory for execution.

Other important onboard features include a watchdog reset, four edge or level sensitive interrupts, a low-voltage reset, and an EEPROM write-protect circuit. The processor board is also equipped with 48 internal general purpose I/O bits and an external processor test port for downloading and troubleshooting software. The processor board has access to additional resources through the expansion bus.

The 1553 board operates as a remote terminal on the spacecraft's 1553 bus. The core component for the bus communication processing is the United Technologies Microelectronics Center SuMMIT chip. The SuMMIT chip communicates over

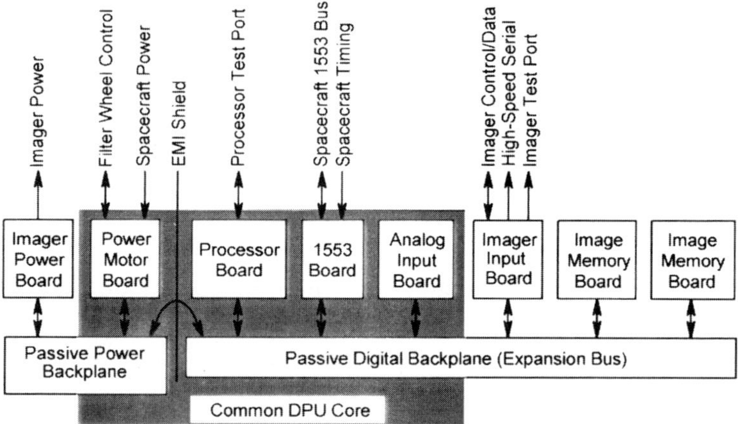

Figure 15. Data Processing Unit block diagram showing the specific boards common to all DPUs on NEAR (shaded), as well as the MSI specific electronics.

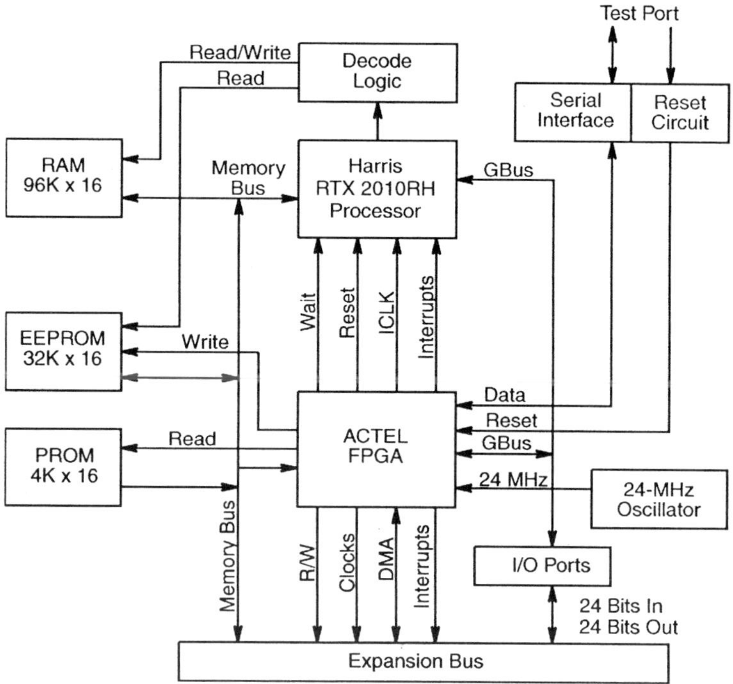

Figure 16. Detailed functional block diagram of the processor board within the DPU.

the spacecraft 1553 bus under the control of the processor. In addition to the 1553 bus capabilities, this board has optically isolated inputs for timing pulses from the redundant CTPs. This signal is used to synchronize many of the imager functions, including image acquisition.

The processor board performs the arbitration necessary to access the 32K words of shared memories and shared internal registers. Data are transferred to and from the processor board through these words in memory and registers in the SuMMIT chip, which are mapped on the expansion bus.

The analog input board contains analog and digital electronics for gathering the imager housekeeping telemetry data. This board digitizes the analog signal measuring the CCD temperature and temperatures measured from both power boards. It also measures the current of the imager's DC to DC converter. The analog to digital converter and input selection on this board are controlled by the processor through the expansion bus.

The imager I/O board provides the control and data interface to the imager, as well as the high speed serial interface to the spacecraft. The data to and from these interfaces are buffered using first-in-first-out (FIFO) memories. Images are retrieved from the input FIFO by the processor and stored temporarily in memory on the image memory boards. After processing, the image data are transferred from the image memory boards to the output FIFO for transmission to the spacecraft via the high speed serial interface. Real-time image display was possible during testing through the use of the imager test port.

The image memory board provides temporary internal storage of the camera's pixel data. These image buffers are used while any image compression is occurring, or until the CTP is ready to accept the data. Each of the two boards contains 1 Mbyte for a total storage capacity of 4 uncompressed images per board. Image buffer memory modules used within the DPU contain four Hitachi chips of $32K \times 8$ bit words making each module $128K \times 8$ bits. In order to make the most efficient use of memory and physical space, the output of the Thomson TH7866A CCD array, originally 244×550 pixels, was reduced to 244×537 pixels to fit within 128K words. The image memory is mapped onto the expansion bus in two 128K word pages. The remaining six pages are swapped by the RTX processor board through the general purpose I/O bits.

The imager power board provides power to the instrument at $+15$ and -12 V. This board contains an in-rush current limit circuit, an Interpoint hybrid EMI filter, Interpoint hybrid DC to DC converters, and a low-voltage inhibit circuit. A voltage regulator is used to maintain the -12 V supply. Current monitors compatible with the analog input board are implemented using a sense resistor in the input power return, and Analog Devices AD590s are used to monitor the board temperatures.

The power motor board performs two functions for the MSI DPU: It conditions the power from the spacecraft to provide internal DPU voltages at $+5$, $+15$, and -15 V, and also provides the drive to the filter wheel two-phase stepper motor. To minimize power dissipation on the power motor board, the current to each motor phase is regulated using pulse width modulation. As part of the stepper motor control circuitry, interfaces for fiducial sensors are used to monitor the motor position. This board is accessed by the processor through the internal general purpose I/O bits.

4.6. SOFTWARE OVERVIEW

The NEAR instruments' software design maximized the features shared among the DPUs of the other instruments. As described above, much of the DPU hardware was common and only certain specialized boards were MSI specific. This same philosophy was implemented in the instrument software. The common software includes all of the code necessary to communicate with the CTP and over the 1553 bus and to implement a common core of commands and housekeeping functions.

The MSI instrument specific code handles execution of image acquisitions, filter wheel control, data compression, housekeeping data, and packaging the images into a variety of formats. The real-time software is event driven with the highest priority given to the image acquisition process. Second in the priority ladder is delivering the image data to the CTP, followed by servicing of the 1553 telemetry and command systems. Filter wheel processing, automatic exposure control (AEC), and data compression are all lower in priority, with the lowest priority given to memory dumps.

All of the code for the RTX2010RH processor was written in the FORTH language. The primary requirements of the MSI specific code were image collection in 'acquisition sequences' at the exact time commanded to do so and with proper instrument setup. Once an image is acquired, it must be formatted before being sent to the spacecraft CTP.

The NEAR telemetry systems use the standards of the Consultative Committee for Space Data Systems (CCSDS) for data transmission. The most recent release of the CCSDS telemetry protocols is described in document: CCSDS 102.0-B-4, (1995). Using this CCSDS model, NEAR defines the following virtual channels (VC) for sharing the physical downlink telemetry resources:

Virtual Channel 0: packetized data written to the solid state recorders
Virtual Channel 1: packetized playback data from the solid state recorders
Virtual Channel 2: unpacketized transfer frames of video images to the recorders
Virtual Channel 3: packetized data transmitted in real time

Each virtual channel may be supplied by multiple data sources, or 'application processes' in CCSDS parlance. For example, each unique data type from each NEAR instrument sensor is assigned a separate application process identifier (ID) for VC0 and VC3 data. By using these IDs as indicated by the standard protocols, many separate data streams of varying rates are easily multiplexed into a single channel and separated out just as easily by the receiving software in the ground data systems. Similar CCSDS protocols are used for routing commands to individual subsystems from the ground.

Because of the large amount of data associated with every image ($\sim$1.6 million bits), an instrument specific data path, the 2 Mbits s^{-1} high speed serial link, connects MSI with the CTP. The nominal method for acquiring images from MSI is to send them as imager specific VC2 transfer frames over the high speed link enabling an entire image to be written to the recorder in only 1 s. An alternative method

Table VI
Image sequence data structure

Image sequence ID	1–30	Sequence identification
Images in sequence	1–8	Number of images to acquire
Repeating sequence	Yes/No	
Time between images (s)	1–255	
Lossy table	0: None	
	1: Identity	
	2: Linear-Logarithmic	
	3: CME	
	4: Square root	
	5: Shot noise limited	
	6: Linear	
	7: Square	
Differential PCM	On/Off	
Lossless algorithmn	None/FAST/RICE	
Pixels per block	16	
Exposure mode	MAN/AUTO	
Image No. 1	Exposure and filter position	
Image No. 2	Exposure and filter position	
Image No. 3	Exposure and filter position	
Image No. 4	Exposure and filter position	
Image No. 5	Exposure and filter position	
Image No. 6	Exposure and filter position	
Image No. 7	Exposure and filter position	
Image No. 8	Exposure and filter position	

of transferring MSI data to the CTP makes use of either VC0 or VC3 packets. However, this is comparatively slow and is not the normal mode of operation.

Commanding of MSI is accomplished through a data structure known as an image sequence. A sequence provides a simple way to set up a particular observation with as many as eight images. Table VI summarizes the image sequence data structure, and the possible parameters. The software determines the amount of time needed to execute a multispectral image sequence, taking account of the requested time between images, exposure times, and the time for the filter wheel to lock into the next position. If the imager were commanded to take pictures with an insufficient amount of time between images, the software would automatically delay the execution of the image until the filter wheel has stopped its motion.

MSI's software has been optimized to facilitate taking images as quickly as possible. The maximum frame rate is one per second and is achievable only if a single filter is used or if short exposures are combined with short wheel rotations. With this realization, filters normally used together (e.g., for 'true color') were

Table VII

Filter correction factors used in automatic exposure algorithm

Position (0–7)	Passband center wavelength (nm)	DN of solar spectrum in passband	% of light reflected by Eros	Relative sensitivity = DN × % reflected
0	700	20770	0.207	4299
1	550	2460	0.182	448
2	450	717	0.153	110
3	760	2077	0.215	447
4	950	1053	0.197	207
5	900	1680	0.197	331
6	1000	669	0.199	133
7	1050	527	0.203	107

placed adjacent to each other on the filter wheel. Due to the finite time required for the filter wheel to move to another position, three-color requires at least 5 s.

The camera electronics reads an image from the CCD through the DPU's 512 word imager input FIFO every second. Readout of an entire image takes approximately 904 ms. The FIFO input and its data-ready signal are reset by hardware at the 1 Hz timing pulse from the spacecraft. The first pixel of an image is the first data word written into the FIFO after the 1 Hz pulse, and the software determines the end of the image by counting 131,028 pixels.

The image data may be mirrored to the test port, allowing the test ground-support-equipment (GSE) to analyze the data as it is being read in from the camera, if a special control bit is set by issuing a command to the DPU. After completing an image readout, the DPU software disables the test port to ensure that the test GSE only receives one image at a time.

Figure 17 is a flow diagram of the MSI software structure. Within the code, several types of 'ticket' data structures are defined. Like a job shop work request, each ticket contains data that defines the work to be done, and describes the state of the work from the request through completion. There are two ticket types: sequence tickets and image tickets. Each sequence ticket describes an entire sequence of 1–8 images to acquire and process. Under control of the sequence ticket, image tickets are produced which direct the processing of each individual image in the sequence – from acquisition through data compression and telemetry transmission. At the beginning of an image acquisition sequence, the exposure is set and the filter wheel positioned nominally 10 s prior to the acquisition time for the first image. The code then waits until the requested time for the image to be taken. When acquisition of each image is complete, the software sets up and then waits for the next image acquisition time. In the event that the commanded acquisition time is earlier than

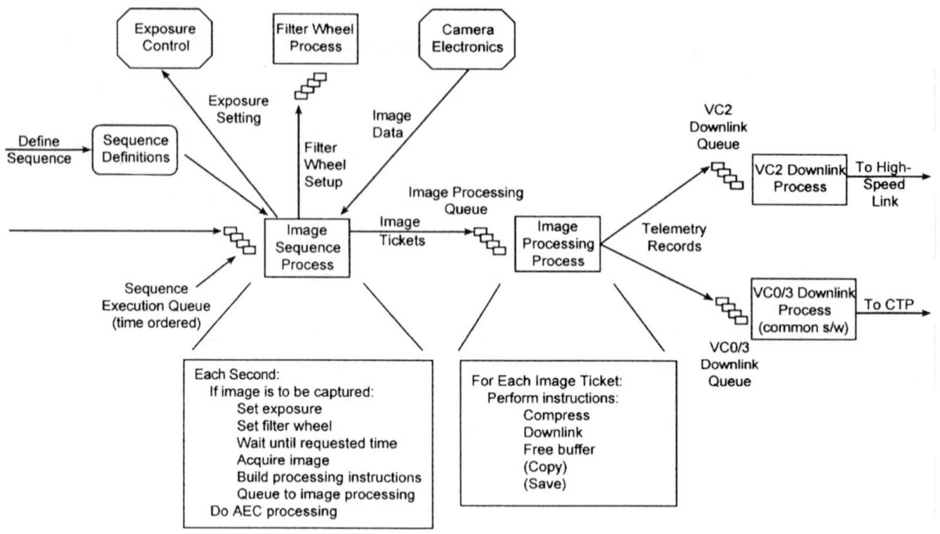

Figure 17. MSI software structure.

the current time, image acquisition will occur as quickly as possible in an effort to 'catch-up'.

The MSI specific code accommodates 20 command types and has buffer space for up to 30 time-tagged image execution sequences to be waiting for execution simultaneously. Both CTP and DPU macros may be used, allowing many more commands and sequences to be pre-loaded to the instrument.

MSI downlinks image science records only in response to specific commands, command sequences, or macros directing it to do so. When it is not collecting images or preparing for downlink, the DPU measures and reports the maximum brightness of each image and performs housekeeping data collection and reporting.

4.6.1. *Compression Routines and Look-Up Tables*

Figure 18 shows the structure of the MSI data compression processing scheme, which consists of up to 3 parts, each of which can be bypassed. The first part allows the selection of lossy compression using one of seven 12 to 8 bit look-up tables (LUTs). The term lossy implies that after the images are compressed, they may not be fully restored without the loss of data accuracy. Each of the seven tables is a vector of 4096 elements, one element per least significant bit of the 12-bit analog to digital converter. The value in each element of the LUT stores an 8-bit number, which then replaces the 12-bit number in the original data.

The seven tables available are Identity, Linear-Logarithmic, CME, Square-Root, Shot-Noise-Limited, Linear, and Square. All tables incorporate a small offset so that the minimum expected signal corresponds to a very low 8-bit value. The Identity table preserves the lower pixel values by simply dropping the upper bits after a specified offset. The Linear-Logarithmic table combines the features of the

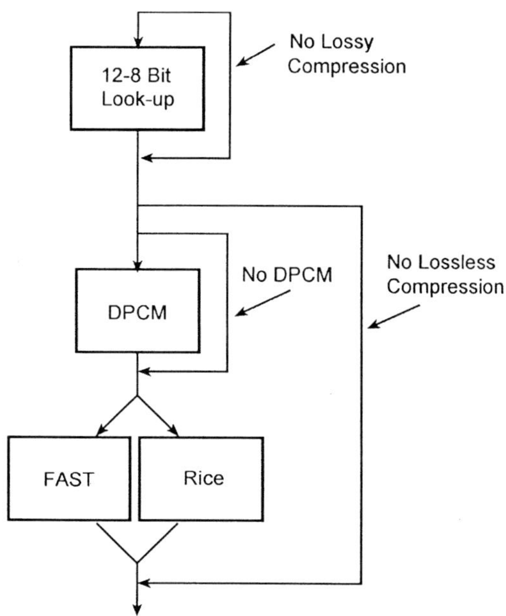

Figure 18. Flow for MSI data compression processing.

Identity table for the lower bits, but compresses the upper bits using a logarithmic scale. CME or the Constrained Maximum Error table uses a technique to minimize the maximum percentage error. The Shot-Noise-Limited and Square-Root tables are very similar and both are designed to utilize the known variation of shot noise with signal level. As the name implies, the Linear table linearly converts from the 12 to 8 bits. Finally, the Square table is designed to enhance small contrast between bright pixel values.

Uncertainty in measured brightness is approximately proportional to the square root of the total detected photons. The CME, Shot-Noise Limited, and Square Root tables effectively spread the loss of information by compression evenly over MSI's 12-bit range. Assuming a signal-to-noise of ~500 at the maximum signal level, compression to 8 bits effectively decreases data quality by a factor of ~2. Identity and Linear-Logarithmic compression tables preferentially preserve information content of the data in the low-brightness part of the dynamic range (e.g., shaded slopes on the asteroid), whereas the Square table preferentially preserves information in the upper part of the dynamic range (e.g., bright fresh craters).

The second part of the compression process permits Differential Pixel Code Modulation or DPCM encoding which exploits pixel to pixel correlations. The final level of compression uses either of two lossless algorithms, FAST or RICE. The FAST algorithm looks at a block of data, selects the maximum value in that block, and then represents every pixel in that block with only that number of bits. Thus, for dark scenes a much larger compression factor is possible than for brighter scenes, although if DPCM is also implemented, then bright uniform scenes may also have a large compression factor.

A slightly more complicated routine, the RICE algorithm, is still computationally simple and requires only a single pass over the data. Like FAST, it is also a block-oriented algorithm, although it may use a different number of bits to represent each value in a block. RICE has limited memory requirements and works well with the NEAR telemetry system which is based on data packets. Compression factors are generally slightly higher than FAST with DPCM; however, RICE requires more processor time.

4.6.2. *Exposure Time and AEC Processing*

One of the optional parameters which may be selected for any sequence is to use automatic exposure control (AEC). When a sequence uses AEC, the MSI software acquires and processes a test image before the start of the desired acquisition sequence in order to determine the exposure times to be used for each of the science images in the sequence. A command argument specifies the number of seconds the test image should be taken prior to the desired first science exposure. As with any image sequence acquisition, the software commands the exposure time and positions the filter in order to prepare for the test image, nominally 10 s ahead.

When the test image exposure is complete, the software reads out the test image into one of its eight image buffers and executes the autoexposure algorithm. The intent is for the test image to be underexposed, and for the algorithm to analyze the image to determine an exposure time (greater than that used for the test image) that would result in a 'properly' exposed image.

The algorithm first forms a 128-bin histogram of the pixel values in the image, with 32 DN levels per bin. It then uses a saturation parameter to determine if the image is underexposed. If more than the allowable number of pixels fall in the last (brightest) bin, then the image is considered overexposed. In this case the software scales the commanded exposure time used for the test image by a constant less than unity. Otherwise, the test image is underexposed as intended. The software integrates under the histogram, starting at the last (brightest) bin and working downward, until the desired saturation level is reached. The histogram bin at which this occurs is a control point. To determine the proper exposure time, the software multiplies the exposure time used for the test image by the ratio of the target saturation DN level to the DN level of the control point histogram bin.

Finally, to determine the exposure time for each image in the sequence, the software adjusts the resulting exposure time for the sensitivity difference between the filter used for the test image and the filter to be used for the science image, using the relative sensitivity values shown in Table VII. The resulting exposure time is the previously calculated exposure time, multiplied by the ratio of the relative sensitivity of the test image filter to the relative sensitivity of the science image filter. The time required to compute the autoexposure time from the test image depends on the test image data, but nominally completes within 4 s, including the overhead of other concurrent tasks in the processor.

4.7. ENVIRONMENTAL TESTS

As with all flight instruments, an extensive amount of testing was performed on MSI to simulate the harsh environment of space. The environmental tests included many hours of thermal vacuum testing at the subsystem level. To expedite the development of the instrument, MSI's camera and DPU were tested separately. In addition to the sensor level tests, the integrated spacecraft spent more than 450 hours in the large Space Environment Simulator in vacuum at the Goddard Space Flight Center. Operation of the instrument during these tests was verified through the use of an external stimulator attached to the light baffle cover. This stimulator used a small incandescent lamp, a mylar diffusing sheet, and a US Air Force resolution pattern at the focus of a simple lens. This pattern was projected into the camera's telescope, and the images were monitored throughout the tests.

In order to qualify the structural design of MSI and characterize any changes in image quality which might occur during launch, the instrument was vibrated in three axes to levels exceeding those expected during launch. No mechanical or electrical anomalies in either the DPU or camera were detected. Changes in boresight of the instrument were characterized by mapping the shift in the position of the CCD with respect to the optical alignment cube on the camera using a theodolite and a large flat reference mirror. The center pixel of the CCD (as viewed through the MSI optics by the theodolite) defined the boresight of the instrument. A shift in any of the components (lenses, filters or the CCD itself) would contribute to a shift in the measured boresight position. To quantify the boresight position, the angular position of the center pixel of the MSI CCD was measured relative to the normal of the front face of the MSI alignment cube. The magnitude of the shift was determined by a comparison of the boresight position (relative to the alignment cube) prior to and after the vibration of the instrument.

The theodolite established the horizontal (azimuth) and vertical (elevation) angular coordinates in which the position of the CCD and the MSI alignment cube were mapped. When focused at infinity, the theodolite projected a cruciform image into the camera. Focus of the theodolite was established by autocollimating the cruciform off a flat reflective surface which was the front surface of MSI's alignment cube. Autocollimating off the alignment cube also established the coordinates of the normal vector of the cube which was used as the reference for camera boresight, determined by projecting the image of the cruciform into the telescope and aligning it to the center camera pixel. This was observable using the MSI GSE. Angular measurements made with this particular theodolite provided arcsecond accuracy.

Since the physical distance between the alignment cube and the MSI optical axis exceeded the field of view of the theodolite, the theodolite had to be repositioned between measurements. Moving the theodolite changed the coordinate system

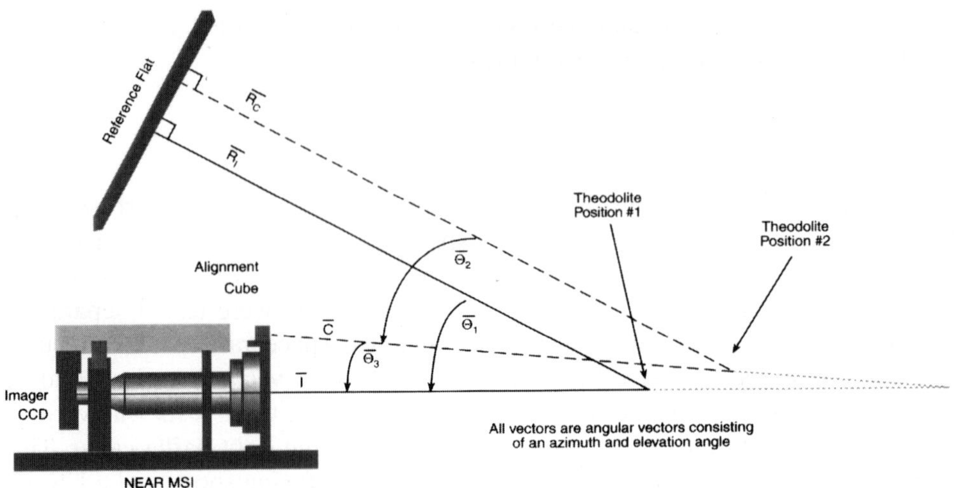

Figure 19. Boresight measurement setup using MSI and a theodolite. Measurements were made between each vibration axis in order to quantify any shift in boresight.

used for measuring the angular position of the alignment cube and CCD, hence it was necessary to use a reference surface to compare the two measurements. The reference surface used here was a large, high quality, flat mirror. The measurement setup is shown in Figure 19.

The mapping results indicated that an azimuthal (horizontal) shift of less than 3 arc min ($<$9 pixels) occurred during the vibration testing, well within the budgeted allowance of 6 arc min of shift. There was a small shift in elevation (vertical) mapping of the CCD to the alignment cube of approximately 5 arc sec ($<$0.2 pixels), which falls within the bounds of measurement error. The small shifts seen in the image result from machine tolerances on the hardware mounting the CCD heatsink to its housing. These detailed boresight measurements were used in determining the overall coalignment of MSI to the other instruments on NEAR. Coalignment between the Near-Infrared Spectrograph (NIS) and MSI is known to better than 8 arc sec and the difference in alignment of the boresights of MSI to the NEAR Laser Rangefinder (NLR) was determined to be 0.218°. The boresight alignment difference between MSI and NIS is somewhat better than NLR, however, due to uncertainties in the NIS alignment cube an accurate figure is not presently known (A. C. Sadilek, private communication). However, these uncertainties are small compared to the inherent uncertainties in the measurements due to thermal and gravitational effects.

By definition, the imager points in the direction of the spacecraft's positive X'-axis – the common boresight of the science instruments. The orientation of the CCD is such that the long axis (2.93°) is parallel to the spacecraft Z-axis (the direction of the high gain antenna). This CCD orientation was selected so that the long axis is parallel to the dispersion direction of NIS and the MSI field of view can

Table VIII

Calibration measurements made on the MSI camera

Test name	Inflight	Ground-based
Spatial resolution	Yes	Yes
Absolute radiometric calibration	Yes	Yes
Off-axis signal rejection	Yes	Yes
Out-of-band rejection	No	Yes
Spectral response	Yes	Yes
Response uniformity	Yes	Yes
Signal versus exposure setting	Yes	Partial
Thermal response:		
Dark signal level	Yes	Yes
Calibration ractors	No	Yes
Field of view and field distortion	Yes	Yes
Calibration of known mineral specimens	No	Yes
Small step response	Yes	Flight spare only

be sampled by three steps of the NIS scan mirror. Images displayed on a standard raster monitor will be oriented so that the spacecraft's Z-axis points to the right, and the Y'-axis is directed to the top of the display. From the viewpoint of an observer on the NEAR spacecraft sitting on the high gain antenna, 'up' is to the right on the monitor.

5. Calibration

Calibration of MSI was designed as a three-tiered procedure in which each successive level builds on preceding ones. The first level characterized the individual components, both optical and electro-optical. These measurements were made prior to assembly of the camera. The second level of calibration commenced once the camera was fully assembled and a detailed ground-based calibration was performed. The third level of the instrument calibration consists of a variety of inflight observations. In addition to verifying the ground-based calibration, the inflight measurements will permit a detailed characterization of the instrument throughout the mission.

The objectives of the MSI calibration are to characterize the performance and to provide calibration factors for use in converting the data into physical units. Table VIII lists the specific areas in which measurements were made to meet these objectives. As shown in this table, most of these areas were covered during the ground-based calibration.

5.1. COMPONENT LEVEL CALIBRATION

The FPD calibration was performed in a thermally controlled vacuum chamber specifically designed to calibrate small detectors. The chamber itself contains a small quartz-halogen lamp, a variety of small apertures, and two filter wheel assemblies for spectral and neutral density filters. Detector level calibration included a variety of tests designed to characterize the linearity of the electronic exposure control, CCD response linearity, spectral response, and temperature effects on background and response. All tests were performed at four temperatures: room ($\simeq 20$ °C), 'high' (-20 °C), 'medium' (-30 °C), and 'low' (-40 °C).

Calibration data for the camera components – CCD, lenses, and filters – were obtained from the manufacturers or measured directly prior to assembly. The tests performed at the component level calibration confirmed the spectral responsivity of the CCD as indicated by the manufacturer, the expected temperature dependence of dark current, and showed that the CCD response is approximately linear. In addition to these results, minor blemishes in the anti-reflective coating of the CCD were mapped. Although these blemishes were apparent under the pinhole illumination of the CCD during the detector calibration, they are completely negligible (see Section 5.2.1.8 below) once the imager is completely assembled because the cover glass of the CCD is out of focus.

The spectral response of each filter was only measured by the manufacture in order to expedite the development schedule of the camera. However, during the ground-based calibration of the camera (see below) it became apparent that there was an error in one set of measurements provided by the manufacturer. The red filter (760 nm) had a small shift in its central wavelength toward the blue by 4 nm.

5.2. CAMERA LEVEL CALIBRATION

5.2.1. *Ground-Based Calibration*

All of the ground-based calibration tests were undertaken in a large vacuum system in which the optical sources, viewing geometries, and thermal environment were carefully controlled. Most of the calibration was performed under expected operating conditions (i.e., the CCD nominally at -30 °C and the telescope at $+20$ °C); however, additional thermal configurations were also utilized in order to bracket the behavior of the instrument under possible flight conditions.

5.2.1.1. *Test equipment.*   All calibration tests were performed in the Optical Calibration Facility (OCF) of The Johns Hopkins University Applied Physics Laboratory consisting of large, linked vacuum systems. The largest of these chambers, the instrument chamber, has an internal diameter of 1.3 m and a length of 2 m and permits mounting the instrument under test on a two-axis motion stage. The motion stage can rotate the instrument in azimuth and elevation through a range limited only by the instrument cabling and mounting hardware. The interior and ends of

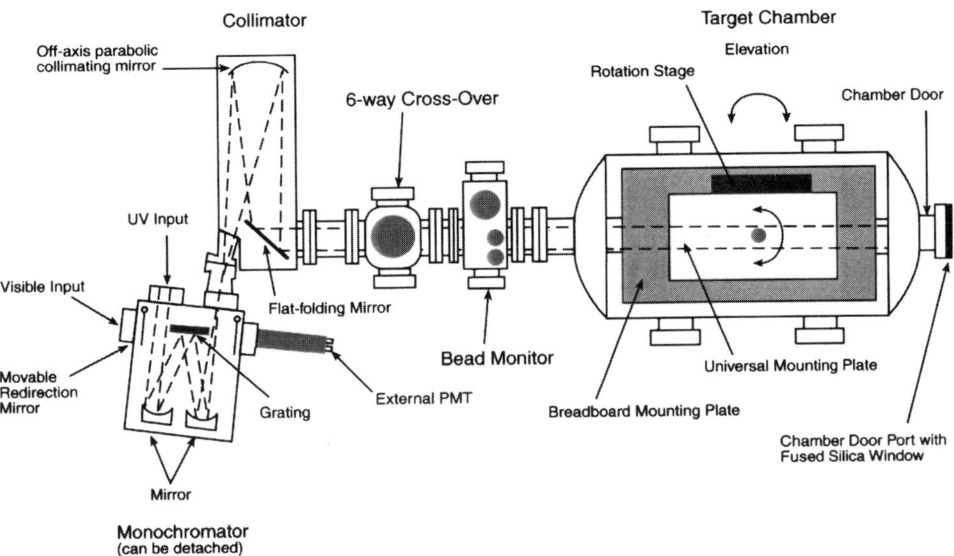

Figure 20. The Johns Hopkins University Applied Physics Laboratory Optical Calibration Facility.

the instrument vacuum chamber are surrounded by cold walls in which an external refrigerator circulates a cooling fluid through coils on the cold walls to reduce the internal temperature to $\simeq$40 °C. Figure 20 provides a schematic overview of the OCF.

Inside the instrument chamber, there are two sources of illumination. In one orientation the instrument can face a quartz chamber window which has a diameter of 203 mm. Various sources can be mounted in air outside the window. With a rotation of 180° the instrument looks along a beam tube into an off-axis parabolic collimating mirror. A vacuum monochromator is the typical source at the focus of the collimator.

The monochromator can be illuminated by an incandescent lamp with quartz optics for spectral calibration. At the input slit is a set of neutral density filters to attenuate the light by known amounts. At the exit slit a set of long pass filters is used to remove higher orders (i.e., shorter wavelengths) when the system is used at longer wavelengths. The wavelength can be changed manually or can be set to scan under computer control. The exit slit of the monochromator is located at the focus of the collimator which has a focal length of 1.43 m and operates at $F/7$. The collimated beam is folded into the beam path of the vacuum chamber by a plane folding mirror. Other sources can also be placed at the focus of the collimator and the light transmitted into the chamber through an intervening window. These sources include a point source (pinhole), a resolution target, a radiance source, or test targets.

5.2.1.2. *Dark current.*   The first attribute of MSI to be considered in calibration is the background level which must be removed from all images before further analysis. The background is the combined effect of an electronic offset which defines the zero of the analog to digital converter, and any additional signal due to thermal electrons (dark current). The background level was measured frequently in the OCF, by observing either the focus of the collimator or an integrating sphere with all light sources off. The observed behavior is as expected. The accumulation of dark current at a given temperature will be higher for longer exposure times because more thermal electrons can accumulate. As mentioned in Section 4.3.2, the dark current contribution is not spatially uniform and appears as a gradient across a dark image (Figure 21). This is a result of the way the image is read out of the detector (cf. Figure 9). Rows near the top of the CCD (bottom of the image) remain in the CCD longer and therefore accumulate more dark current than rows near the top of the image. In a frame transfer device, after the image photoelectrons have accumulated for the desired integration period, the entire active region is quickly transferred to the memory region. Dark current still accumulates in the remaining image stored in the memory region, while it is read out line by line into the serial readout register. As shown in Figure 22, higher temperatures lead to more thermally generated electrons and thus a higher background level. There is also a greater dependence on exposure time and a stronger gradient across the image. Over the range of temperatures at which measurements were acquired ($-17\,^{\circ}$C to $-43\,^{\circ}$C) the mean background level was observed to vary between 80–90 DN out of 4095 DN.

Figure 22 also shows that for a given temperature, the background is a linear function of exposure time. Therefore a short and a long exposure dark frame may be used to model the background level at any arbitrary exposure time at the same temperature. Testing of this model using the dark frames acquired during the ground-based calibration showed that it is accurate to $\sim$1 DN.

5.2.1.3. *Spatial resolution and focus determination.*   To focus MSI, the region just below the surface of the silicon chip of the CCD must coincide with the imager focal plane as defined by the optics. Since MSI's optics are fixed and focused at infinity, focusing the camera entails correctly positioning the CCD along the optical axis.

Focus measurements conducted at room temperature provided the data needed to fabricate a preliminary set of four high precision titanium standoffs which mount the CCD to its housing. These were installed in the camera prior to the final focus measurements, conducted under MSI's operating conditions in the OCF. The change from room temperature to the instrument's operating temperatures thermally contracts the imager and causes a small focus shift (Figure 23). By experimentally determining this shift due to thermal contraction, new standoffs sized to correct for this shift could be fabricated and installed, positioning the CCD into the imager focal plane.

Figure 21. Raw, contrast-enhanced, dark-field images acquired at an exposure of 500 ms and temperatures of (a)$\sim$ +20 °C and (b) −38 °C. Note the gradient in the room temperature image due to dark current, which extends from 212 to 556 DN. The odd/even effect in (b) is about 5 DN.

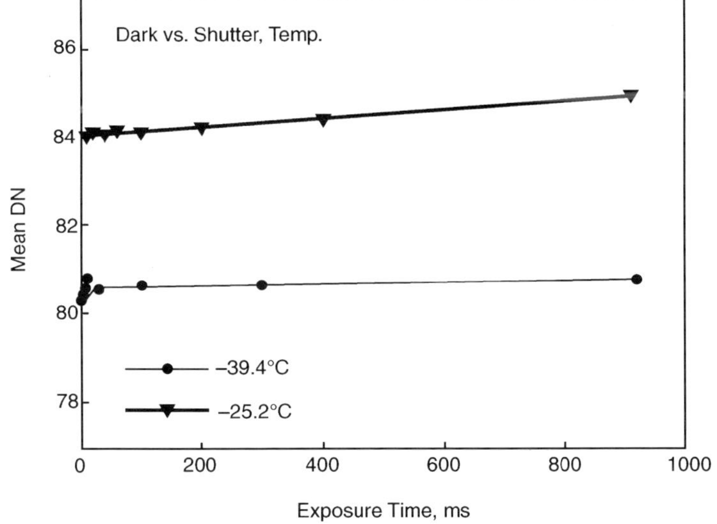

Figure 22. Mean dark-field DN as a function of exposure time at two temperatures. Note that both dark current and its dependence on exposure time increase with increasing temperature.

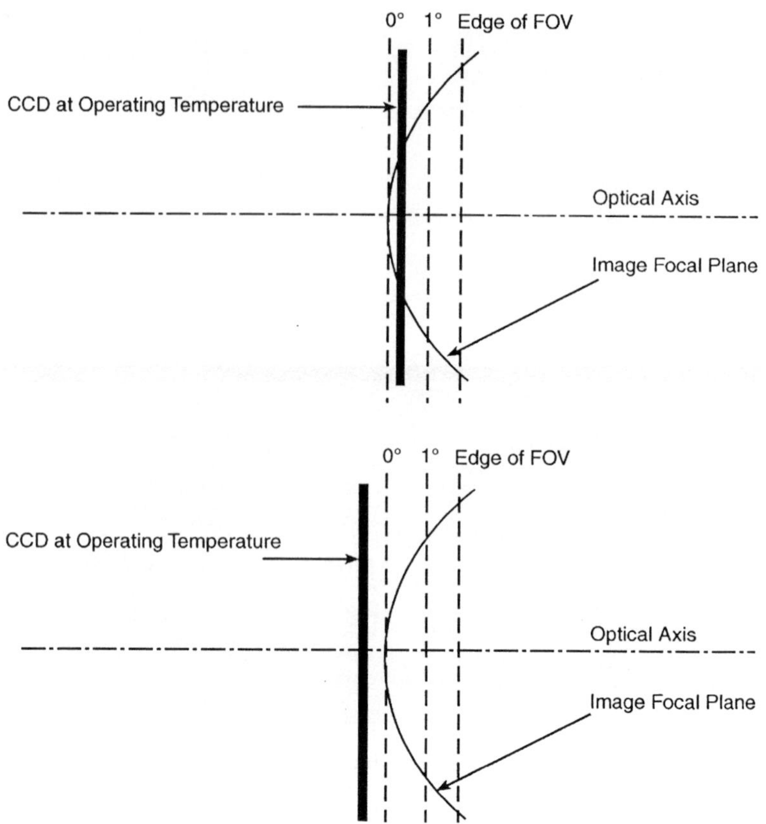

Figure 23. Curved focal plane of MSI. The effect of thermal contraction is to shift the detector in the direction shown, effectively moving the focal plane beyond the CCD.

Unlike tests conducted at room temperature and ambient pressure, the CCD was not accessible for mechanical translation while in the OCF chamber. Instead, focus was determined by moving the image plane while the detector itself remained stationary. Moving the image plane was done by adjusting the focus position of a pinhole source relative to the collimator focus. The collimator and pinhole were used to project an image of the pinhole into the camera. A longitudinal translation of the source results in a longitudinal translation of the image. Because the longitudinal magnification of an optical system is proportional to the square of the lateral or transverse magnification, a very accurate determination of the camera's best focus could be determined by moving the pinhole source in small steps. With a collimator focal length of 1.431 m and the MSI focal length of 0.167 m, the optical magnification is -0.117; hence an easily adjusted 2 mm shift at the source produced a finely adjusted -27 μm shift of the image. Focus was determined by maximizing the percent of the total radiometric energy of the image on a single CCD pixel.

Table IX

Point source image energy distribution: percent of the total radiometric energy falling on a single pixel (note filter 2 shown below was not used in flight)

Camera Temperature	Position (AZ, EL)	Filter number (center wavelength, nm)							
		0 (700)	1 (550)	2 (450)	3 (760)	4 (950)	5 (900)	6 (1000)	7 (1050)
Cold	$(0°, 0°)$	45%	64%	50%	47%	41%	43%	42%	41%
Operating	$(0°, 0°)$	43%	64%	28%	54%	48%	52%	47%4	4%
	$(0.71°, -0.71°)$	49%	61%	22%	52%	50%	49%	42%	39%
Ambient	$(0°, 0°)$	46%	48%	14%	38%	36%	34%	33%	33%
	$(0.71°, -0.71°)$	46%	32%	17%	35%	39%	37%	32%	26%
	$(0.95°, -1.4°)$	31%	32%	23%	37%	30%	38%	28%	27%

Point source images were recorded at 2 mm increments about the collimator focus. MSI's green filter (550 nm) was used for the focus measurements because preliminary tests showed this filter to be representative of the focus of all the other filters, and ray trace predictions showed the image spot size with green filter to be the smallest of the eight filters. Because the image plane of the MSI telescope is curved (cf. Figure 23) and the CCD is flat, the best focus position for the CCD changes as a function of position within the field of view. Therefore, images were recorded at the instrument's boresight, one degree from boresight, and at a corner of the field of view.

Based on these measurements, new CCD standoffs with the corrected lengths were fabricated and installed. The imager was returned to the OCF and thermal operating conditions were attained. Focus quality for the imager was then verified by placing the point source at the collimator focus. Subsequent measurements were recorded for all eight imager filters at various points in the instrument's field of view.

Table IX lists the percent of the total radiometric energy focused on a single pixel for each filter relative to an 11×11 pixel region centered on the spot, taken as part of an initial set of focus measurements. The first column describes the thermal conditions under which the measurements were made, where 'Cold' is defined as the entire instrument at $-30\ °C$ in vacuum, 'Operating' is the telescope at room temperature and the CCD at $-30\ °C$ in vacuum, and 'Ambient' is room temperature, in air. The location of the measurement within the field of view is indicated in the second column showing the azimuth (AZ) and elevation (EL) relative to the instrument boresight. These data indicate that all filters except the blue filter (filter 2), are well focused. The focus degradation with filter 2 was largely due to chromatic aberrations. Therefore, filter 2 (450 ± 50 nm) was replaced with a narrower bandpass filter, 450 ± 25 nm, which improved the focus through this filter. Figure 24 shows spot profiles for each of the eight filters. These profiles were measured at the center of the field of view, with the a cut through each spot along

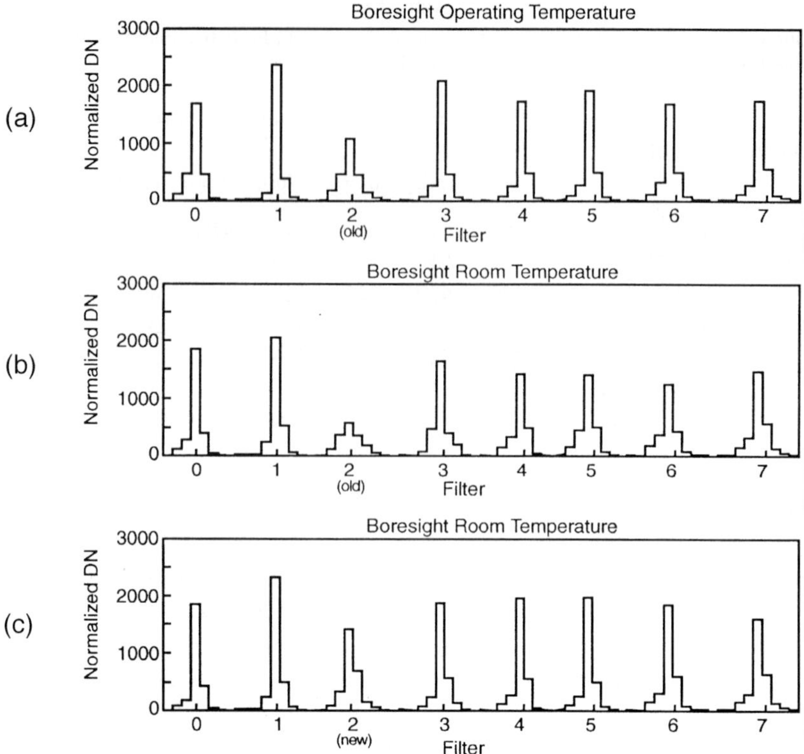

Figure 24. Measured spot profiles from each filter measured along the instrument boresight and across
the narrow pixel dimension. (a) Profiles under operating conditions. Filter 2 shows the response prior
to changing out this filter to a narrow passband. (b) Same as (a) but with the telescope and CCD at
room temperature. (c) Spot profiles measured at room temperature, but after changing filter 2 from
the broader passband, 450 ± 50 nm, to the narrower passband, 450 ± 25 nm.

the narrow pixel dimension. The top panel shows the spot profiles under operating
conditions before changing out filter 2. The middle panel shows the same data but
under ambient conditions. The bottom panel also shows the spot profiles under
ambient conditions, after changing filter 2 with the narrower passband. Comparing
the bottom and middle panels of this figure, one sees the significant improvement
in the 450 ± 25 nm filter.

The rectangular shape of individual pixels introduces an additional complication
to issues of focus and spatial resolution. Because of the unequal angular dimensions
of a pixel's instantaneous field of view, the ability to resolve elongate objects near
the limit of resolution depends on objects' orientations. For example, a narrow
linear feature will be better resolved if it trends perpendicular to the pixels' long
dimension, so that its width is sampled by more pixels. (In the preceding discussion,
the spot profiles are all measured in this direction to better resolve them.) This
peculiarity in no way affects radiometric responsivity to an extended radiance
source like Eros.

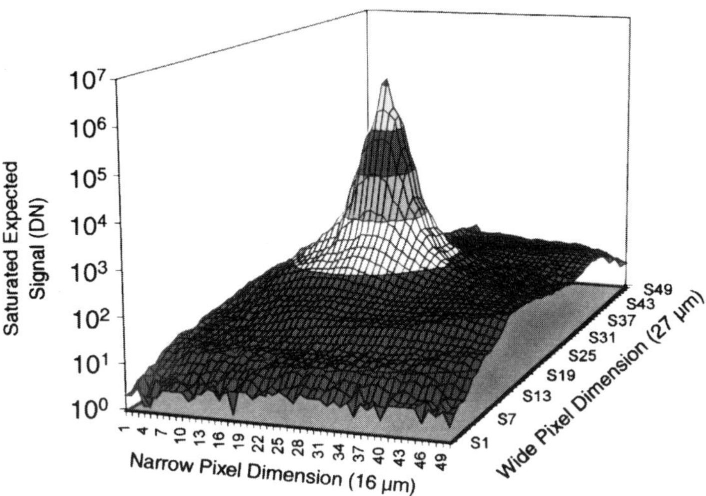

Figure 25. Scatter plot of 50 × 50 pixel array through the 'clear' filter (700 nm) and 917 ms exposure. Data show the logarithmic off-axis response of MSI.

5.2.1.4. *Off-axis light rejection.*   The presence of illuminated areas and deep shadows will produce a wide range of brightness in a single image of Eros. It is therefore essential that scattering from bright areas to faint ones be as small as possible. Several tests were performed in the OCF to measure the point source rejection ratio (PSRR) of the camera. This is the ratio between the point source axial intensity to its off-axis intensity.

As the magnitude of scattering was expected to be very low, a wide dynamic range was essential to measure the PSRR, and scattering from the test facility had to be avoided. A pinhole point source was used at the monochromator exit slit (located at the focus of the collimator). After demagnification by the ratio of the MSI and collimator focal lengths, the source had an effective diameter of 17 μm at the detector – comparable to the 16 × 27 μm pixels of the CCD. Since the monochromator slit is at the focus of the collimator with no intervening optics, the only spurious scattering sources are from the collimator and folding mirrors. The monochromator was illuminated with an Oriel incandescent lamp and the monochromator was set for zero order to transmit all wavelengths. Other sources of light may also contribute to the background level. To extend the dynamic range we started data collection using a short integration time and then increased it in stages to the maximum. As the exposure increased, the central signal saturated making it feasible to study the stray signal level out to larger angles. The background signal was subtracted on a pixel-by-pixel basis from each image to correct for the background signal level and all signals were corrected to the same exposure.

The background subtracted signal distribution for the 700 nm filter, scaled to maximum integration time is shown in Figure 25 for a 50 × 50 pixel sample. The peak of the distribution is almost 1.5×10^6 DN – well over the maximum

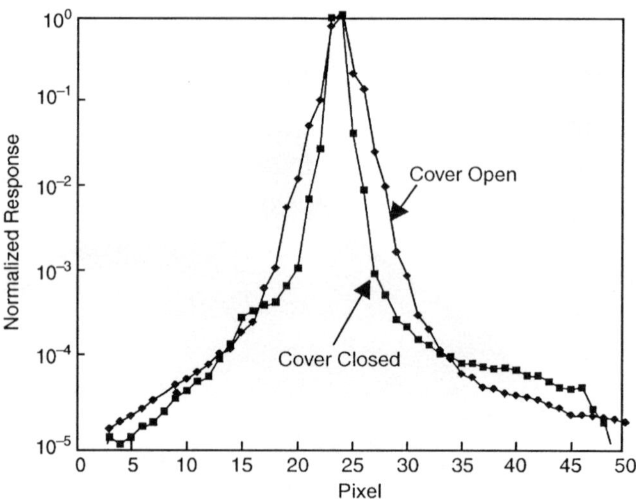

Figure 26. Normalized response of a 17 μm diameter source as imaged on the detector for two cases: (a) through reduced aperture viewport in cover, and (b) with full aperture of instrument.

measurable value of 4095 (12 bits). To show the distribution of stray light the data are plotted logarithmically. At 25 pixels from the center the signal has dropped to a few DN, the limit of the accuracy of the measurement. Figure 26 shows the profile through the peak of the distribution for two conditions: the light baffle protective cover closed and cover open. With the cover closed the aperture is approximately 24 mm in diameter compared with a diameter of 49 mm with the cover open. The wider distribution with the door open may be a consequence of the larger optical scattering area exposed. Extending the dynamic range has made it possible to follow the stray light down to five orders of magnitude.

The PSRR was determined by comparing the energy on a single pixel to the total energy in the image. The angular distance of each pixel from the position of the point source was then calculated assuming that a pixel has angular dimensions of $0.0054 \times 0.0092°$ (95×161 μrad). For a 50×50 pixel matrix this produces a lot of points, a sample of which are plotted for the broadband filter in Figure 27.

The scattered light falls off very rapidly out to $0.04°$ ($\sim$5 pixels) before decreasing with a shallower slope. At an angle of $0.1°$ ($\sim$14 pixels) the scattered PSRR has fallen to 1.0×10^{-5}. (These data also include scattering from the two mirrors in the OCF path so they are a worst case.) The measurements show that stray light is not likely to be a problem in the study of a target with large brightness variations. Only scattering measurements for the 700 nm 'clear' (broadband) filter have been considered in this analysis, and data for the other filters are still to be examined.

5.2.1.5. *Field of view and field distortion.* The CCD pixel array and the focal length of the telescope define the instrument's field of view. The 537×244 pixels correspond to $2.93 \times 2.25°$, and the physical pixel dimensions of 27×16 μm define

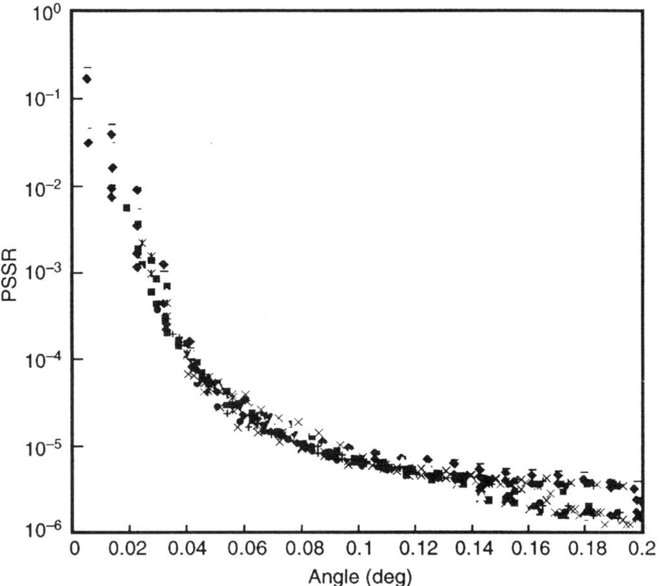

Figure 27. Point Source Rejection Ratio measured through the 700 nm filter, full aperture, and a variety of exposures.

the pixel field of view of $5.46 \times 10^{-3} \times 9.22 \times 10^{-3}$ deg ($95.2 \times 160.9\ \mu$rad). A variety of factors may lead to distortions within the field of view which may be characterized by mapping the response of an object throughout the field.

This mapping was performed under the instrument's operating conditions in the OCF with the camera mounted on the two-axis rotation stage. Field of view distortion measurements were made by using an object plate consisting of four thin aluminum discs with pinholes of diameter 100, 200, 400, and 800 μm arranged as shown in Figure 28. This plate was back illuminated and fixed at the focus of the OCF collimator to project an image into the camera. The image of the four pinholes at the CCD was demagnified by a factor 0.117 and provided a range of spot sizes from subpixel to several pixels in diameter. This selection of sizes helped minimize the effect of the less responsive regions on the CCD (viz., the 8 μm band between alternate pixels due to the anti-blooming gate). By rotating the motion stage, the four pinhole object was moved throughout the field of view and provided a fixed pattern with a known angular separation. By measuring accurately the angular rotation of MSI and computing the pixel location of images formed at the CCD for each position, the extent and type of distortion can be categorized and corrected to yield an undistorted final image. For each of the eight MSI filters, a boresight image was obtained, then the motion stage was stepped in azimuth and elevation in a 7 by 9 grid that covered the field of view. The azimuthal and elevation scans took place with angular steps of $0.38°$.

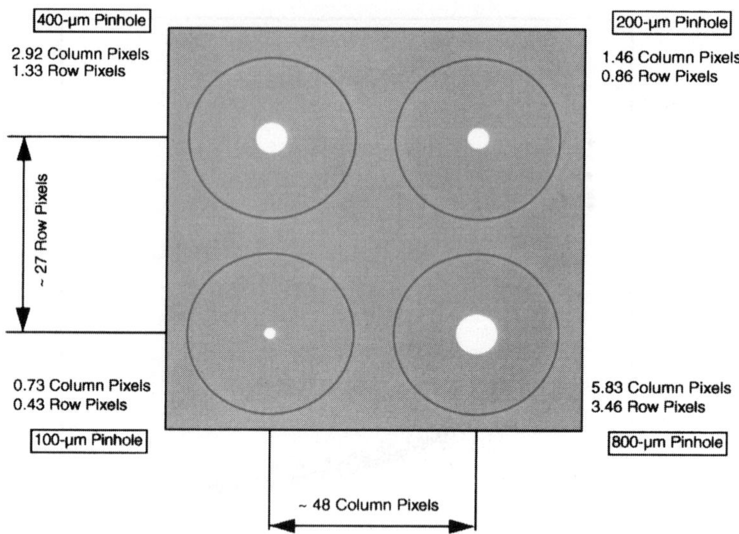

Figure 28. Four pinhole target used to map field distortions. The fixed pattern and angular separation produced a known spot pattern at the MSI focal plane with spot sizes varying from subpixel to several pixels in diameter.

The data acquired during this test were reduced by separating out each pinhole and generating a map across the entire pixel array. Figure 29 shows a sample field map for the 100 μm pinhole. Preliminary analysis of these data suggests that distortions are of the order of 1 pixel or less. However, some discrepancy between predicted and measured locations of the pinholes appears to be related to the pattern of stepping of the pinhole pattern, so a more definitive assessment of the small level of distortion requires inflight calibration images of extended star clusters.

5.2.1.6. *Spectral response.* The spectral response of MSI was measured in the OCF under operating conditions (i.e., the CCD nominally at -30 °C and the telescope at $+20$ °C, in vacuum) and used the monochromator. The objective of this test was to measure the half-width of MSI's filters rather than determine the absolute intensity response. Figures 30(a) and 30(b) show two typical response measurements and the calculated response of the instrument.

Figure 30(a) shows the data of the blue filter supplied by the manufacturer, along with the response measured using the OCF monochromator. The dashed line in this figure is the calculated response derived from the product of the lens transmission, the filter transmission, and the detector response. Generally, there was good agreement between the calculated and measured responses. As mentioned previously, only one filter (760 nm) showed any serious discrepancy in the passband between the measured response and the manufacturer's data. Similarly, the responses of filter 7 (IR4) are shown in Figure 30(b). This figure provides a good example of how the spectral response appears significantly different from the

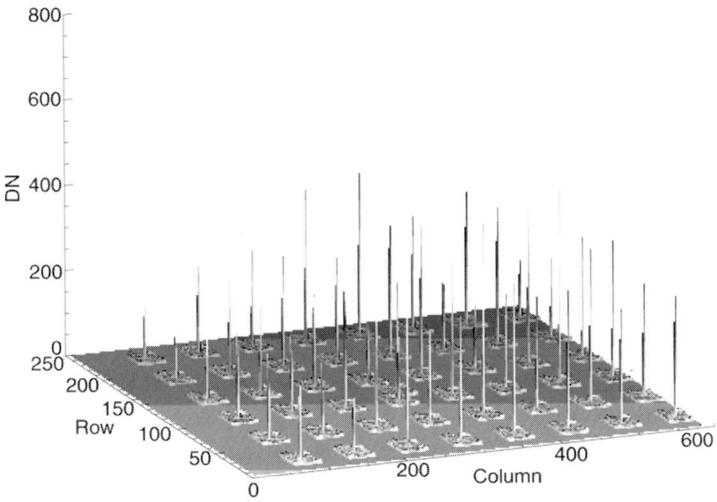

Figure 29. Composite image of the 100 μm pinhole over the field of view. A similar map exists for each of the four target pinholes which were stepped in a 7–9 grid over the field, resulting in a 6–8 distortion map because at the extreme edge of the field, two of the four pinholes were outside the field of view.

manufacturer provided filter transmission data alone. The difference results from the rapid fall-off of the CCD response at these long wavelengths (cf. Figure 10).

5.2.1.7. *Out-of-band rejection.*   The spectral range of each filter was scanned in detail and the full response range of the detector was scanned at intervals of 25 nm for each filter. The out-of-band range was scanned at increased sensitivity by using the maximum integration period, increasing the monochromator slit width, removing a neutral density filter from the beam, or by a combination of these techniques. This procedure allowed an increase in the sensitivity of two to three orders of magnitude. Measurements on the widest filter, 700 nm (filter 0), showed that the out-of-band signal was less than 0.1% of the in-band signal. All of the filters were checked for spectral leakage and in all cases the results were similar to those for filter 0. In all cases, the measured out-of-band signal was less than 0.1% and was probably due to stray light from within the monochromator.

5.2.1.8. *Response uniformity.*   Radiometric calibration of MSI images requires removal of nonuniformities in the imager response across the field, i.e., the 'flat field'. Nonuniformities in imager response can arise from a variety of sources: optical vignetting, the $\cos^4 \theta$ fall-off of image illumination, nonuniform transmission of the lens cover, the CCD window, any of the optical elements of the telescope, or nonuniformity in the responsivity of the CCD itself. In MSI, vignetting is not present and the $\cos^4 \theta$ fall-off will reduce the irradiance at the corner of an image relative to the center (an angle of 1.84°) by only 0.2%, so the major sources of

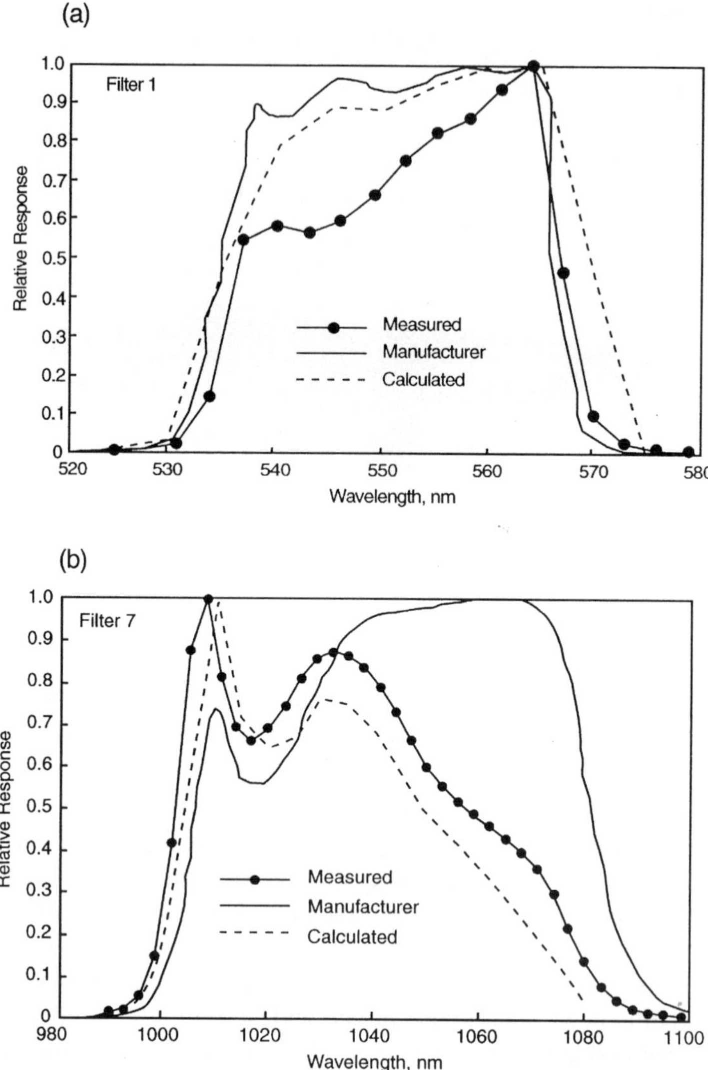

Figure 30. (a) Spectral response of the MSI blue (450 nm) filter. (b) Spectral response of the MSI IR4 (1050 nm) filter. The solid line corresponds to data supplied by the filter manufacturer, the line with individual data points is the measured response in the OCF, and the dashed line is the calculated response of all of the optical and electro-optical efficiencies.

nonuniformities in the flat field will likely involve optical elements or the CCD itself.

An additional source of nonuniformity in an acquired image, unrelated to the optical train or responsivity of the CCD itself, is CCD smear – an artifact inherent to frame transfer CCDs. CCD smear occurs because of the finite time required to shift an image from the active zone to the memory zone of the CCD ($\sim$0.9 ms).

Because there is no mechanical shutter in MSI, the active region is always exposed, and as the image is transferred to the memory zone, any illuminated part of the shifting image will continue to create photoelectrons. This shifting of the image from the active region to the memory region is known as frame transfer. During the frame transfer period, the leading edge of the image moves immediately out of the active zone (cf. Figure 9) and is exposed only for the commanded exposure time. However, the trailing edge receives an additional 0.9 ms of exposure because of the finite time for frame transfer. Intermediate regions receive additional exposure time that is a linear function of position in the frame.

Two different sets of measurements were used to characterize the MSI flat field. Both were collected with the CCD nominally at $-25\,^\circ$C and the telescope body at $+20\,^\circ$C. The first set of measurements were collected using the large integrating sphere as an extended, field and aperture filling, uniform source. The sphere was located outside the rear chamber window, so MSI viewed it with no intervening optics other than the chamber window. These data rely on the uniformity of the sphere, which has been calibrated by Labsphere, Inc., and showed a uniformity of better than 99.5% over most of the aperture. However, due to the brightness of the sphere, instead of using all four internal lamps, either only one or two lamps were used or the sphere was illuminated with an external lamp. Because of these steps to reduce illumination of the interior of the sphere, the original calibration of uniformity may have changed. Images of the large sphere through the rear window were acquired both with the lens cover closed and opened.

The second set of measurements was made to test the uniformity of the large sphere. To do this, the large sphere was positioned at the focus of the collimator, and MSI was rotated on the motion stage so that five different regions of the field of view were sampled: the center and the four corners. Since the source itself did not move during this set of measurements, any variation should be directly attributable to MSI.

The effect of CCD smear can be removed using a recently developed algorithm (P. Murphy, private communication). The smear is scene dependent, and is modeled as the result of the shifting of collected photoelectrons over an image on the active zone of the CCD for an arbitrary exposure time and a constant frame transfer time of 0.9 ms. Figure 31 shows images of the large sphere through the collimator, at exposure times of 1 to 100 ms, before and after the correction algorithm was applied to remove the smear. This scene is particularly useful because it shows a bright, uniform object on a dark background. At 100 ms (before correction) the smear is barely visible; at exposure times less than 30 ms and especially below 10 ms the smear substantially affects measurement of the scene radiance. However after correction, the smear is effectively removed from images with an exposure time as short as 1 ms. The remaining signal represents the actual responsivity of MSI.

Figure 32(a) shows highly contrast-enhanced, smear-corrected flat fields for each filter, obtained using the large sphere through the rear window with the lens

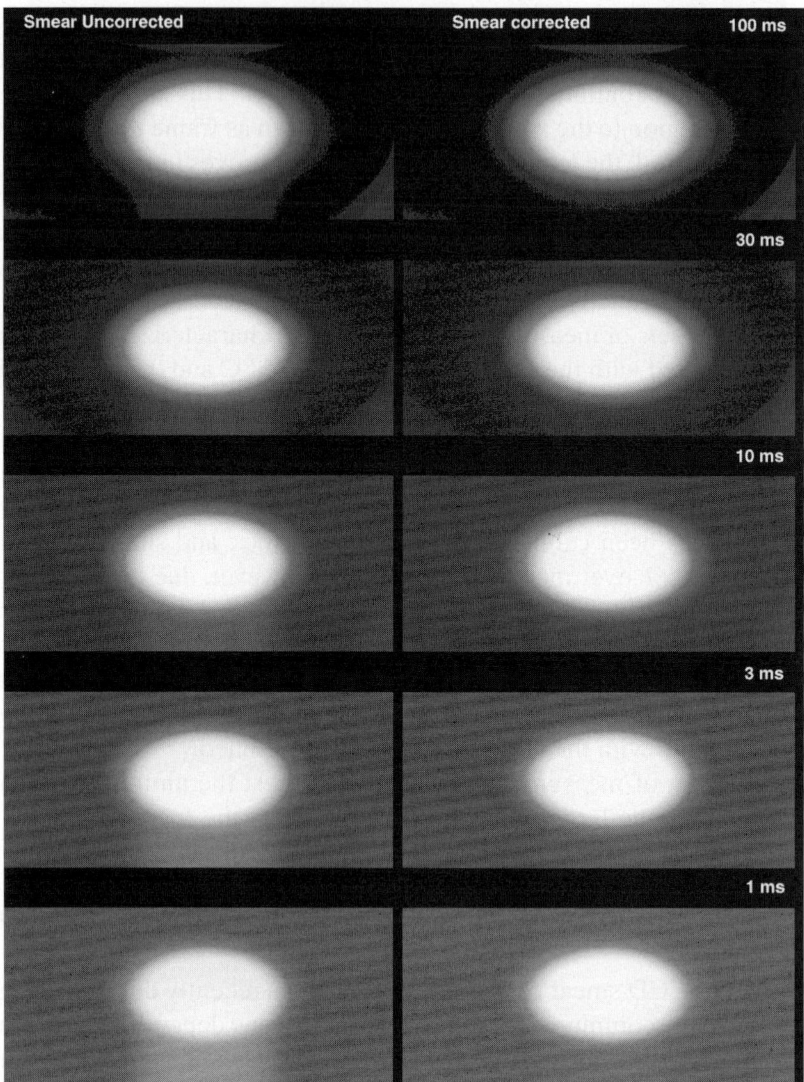

Figure 31. Images of the large integrating sphere viewed through the aperture to the collimator, at exposure times of 1 to 100 ms. The left column contains raw images with only the dark-field removed. Note that readout smear is increasingly evident at shorter exposure times, as the time for image readout (∼0.9 ms) is proportionally larger compared to the exposure time. The right column contains the same images with the scene-dependent smear removed. Images have been highly contrast-enhanced. The diagonal banding is periodic noise from the experimental setup and in not internal to MSI.

cover off. These are uniform to approximately ±2.2%. There is little dependence on filter position, except at the longest wavelengths (1000 and 1050 nm) where responsivity is slightly enhanced at the edges of the field. Most of the nonuniformity present represents pixel-to-pixel variations in responsivity. Figure 32(b) shows enlargements of the center of each flat field, again highly contrast-enhanced. Vari-

Figure 32a.

Figure 32a–b. (a) Flat fields for each MSI filter. Images have been highly contrast-enhanced. (b) Enlargement of the central portion of Figure 32(a), showing that that pattern of more and less responsive pixels is highly reproducible between the flat fields for different filters. Images have been highly contrast-enhanced.

ations in pixel-to-pixel responsivity are highly reproducible filter-to-filter, indicating that the CCD itself is the largest contributor to flat field nonuniformity with the lens cover off.

Images of the large sphere through the collimator at the center and four corners of the field yield measurement of flat field nonuniformity which agree closely with the above data, indicating that the interior of the sphere is essentially uniform when illuminated with either one or two internal lamps or with the external lamp.

Having the lens cover closed creates a small but significant additional source of flat field nonuniformity. Figure 33 shows contrast-enhanced flat fields for the 550 nm filter with the cover on and off. The cover-on flat field contains small nonuniformities that are either subdued or absent in the cover-off flat field. Similar behavior is observed in the other filters. The locations of the nonuniformities correspond with locations of blemishes in the anti-reflective coating on the CCD

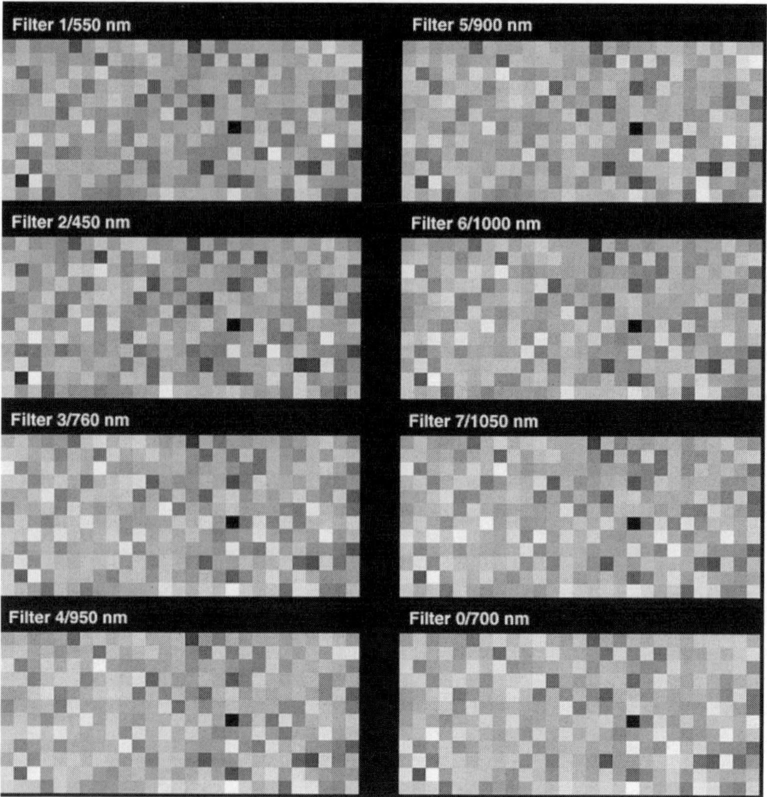

Figure 32b.

coverglass that were detected during component level CCD calibration. Imaging with the lens cover on reduces the optical aperture of the imager. The rays entering the reduced aperture optical system are more paraxial, and therefore enhance the effect of any blemishes on the CCD window. Although these nonuniformities are visible in the highly contrasted image of Figure 33, they add much less that 1% to the overall nonuniformity of the flat field, and are therefore not major effects.

5.2.1.9. *Absolute radiometric calibration.*   A radiometric calibration is most accurate if it is performed in the manner in which the instrument will be operated. This was possible in the case of MSI with the use of a field and aperture filling radiance source. The instrument viewed the source directly through the 203 mm chamber window. A correction was made for the vacuum chamber window, and a calibration factor was obtained which converts instrument signal (DN level) to units of radiance at the center wavelength of each filter.

Two sources were used with MSI, a large (203 mm aperture) integrating sphere, and a smaller (102 mm aperture) gold coated integrating sphere. The spheres were located outside the external window of the vacuum chamber as just described, or

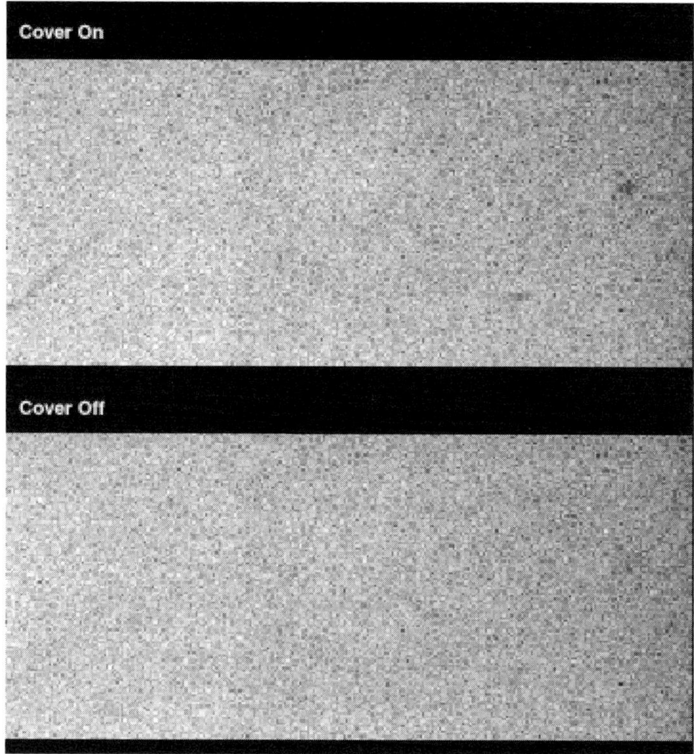

Figure 33. Comparison of the flat field for the 550 nm filter with the lens cover on and off. Images have been highly contrast-enhanced.

alternatively at the focus of the collimator. The former has the advantage of the most direct calibration but the brightness of the source requires low exposures. This makes the measurement less representative of typical operation, probably increasing the calibration error.

With the source at the collimator focus there are two reflections that must be corrected for, but the source can be easily attenuated with calibrated neutral density filters so long exposure values can be used. Both techniques were used and the results given in Table X are considered to be the best estimates of calibration factors for MSI based on the calibration sources used onground for an exposure time of 100 ms.

The signal falling on the detector is the integrated response of the spectral bandpass of the filter and the spectral distribution of the source. Thus, the calibration factors given in Table X will vary depending on the spectrum of the source. These calibration factors reflect the spectral distribution of the sources used and have not been corrected to correspond to a solar spectrum. For objects observed inflight, the calibration factors must be corrected for the spectrum of the source by convolving it with the spectral response of the filters (shown in Figure 2).

Table X
Radiometric calibration of MSI at 100 ms exposure

Filter number	Center wavelength (nm)	Calibration factor DN/(W m^{-2} sr^{-1} μm^{-1})
0	700	4578.0
1	550	509.1
2	450	148.3
3	760	503.4
4	950	291.2
5	900	470.7
6	1000	159.5
7	1050	64.4

5.2.1.10. *Variation of responsivity with temperature and lens cover status.* The absolute calibration of MSI is expected to vary with CCD temperature, primarily because of decreasing sensitivity of the CCD at wavelengths 900 nm at lower temperatures. In addition the filters may exhibit temperature dependent variations in transmissivity. The temperature dependence of responsivity was determined by comparing images of the large integrating sphere acquired at temperatures from −33 to +25 °C. Figure 34 shows the variations in sensitivity, fitted empirically and normalized to −29.6 °C – the nominal inflight operating temperature of the CCD – which is within one degree of the temperature at which the accuracy of the calibration coefficients in Table X has been tested using inflight data (see Section 5.3.2). There is a very regular pattern of increased responsivity of shorter wavelength filters and decreased responsivity of longer wavelength filters with decreasing temperature. At 700 nm – the center wavelength of the broadband filter – responsivity is nearly independent of temperature.

The status of the lens cover, open or closed, also affects the absolute calibration because of its attenuation of incoming light due both to the decreased aperture and the transmissivity of the fused silica. The attenuation due to the lens cover was measured by observing the integrating sphere through all filters alternately with the cover open or closed. No wavelength dependence was measured, and the incoming light is attenuated by 0.236 ± 0.001.

5.2.1.11. *Test of calibration on known mineral specimens.* The ultimate test of the radiometric and wavelength calibration of MSI is that it yields multicolor images from which accurate brightness and spectral data can be extracted, and in which spectrally different materials can be resolved spatially. To this end, color images of rock and mineral samples analogous to plausible surface materials on S asteroids were acquired in the OCF. The relative spectral features of the specimens are well known but the absolute reflectance levels depend on variables such as sample texture. A specimen of packed halon powder (the widely used spectral

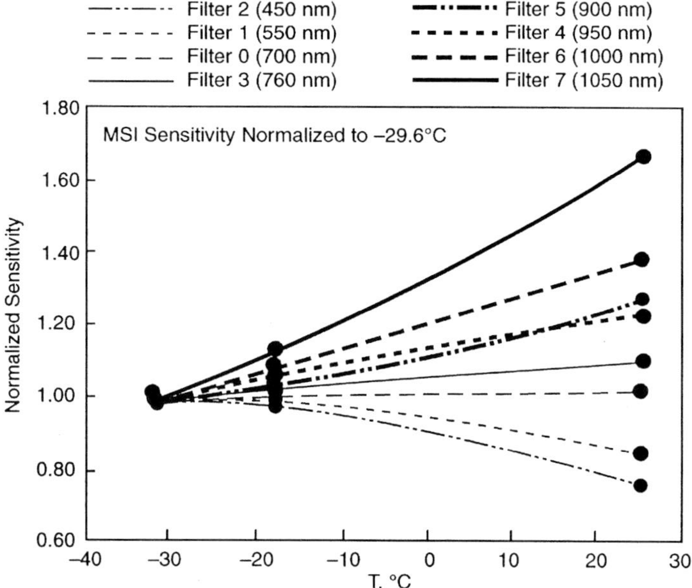

Figure 34. Temperature dependence of responsivity of MSI through its eight filters. Responsivity is normalized to typical inflight operating temperature, −29.6 °C.

standard for laboratory rock and mineral spectroscopy) was also measured for reference. The spectra of the same specimens were also measured by NIS (see Warren et al., this issue).

Figure 35 shows overlaid MSI and NIS reflectance spectra of a sample of nontronite, a ferric iron-containing clay. The NIS spectrum shows the characteristic Fe absorption feature near 900 nm, as well as longer wavelength features attributable to OH and H_2O in the mineral crystal lattice which are critical to mineral identification. The MSI spectrum exhibits excellent agreement in the wavelength region of overlap, and accurately measures both the shape and the depth of the Fe absorption.

The importance of the higher spatial resolution of MSI in extrapolating NIS measurements to smaller spatial scales is illustrated in Figure 36. This image represents a fragment of the Allende carbonaceous chondrite meteorite, an assemblage different from those expected on S asteroid surfaces, but one which does contain the same major minerals (olivine and pyroxene). The spectral properties of the individual grains are indicative of their different mineralogies. A NIS measurement of the same sample would contain only ten resolution elements, sufficient to determine the average mineralogy, but not adequate to detect the small-scale compositional variations that are evident in the MSI imaging.

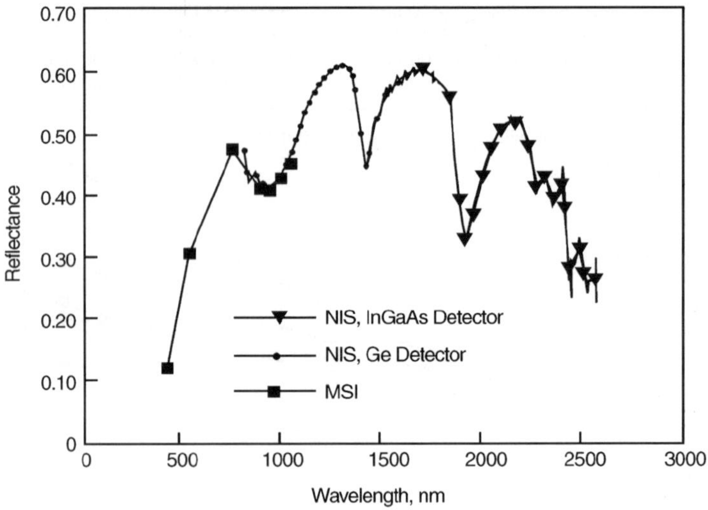

Figure 35. Overlaid MSI and NIS spectra of the Fe-containing lay mineral nontronite. The MSI spectrum accurately measures both the shape and the depth of the Fe absorption in the region of wavelength overlap of the two instruments.

5.2.2. *MSI calibration equation*

An important result of the tests outlined above is an equation which converts DN level for a pixel in an arbitrary MSI image to the physical units of radiance, $W\ m^{-2}\ \mu m^{-1}\ sr^{-1}$. This has the form:

$$\text{Radiance}_{0x.y.f.T.t.c} = \frac{\{[\text{DN}_{x.y.f.T.t.c} - \text{Dark}_{x.y.t.T}] - \text{Smear}_{x.y.t}\} \cdot 100}{\text{Flat}_{x.y.f.c} \cdot \text{Coef}_f \cdot \text{Resp}_{f.T} \cdot \text{Atten}_c \cdot \text{Exp}_t}, \tag{1}$$

where $\text{DN}_{x.y.f.T.t.c}$ is raw DN measured by the pixel in column x, row y through filter f at exposure time t and temperature T with the cover status c open or closed. $\text{Dark}_{x.y.t.T}$ is the dark level modeled for this pixel at exposure time t and temperature T, derived from a short and long dark field exposure at the same temperature. $\text{Smear}_{x.y.t}$ is the scene dependent smear for the pixel at exposure time t. $\text{Flat}_{x.y.f.c}$ is the flat field for filter f with the cover status c open or closed. Coef_f is the calibration coefficient for filter f, and $\text{Resp}_{f.T}$ is the relative responsivity for this filter at temperature T. Atten_c, if appropriate, is the attenuation with the cover closed. Exp_t is exposure time in milliseconds, and the constant 100 is the exposure time in milliseconds for which the calibration coefficient Coef_f is applicable.

5.3. INFLIGHT CALIBRATION

5.3.1. *Objectives*

Inflight calibration of MSI outlined in Table I has three major objectives: (a) to verify onground calibrations; (b) to track any changes in instrument radiometric response, dark current, field of view, alignment, or point spread function that occur inflight; and (c) to measure attributes of the instrument that cannot be well characterized onground. Objectives (a) and (b) represent verification, and if necessary updating, of the two earlier layers in instrument calibration; most of the instrument's characteristics can be rederived if necessary from the inflight data. Objective (c) includes measurements of distortions using star clusters, careful measurements of light scattered from a source within the MSI field of view, and stray light within the MSI field of view that originates from out of field sources. Scattered and stray light have proven to be significant problems in determining multispectral properties of high contrast planetary surfaces (e.g., Gaddis et al., 1995), such as Eros will represent at the moderate phase angles viewed by the imager. Accurate measurement of these effects is difficult onground.

MSI is body-fixed and has no mechanical shutter, so there is no onboard calibration target. The primary inflight calibration target will be highland regions of the Moon imaged during the Earth swingby, specifically the Hevelius Formation (ejecta of the Orientale basin, on the western limb of the near side). This region is identical spectrally to Apollo 16 samples, which are well-studied in the laboratory (Pieters et al., 1993b). The swingby occurs nearly one year prior to arrival at Eros; stability of the imager over this time period will be ascertained and tracked in two ways. First, Canopus will be imaged both at the time of swingby and during approach to Eros (as well as throughout the mission) to quantify and track any changes in instrument response characteristics. Second, the Moon will be imaged twice at closely similar photometric geometries nearly two years apart. The basic equation describing calibration of Eros to the Moon is:

$$\frac{Eros}{Sun} = \frac{Eros}{Canopus} \frac{Canopus}{Lunar\ highland} \frac{Lunar\ soil}{Sun} , \tag{2}$$

where Eros/Canopus and Canopus/Lunar highland are determined inflight, and Lunar soil/Sun is known from ground-based studies.

Other characteristics of MSI will be measured at different times during cruise and at Eros. Geometric distortion will be measured on several occasions, each time using multiple images of a star cluster in different parts of the field of view. The point spread function will be measured repeatedly using the star clusters, as well as bright stars employed for tracking radiometric calibration. The flat-field character of the imager and in-field scattering of light will be examined both at the Moon and at Eros, by acquiring multiple images of these objects in different parts of the field of view. Out-of-field signal will be quantified by acquiring swaths of images across the limb of each object, and this process will be repeated at different orbits of Eros as the source of stray light (Eros) changes in its size and illumination. Dark frames will be acquired during major imaging sessions, using either the night side of Eros or a region of space free of bright stars as a target.

Alignments of MSI with NIS and NLR will be redetermined at Eros to correct for possible effects of the difference in gravity and temperature from that

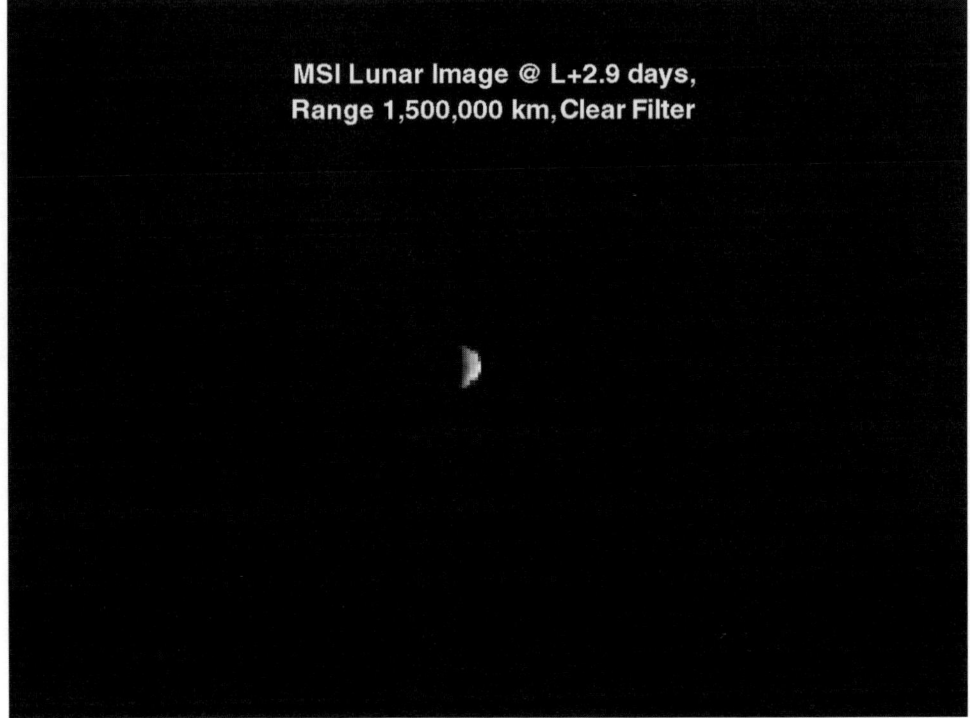

Figure 37. Calibration image of the moon acquired shortly after the launch of NEAR, through the clear filter with an exposure time of 8 ms. The image has been fully calibrated using the MSI calibration equation, and has been aspect-corrected for the rectangular pixels.

during onground measurements. Coalignment with NIS will be determined during approach to Eros when the asteroid is ~0.1° in size, by scanning swaths of measurements from both instruments across the disk along both axes of the imager field of view. Correspondence in the signal levels from NIS with the disk's position in the MSI frame will show the relative position of the two instruments' fields of view. Coalignment with NLR will be measured by firing the laser transmitter in 8 Hz burst mode at the night side of Eros, and determining the location of the illuminated spot in simultaneously acquired MSI frames.

5.3.2. *First Results*

The first set of lunar images was acquired on 20 February 1996 from a range of 1.5 million km at a phase angle of 99°, and included tests of instrument linearity and responsivity. A typical image (clear filter, 8 ms exposure) is shown in Figure 37, and covers the eastern far side. This restricted set of images has already proven the utility of inflight calibration.

Results for instrument linearity are shown in Figure 38, using a sequence of images acquired using the 550 nm filter at exposure times of 3 to 80 ms. The total

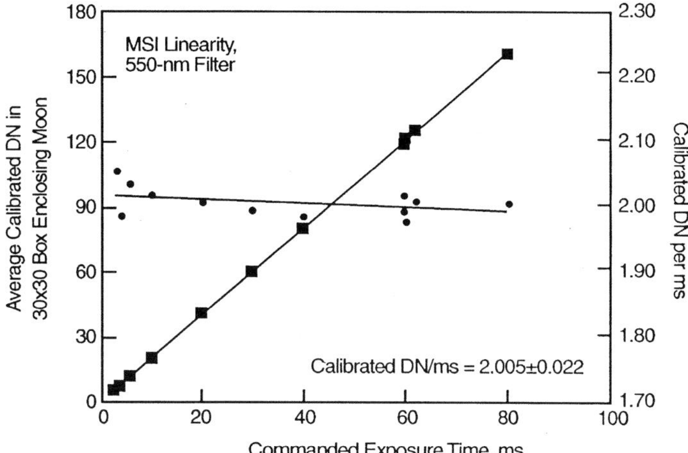

Figure 38. Linearity of MSI derived from a sequence of 550-nm lunar images with different exposure times. The left axis shows the average DN level in a 30 × 30 pixel box enclosing the moon; the right axis shows the mean DN accumulation rate per millisecond.

signal within a 30 × 30 box enclosing the Moon is a linear function of exposure time and is expressed in the figure for each image as the signal collected per millisecond. The responsivity is constant to within ±1%, and individual pixels encompass nearly the full dynamic range of MSI.

The absolute calibration of MSI was also tested by deriving calibration coefficients independent from onground coefficients, applying both sets of coefficients to the sample of pressed halon measured in the OCF and comparing the resulting calculated radiances. For this procedure, the illuminated highlands region of the Moon was used as a calibration standard. The expected radiance was calculated assuming spectral reflectance and phase angle dependence of reflectance like that of Apollo 16 soil, convolved with the solar spectrum and the spectral response of each MSI filter (Figure 2). Using the measured DN levels and known instrument parameters (cover on, exposure times, etc.), the MSI calibration equation (1) was solved for filter-dependent calibration coefficients. These coefficients were then applied to images of halon measured onground, and the derived radiances were compared with radiances of halon calculated using the onground calibration coefficients. In Figure 39 the halon spectrum derived from inflight calibration is expressed relative to the spectrum derived using the onground calibration coefficients. The overall agreement is excellent; in any given filter the uncertainty in absolute calibration is approximately ±3%.

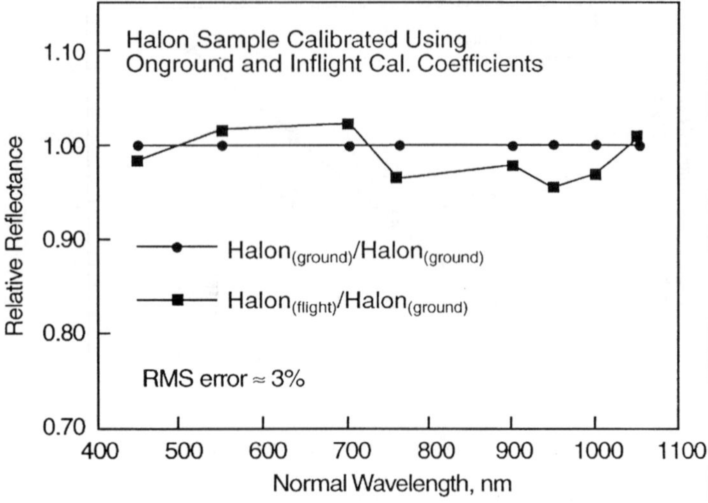

Figure 39. Spectra of a sample of pressed Halon measured onground, calibrated using radiometric calibration coefficients derived from onground and inflight measurements. The two absolute calibrations agree to within approximately 3%.

6. Observation Strategy

6.1. MEASUREMENT PLAN AT EROS

The observation plan at Eros, reviewed by Veverka et al. (1997), can be divided into three periods: approach to the asteroid, imaging from high orbits, and imaging from low orbits. Observations of Eros will commence approximately 8 days prior to encounter. During this time, while the spacecraft is still thousands of kilometers away, MSI will be used to image the sphere of influence of the asteroid (∼500 km radius) several times to search for natural satellites. The broadband 'clear' filter will be used for this purpose. With the multiple mosaics obtained during this period, if a satellite is identified in one mosaic, its positions in the other mosaics can be used to determine its orbit. Given the sensitivity of MSI through the clear filter determined from onground calibrations (detection of an object with apparent magnitude +10.5 is expected at the 3σ level), it is predicted that the imager could detect a satellite with Eros' reflectance properties less than 15 m in size (Veverka et al., 1997).

Near the middle of the approach observations, the range to Eros will be ∼1000 km and the asteroid will fill the imager's field of view with a spatial resolution ∼130 m. At this time, the southern hemisphere and equatorial region of Eros will be illuminated. (Eros' spin axis is inclined ∼90° to the orbital plane; this period of time is southern-hemisphere 'summer'.) The phase angle of the images (sun-asteroid-spacecraft angle) will be 76°–90°. Images will be acquired every 10° of the asteroid's rotation (every 9 min), and 7 color mosaics will be acquired every

20° of asteroid rotation. These data will form the basis for the first global map and shape model of Eros.

NEAR will pass between Eros and the Sun and enter retrograde orbit on 6 February 1999. The orbit will be lowered stepwise from a radius of 1000 km to 400, 200, and then 100 km. MSI will acquire a 7 color mosaic of the entire illuminated surface from each altitude; with each stepwise lowering of the orbital radius, spatial resolution will improve by a factor of ∼2. At the 100 km orbit spatial resolution will be ∼13 m, and the phase angle of the imaging will be approximately 90°.

With one notable exception described below, from March to December 1999, NEAR will be maintained in orbits from 35 to 50 km in radius, at which time the NLR, XGRS, and the Magnetometer will obtain their primary measurements of Eros. In the low orbits (period ∼14 hr) through August, image strips parallel to the spacecraft groundtrack will be acquired. Slewing of the spacecraft will allow off-nadir pointing, such that all illuminated latitudes are accessible for imaging. The subsolar latitude will still be in the southern hemisphere, so the north polar region will not yet be illuminated. The spatial resolution of MSI images at these times will be as high as 3 m, and the imaging most critical to understanding small scale surface structure and processes will be obtained. Areas of particular interest identified in the earlier image sequences, such as young craters, grooves, and geologic boundaries, will receive the highest priority for coverage.

During August 1999 the orbital radius will be increased for several days to 100 km and the orbital plane will be flipped 180° to maintain retrograde revolution. At the time of this maneuver, the subsolar latitude will cross from the southern to northern hemisphere (Eros' 'vernal equinox') and all latitudes will be illuminated. A 13 m resolution global mosaic is planned for this maneuver.

From September–December 1999, the spacecraft will return to low orbits, and the northern hemisphere will receive the highest priority for coverage in 'push-broom' mode. End-of-mission will occur on 6 February 2000. Throughout all phases of the mission at Eros, imaging with MSI will be coordinated with spectroscopic observations from NIS to build up continuous 400–2600 nm wavelength coverage of Eros' surface.

6.2. IMAGING OF MATHILDE

En route to Eros, on 27 June 1997 NEAR will pass about 1200 km from the 61 km diameter, C-type asteroid 253 Mathilde. C asteroids are low albedo bodies thought to be composed of carbonaceous chondrite-like material, which dominate the outer part of the main asteroid belt. MSI may be commanded to acquire both monochrome and color images of Mathilde, at a best resolution of ∼200 m. This flyby would represent the first spacecraft investigation of this class of low albedo asteroids.

7. Conclusions

The NEAR mission will provide the first detailed study of an asteroid, and the Multi-Spectral Imager is an integral part of this investigation. MSI will provide the optical navigation essential to the rendezvous mission. Through its multicolor imaging, the data acquired by the camera will be used to study the surface topography, composition, and evolution of Eros. The mission science objectives were achieved in MSI by adapting existing designs, concentrating on improving the most important features. For NEAR, imaging performance and robustness were more important than low mass or power consumption.

Based on the experience of MSI, one might generalize that accelerated timescale programs inhibit the flying of new instrument designs. To use such new instrument technology in a fast-paced program like NEAR, a prototype to demonstrate feasibility would be required before the program start date. However, incremental improvements in existing designs can be made within these tight schedules.

The final mechanical assembly which includes optical alignment and focusing, and the calibration of the complete camera, are time consuming but essential parts of producing a flight instrument. For MSI, time and effort were expended early in development to increase the efficiency of these late stages of production. For example, the largely athermal design and determination of best focus by changing the object plane enabled the final flight configuration to be determined in one pass instead of a traditional iterative approach. This significantly shortened the time required in the OCF. Optical filters with different thicknesses corrected for wavelength dependent focus changes and also reduced any residual chromatic aberration. Adopting an existing GSE helped to reduce the time needed to develop appropriate test equipment. No doubt with increased innovation and flexibility of approach, more such methods will be developed for future missions.

Acknowledgements

The authors would like to express their appreciation to all of the MSI team members. Their dedicated efforts were directly responsible for the success of the Multi-Spectral Imager. In particular we thank C. J. Kardian, the mechanical designer of MSI. His patience, innovative ideas, and experience proved invaluable to the success of this instrument. We thank S. J. Krein, MSI's thermal engineer, for his expertise and wit, K. Strohbehn for the design of the FPD digital electronics, and S. E. Schlemmer for the FPD board layouts. We also wish to express our gratitude to all of the members of the Technical Services Department of The Johns Hopkins University Applied Physics Laboratory, the members of the NEAR spacecraft I&T team whose patience permitted the timely integration and comprehensive testing of the instrument, the members of the NEAR Mission Operations team, and the NEAR program manager, T. B. Coughlin.

References

Adams, J.: 1974, *J. Geophys. Res.* **79**, 4829.

Allen, C., Morris, R., and McKay, D.: 1994, *J. Geophys. Res.* **99**, 23173.

Bell, J., Davis, D., Hartmann, W., and Gaffey, M.: 1989, in R. Binzel, T. Gehrels, and M. Matthews (eds.), 'Asteroids: The Big Picture', *Asteroids II*, University of Arizona, Tucson, pp. 921-945.

Belton, M., Veverka, J., Thomas, P., Helfenstein, P., Simonelli, D., Chapman, C., Davies, M., Greeley, R., Greenberg, R., Head, J., Murchie, S., Klaasen, K., Johnson, T., McEwen, A., Morrison, D., Neukum, G., Fanale, F., Anger, C., Carr, M., and Pilcher, C.: 1992, *Science* **257**, 1647.

Belton, M., Chapman, C., Veverka, J., Klaasen, K., Harch, A., Greeley, G., Greenberg, R., Head, J., McEwen, A., Morrison, D., Thomas, P., Davies, M., Carr, M., Neukum, G., Fanale, F., Davis, D., Anger, C., Gierasch, P., Ingersoll, A., and Pilcher, A.: 1994, *Science* **265**, 1543.

Carbary, J. F., Darlington, E. H., Harris, T. J., McEvaddy, P. J., Mayr, M. J., Peacock, K., Meng, C. I.: 1994, *Applied Optics* **33**, 204.

Carr, M., Kirk, R., McEwen, A., Veverka, J., Thomas, P., Head, J., and Murchie, S.: 1994, *Icarus* **107**, 61.

CCSDS 102.0-B-4: 1995, Recommendation for Space Data System Standards: Packet Telemetry, Issue 4, November 1995.

Chapman, C., Morrison, D., and Zellner, B.: 1975, *Icarus* **25**, 104.

Cloutis, E. and Gaffey, M.: 1991, *J. Geophys. Res.* **96**, 22 809.

Cloutis, E., Gaffey, M., Jackowski, T., and Reed, K.: 1986, *J. Geophys. Res.* **91**, 11 641.

Cole, T. D. and Cheng, A. F.: 1996, 'The Laser Rangefinder on the Near Earth Asteroid Rendezvous Spacecraft', *Acta Astronomica*, in press.

Drummond, J., Cocke, W., Hege, E., and Strittmatter, P.: 1985, *Icarus* **61**, 132.

Farquhar, R. W., Dunham, D. W., and McAdams, J. V.: 1995, *J. Astron. Sci.* **43**, 353.

Fischer, E. and Pieters, C.: 1994, *Icarus* **111**, 475.

Gaddis, L., McEwen, A., and Becker, T.: 1995, *J. Geophys. Res.* **100**, 26 345.

Gaffey, M., Bell, J., and Cruikshank, D.: 1989, in R. Binzel, T. Gehrels, and M. Matthews (eds.), 'Reflectance Spectroscopy and Asteroid Surface Mineralogy', *Asteroids II*, University of Arizona, Tucson, pp. 98–127.

Gaffey, M., Bell, J., Brown, R., Burbine, T., Piatek, J., Reed, K., and Chaky, D.: 1993, *Icarus* **106**, 573.

Granahan, J., Fanale, F., Carlson, R., Kamp, L., Klaasen, K., Helfenstein, P., Thomas, P., McEwen, A., Chapman, C., Sunshine, J., and Belton, M.: 1995, *Lunar Planetary Sci.* **XXVI**, 489.

Harch, A. P. et al.: 1995, *J. Astron. Sci.* **43**, 399.

Hawkins, S. E.: 1996, 'Overview of the Multi-Spectral Imager on the NEAR Spacecraft', *Acta Astronomica*, in press.

Helfenstein, P., Veverka, J., Thomas, P., Simonelli, D., Lee, P., Klaasen, K., Johnson, T., Breneman, H., Head, J., Murchie, S., Fanale, F., Robinson, M., Clark, B., Granahan, J., Garbeil, H., McEwen, A., Kirk, R., Davies, M., Neukum, G., Mottola, S., Wagner, R., Belton, M., Chapman, C., and Pilcher, C.: 1994, *Icarus* **107**, 37.

Hersman, C. B., Boldt, J. D., Eisenreich, P., Oden, S. F., Temkin, D. K.: 1996, *Modular Design of Data Processing Hardware for Spacecraft Instruments*, 2nd IAA International Conference on Low-Cost Planetary Missions.

Jones, T. and Morrison, D.: 1974, *Astron. J.* **79**, 892.

Lebofsky, L. and Rieke, G.: 1979, *Icarus* **40**, 297.

McFadden, L. A., Tholen, D. J., Veeder, G. J.: 1989, in R. Binzel, T. Gehrels, and M. Matthews (eds.), 'Physical Properties of Atten, Apollo, and Amor Asteroids', *Asteroids II*, University of Arizona, Tucson, pp. 442-467.

McKay, D. S., Swindle, T. D., and Greenberg, R.: 1989, in R. Binzel, T. Gehrels, and M. Matthews (eds.), 'Asteroidal Regoliths: What We do Not Know', *Asteroids II*, University of Arizona, Tucson, pp. 617–643.

Mill, J. D., O'Neil, R. R., Price, S., Romick, G. J., Uy, O. M., Gaposchkin, E. M.: 1994, *J. Spacecraft Rockets* **31**, 900.

Millis, R., Bowell, E., and Thompson, D..: 1976, *Icarus* **28**, 53.

Miner, E. and Young, J.: 1976, *Icarus* **28**, 43.

Morris, R.: 1977, 'Origin and Evolution of the Grain-Size Dependence of the Concentration of Fine-Grained Metal in Lunar Soils: The Maturation of Lunar Soils to a Steady-State Stage', in *Proc. Lunar Planetary Sci. Conf. 8th*, p. 3719.

Morrison, D.: 1976, *Icarus* **28**, 125.

Murchie, S. and Pieters, C.: 1996, *J. Geophys. Res.* **101**, 2201.

Murchie, S., Britt, D., Head, J., Pratt, S., Fisher, P., Zhukov, B., Kuzmin, A., Ksanfomality, L., Zharkov, A., Nikitin, G., Fanale, F., Blaney, D., Robinson, M., and Bell, J.: 1991, *J. Geophys. Res.* **96**, 5925. Ostro, S., Rosema, K., and Jurgens, R.: 1990, *Icarus* **84**, 334.

Pellicori, S. F., Russell, E. E., and Watts, L. A.: 1979, *Applied Optics* **18**(15), 2618.

Pieters, C. M., Gaffey, M., Chapman, C., and McCord, T.: 1976, *Icarus* **28**, 105.

Pieters, C., Fischer, E., Rode, O., and Basu, A.: 1993a, *J. Geophys. Res.* **98**, 20817.

Pieters, C., Head, J., Sunshine, J., Fischer, E., Murchie, S., Belton, M., McEwen, A., Gaddis, L., Greeley, R., Neukum, G., Jaumann, R., and Hoffman, H.: 1993b, *J. Geophys. Res.* **98**, 17 127.

Santo, A. G., Lee, S. C., and Gold, R. E.: 1995, *J. Astron. Sci.* **43**, 373.

Schott Report 7600e–1975: 1975, 'Radiation Resistant Optical Glasses (Cerium Stabilized)', Jenaer Glaswerk Schott and Gen. Mainz, Germany.

Sullivan, R., Greeley, R., Pappalardo, R., Belton, M., Carr, M., Chapman, C., Greenberg, R., Geissler, P., Granahan, J., Head, J., Kirk, R., McEwen, A., Lee, P., Thomas, P., Veverka, J., Morrison, D., and Moore, J.: 1996, *Icarus* **120**, 119.

Veverka, J. and Thomas, P.: 1979, T. Gehrels (ed.), 'Phobos and Deimos: A Preview of What Asteroids Are Like?,' *Asteroids*, University of Arizona, Tucson, pp. 628–651.

Veverka, J., Bell III, J. F., Thomas, P., Harch, A., Murchie, S., Hawkins III, S. E., Warren, J. W., Darlington, H., Peacock, K., Chapman, C. R., McFadden, L. A., Malin, M. C., and Robinson, M. S.: 1997, 'An Overview of the NEAR Multispectral Imager-Near-Infrared Spectrometer Investigation', *J. Geophys. Res.* **102**, 29709.

Veverka, J., Helfenstein, P., Lee, P., Thomas, P., McEwen, A., Belton, M., Klaasen, K., Johnson, T., Granahan, J., Fanale, F., Geissler, P., and Head, J.: 1996, *Icarus* **120**, 66.

Veverka, J., Belton, M., Klaasen, K., and Chapman, C.: 1994, *Icarus* **107**, 2.

Wasson, J.: 1974, *Meteorites: Classification and Properties*, Springer-Verlag, New York.

Weidenschilling, S. J., Paolicchi, P., and Zappala, V.: 1989, in R. Binzel, T. Gehrels, and M. Matthews (eds.), 'Do Asteroids Have Satellites?', *Asteroids II*, University of Arizona, Tucson, pp. 643–658.

Wetherill, G. and Chapman, C.: in J. Kerridge and M. Matthews (eds.), 'Asteroids and Meteorites', *Meteorites and the Early Solar System*, University of Arizona, Tucson, pp. 35–67.

Wisniewski: 1976, *Icarus* **28**, 87.

Zellner, B.: 1976, 'Physical Properties of Asteroid 433 Eros', *Icarus* **28**, 149–153.

Zellner, B. and Gradie, J.: 1976, *Icarus* **28**, 117.

Zellner, B., Tholen, D., and Tedesco, E.: 1985, *Icarus* **61**, 355.

NEAR INFRARED SPECTROMETER FOR THE NEAR EARTH ASTEROID RENDEZVOUS MISSION

JEFFERY W. WARREN, KEITH PEACOCK, EDWARD H. DARLINGTON,
SCOTT L. MURCHIE, STEPHEN F. ODEN and JOHN R. HAYES
Applied Physics Laboratory, Johns Hopkins Road, Laurel, MD 20723–6099, U.S.A

JAMES F. BELL III
*Cornell University Department of Astronomy, 424 Space Sciences Building, Ithaca,
NY 14853–6801, U.S.A.*

STEPHEN J. KREIN
Orbital Sciences Corporation, 20301 Century Boulevard, Germantown, MD 20874, U.S.A.

ANDY MASTANDREA
SSG, Inc., 150 Bear Hill Road, Waltham, MA 02154, U.S.A.

(Received 24 July, 1996)

Abstract. The Near-Infrared Spectrometer (NIS) instrument on the Near-Earth Asteroid Rendezvous (NEAR) spacecraft is designed to map spectral properties of the mission target, the S-type asteroid 433 Eros, at near-infrared wavelengths diagnostic of the composition of minerals forming S asteroids. NIS is a grating spectrometer, in which light is directed by a dichroic beam-splitter onto a 32-element Ge detector (center wavelengths, 816–1486 nm) and a 32-element InGaAs detector (center wavelengths, 1371–2708 nm). Each detector reports a 32-channel spectrum at 12-bit quantization. The field-of-view is selectable using slits with dimensions calibrated at $0.37° \times 0.76°$ (narrow slit) and $0.74° \times 0.76°$ (wide slit). A shutter can be closed for dark current measurements. For the Ge detector, there is an option to command a 10x boost in gain. A scan mirror rotates the field-of-view over a $140°$ range, and a diffuse gold radiance calibration target is viewable at the sunward edge of the field of regard. Spectra are measured once per second, and up to 16 can be summed onboard. Hyperspectral image cubes are built up by a combination of down-track spacecraft motion and cross-track scanning of the mirror. Instrument software allows execution of data acquisition macros, which include selection of the slit width, number of spectra to sum, gain, mirror scanning, and an option to interleave dark spectra with the shutter closed among asteroid observations. The instrument was extensively characterized by on-ground calibration, and a comprehensive program of in-flight calibration was begun shortly after launch. NIS observations of Eros will largely be coordinated with multicolor imaging from the Multispectral Imager (MSI). NIS will begin observing Eros during approach to the asteroid, and the instrument will map Eros at successively higher spatial resolutions as NEAR's orbit around Eros is lowered incrementally to 25 km altitude. Ultimate products of the investigation will include composition maps of the entire illuminated surface of Eros at spatial resolutions as high as $\sim$300 m.

1. Science Objectives and Requirements

The Near Earth Asteroid Rendezvous (NEAR) mission, the first of NASA's new low-cost, rapid development Discovery-class spacecraft, will conduct the first detailed orbital study of an asteroid. The spacecraft was successfully launched from Cape Canaveral on February 17, 1996, and it will approach the asteroid 433 Eros in December 1998 before beginning a year-long detailed examination from low orbit (Farquhar et al., 1995). The NEAR mission has four primary science goals: (1) to

Space Science Reviews **82**: 101–167, 1997.

© 1997 *Kluwer Academic Publishers. Printed in Belgium.*

determine the asteroid's global physical structure, including size, shape, volume, density and spin state; (2) to measure the elemental and mineralogical composition of the surface with sufficient accuracy to enable comparisons with major meteorite types; (3) to characterize the geology and morphology of the surface at the highest possible spatial resolution; and (4) to infer regolith and textural properties of the surface (Cheng et al., 1997). These overall goals drive a stringent set of science requirements for instruments on the mission. Among these requirements are high signal-to-noise ratio (SNR) for all compositional/mineralogic measurements, high spatial resolution global coverage of the surface in order to measure spatial variations in surface properties, and good co-registration and scientific complementarity between the various instruments.

To meet the mission goals and to satisfy the scientific requirements, a diverse and robust set of instruments is included on NEAR (Santo et al., 1995). A key instrument among these is the Near-Infrared Spectrometer (NIS), a grating IR spectrometer designed to measure the spectrum of sunlight reflected from the surface of Eros in the 0.8 to 2.6 μm wavelength range. Reflectance spectroscopy has proved to be a powerful tool for the remote mineralogic characterization of planetary and asteroidal surfaces from both ground-based and spacecraft platforms because many common rock-forming minerals have diagnostic absorption features at visible to near-infrared wavelengths (McCord et al., 1970; Wendlandt and Hecht, 1966; Pieters and Englert, 1993a). A particular advantage of reflectance spectroscopy is the ability to compare remote sensing observations with laboratory spectra of rock, mineral, and meteorite samples (e.g., Gaffey et al., 1993a; Pieters and McFadden, 1994), in order to infer the origin and geologic evolution of solar system objects. In particular, reflectance spectra can be used to infer links between various classes of asteroids and meteorites (Chapman and Gaffey, 1979; Fanale et al., 1992) and to constrain the nature of 'space weathering' or other processes thought to alter asteroid surfaces (Bell et al., 1989; Clark et al., 1992).

Eros is a member of the 'S' class of asteroids, one of several classes that have been defined based on astronomical measurements of albedo, spectral, and polarimetric properties (Chapman et al., 1975). The basic spectral attributes of the class are a moderate albedo, a relatively red spectral continuum, and broad absorptions near 1 μm and 2 μm attributable to ferrous-iron containing silicates. S asteroids dominate the inner main belt and constitute 50% or more of near-Earth asteroids.

Astronomical spectroscopy has been the major technique for identifying the mineral compositions of S asteroids. Transition metal ions in minerals typically have characteristic absorption bands in the visible and near-IR wavelength regions; the exact positions and relative strengths of the absorptions differ between minerals, such that these parameters serve as mineralogic 'fingerprints.' For example, in the Fe, Mg-silicate mineral olivine, a strong, composite absorption is located at 1050 nm. In pyroxene, Fe occurs with Mg and Ca, and creates an absorption that varies in position from 900–1050 nm. Unlike olivine, pyroxene also exhibits

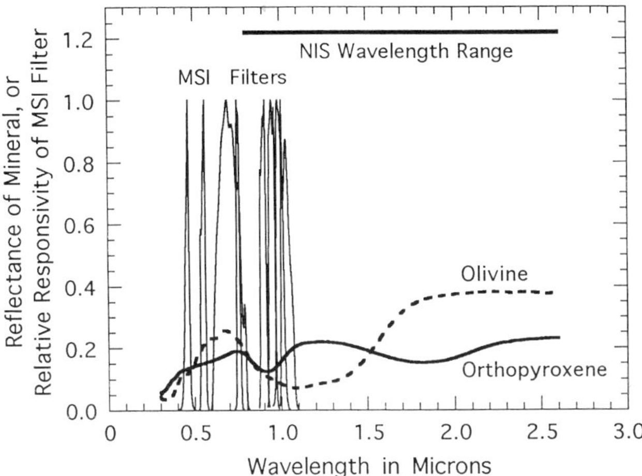

Figure 1. Reflectance spectra of the minerals believed to be the major silicate components of S asteroids, compared with the spectral range of NIS and the seven narrow bandpasses of MSI. NIS covers the wavelength range of diagnostic NIR mineral absorptions. The MSI filter bandpasses are clustered in the wavelength region of the critical 1000-nm mineral absorption, allowing spectral variations detected by NIS to be extrapolated to higher spatial resolution in this wavelength region.

an additional Fe absorption feature at 1900–2300 nm (Figure 1). The wavelength positions of the two pyroxene absorptions vary together systematically as a function of the balance of Fe, Mg, and Ca cations in the pyroxene crystals, allowing detection of differences in pyroxene composition as well as mixing of olivine with pyroxene (Adams, 1974; Cloutis et al., 1986; Cloutis and Gaffey, 1991). Several studies using these techniques (e.g., Gaffey et al., 1989, 1993) have shown that the overall character of S asteroid absorption features is consistent with mixtures of olivine and pyroxene with Ni-Fe metal. However, compositions of S asteroids are highly varied, and the mineralogic heterogeneity within the class is indicative of a variety of rock types on the different S asteroids.

The mineral assemblage on S asteroids is the same type present in meteorite types with highly divergent histories, and substantial debate has arisen regarding the connection between S asteroids and various classes of meteorites. One hypothesis holds that many S asteroids consist of the most common type of meteorites, primitive, undifferentiated ordinary chondrites (e.g., Wetherill and Chapman, 1988). However, S asteroids exhibit a redder spectral continuum and weaker absorption bands than do ordinary chondrites measured in terrestrial laboratories. This difference in spectral properties has been attributed to optical modification of the asteroids' surfaces by processes that affect the fragmental surface regolith, otherwise known as 'space weathering' (Morris, 1977; Clark et al., 1992, Pieters et al., 1993b; Fischer and Pieters, 1994; Allen et al., 1994). An alternative hypothesis is that the spectral contrast between S asteroids and ordinary chondrites occurs because the S-asteroids consist predominantly of different rock types than ordinary

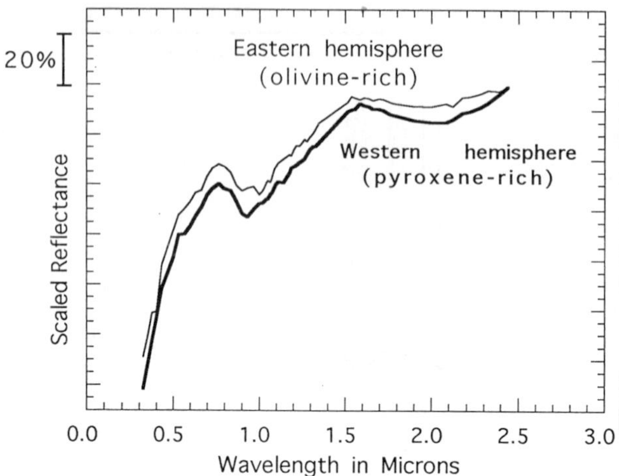

Figure 2. Composite visible-near-infrared spectra of opposite faces of Eros. This presentation highlights differences in strengths and positions of the absorption features due to olivine and pyroxene. The reflectance is scaled to the measured value at 560 nm. The two plots are offset vertically for clarity. The significant features are the different depths and shapes of the absorption bands, not the absolute magnitude of the reflectance.

chondrite (Bell et al., 1989; Gaffey et al., 1989, 1993). The meteorites most closely resembling the bulk of S asteroids spectrally are stony-iron and primitive achondritic meteorites, which have undergone varying degrees of igneous melting and separation of mineral phases. The implications of either model for the origin and evolution of S asteroids, their relationship to meteorites, and the evolution of the early solar system are profound: if S asteroids consist mainly of undifferentiated ordinary chondrite, they would be primitive bodies largely unchanged since accretion; if S asteroids are differentiated, they instead represent collisional fragments of parent bodies that experienced a much greater degree of geologic evolution early in solar system history. In either case, the diversity of S asteroid compositions suggests that there is not a single source for all asteroids of this class.

Spectroscopic analysis shows that the average mineralogy of Eros is in the middle of the range for S asteroids, and at face value is not inconsistent with the Fe mineralogy of ordinary chondrites. However, one hemisphere has a shorter-wavelength 1000-nm absorption and stronger 2000-nm absorption than does the opposite hemisphere (Figure 2). These differences imply an $\sim$20% difference in the abundance of pyroxene relative to olivine, which is comparable to the greatest mineralogic diversity yet recognized on any single S asteroid and greater than the diversity within individual compositional types of ordinary chondrites (Murchie and Pieters, 1996). This degree of mineralogic variation suggests that Eros's surface may contain geologic boundaries that are key to deciphering its formation and evolution.

The primary objectives of the NIS investigation are to map the distribution and abundance of minerals on the surface of Eros and to provide information on the physical and textural properties of these surface minerals (Veverka et al., 1997). NIS is an important part of a complementary instrument strategy involving higher spatial resolution observations of Eros to 1050 nm with the Multi-Spectral Imager (MSI), and lower spatial resolution elemental abundance measurements with the X-ray/Gamma-ray Spectrometer (XGRS). MSI data will allow mineralogic measurements from NIS to be extrapolated down to much finer scales, revealing details of the physical interrelationship of discrete materials identified by NIS. Elemental abundances from XGRS, together with NIS measurement of the distribution and abundance of minerals, will support a more accurate assessment of the rock types composing Eros than would either data set alone. This information together will be used to constrain the history of geologic processes that formed Eros, and determine how the materials on Eros may be related to meteorites. Achieving these NIS objectives will satisfy a major subset of the overall NEAR mission goals, and will also provide a wealth of important new information on the origin and evolution of asteroids and on the possible links between asteroids and meteorites.

Four key measurement and calibration requirements were developed which reflect past scientific experience in this field, plus the realization that compositionally significant variations in Eros's spectral properties may generally be quite subtle. The first requirement was to obtain high signal-to- noise ratio (SNR) using an operationally acceptable integration time. As discussed in the mission overview paper of this issue (Cheng, 1996b), NEAR will be in orbit at a very low velocity relative to the asteroid, so significant integration times to improve SNR can be used. Second, to use high SNR data effectively, the responsivity of the instrument as a function of wavelength must be well characterized. A comprehensive pre-launch calibration effort was therefore required. Third, because the instrument is operating with DC signals at very low levels, the possibility of channel to channel drifts in radiometric calibration over three years in flight is a real concern. An in-flight optical calibration capability had to be included. Fourth, separate spectral channels need to be accurately coaligned (looking at the same field of view). This is particularly important for spectrally adjacent channels, but all channels need to be relatively well aligned. Since the spectral features that are observed vary only slightly, it might appear that this is not very critical. However, with the relatively large field of view of NIS, observations frequently will be made with highly variable illumination within the field of view. Channels which have a different field of view would observe different illumination levels, which could be misinterpreted as spectral features. The most dramatic example is viewing the limb, where part of the field of view is illuminated asteroid and part is dark space.

NIS was also required to obtain data over a wide range of phase angles, including zero. The NEAR spacecraft concept calls for all instruments to be body fixed and to operate in the vicinity of 90° phase angle, which is the typical viewing geometry of Eros during the orbital phase of the mission. Because the higher illumination

Table I

Near infrered spectrometer characteristics

Spectral range	
First order (InGaAs detector)	1.5 to $\approx$ 2.4 μm
Second order (Ge detector)	0.8 to 1.5 μm
Spectral resolution with narrow slit	
First order	44 nm
Second order	22 nm
Field of view	
Narrow slit	$0.37° \times 0.76°$
Wide slit	$0.74° \times 0.76°$
Field of regard	140°
Scan step size	0.4°
Collecting area	5 cm^2
Detector array size	1×16 mm
Number of elements per array	32
Size of detector elements	1×0.5 mm
Mass	14.0 kg

and lack of shadows near zero phase angle is critical to reflectance spectroscopy, a scanning capability was specified for NIS that included the ability to observe at zero phase during the initial flyby and possibly during the plane flip maneuver later in the mission. NIS is the only NEAR instrument with this capability.

One critical requirement was neither technical nor scientific. In order to allow the spacecraft to launch to Eros in the only available launch window, the instrument had to be delivered to the spacecraft fully qualified and calibrated by the end of July 1995. The total time for development, qualification, and calibration was just under 20 months.

2. Instrument Design

2.1. OVERVIEW

Figure 3 is a photograph of the NIS and associated components taken during final instrument level testing. The photograph is labeled to show the optics housing, the Ge and InGaAs detector modules which are bolted to the housing, the mounting bracket that mounts NIS to the spacecraft at the appropriate angle, the support electronics, and the Data Processing Unit (DPU). In this photograph all of the intra-instrument harness is in place, and the heaters and thermostats have been installed, but no thermal blanketing is in place.

The mechanical and optical design of the NIS is based on an instrument called a Special Sensor Ultraviolet Spectrographic Imager, or SSUSI, developed for the

Figure 3. Photograph of NIS taken during integration testing. The NIS instrument is mounted on the bracket that is used to properly position it on the spacecraft. Thermal blankets are not in place in this photo. Relative to the instrument, the nominal direction to the Sun when mounted on the spacecraft would be into the floor. During the flyby, the direction to the asteroid would be overhead in this photo. During orbital operations the direction to the asteroid would be approximately towards the camera in the right photo.

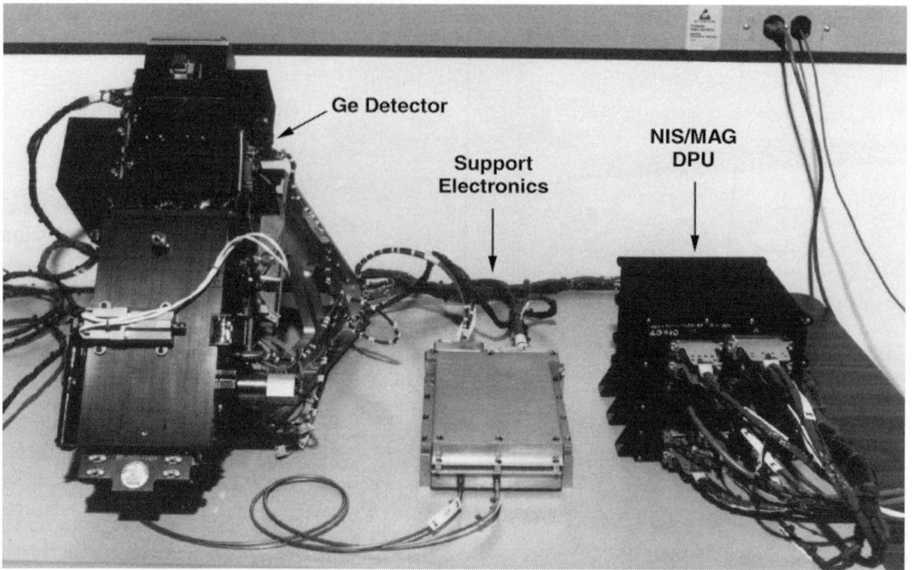

Figure 3. Continued.

Defense Meteorological Support Program (DMSP) satellites. The modification of a heritage design allowed the aggressive NEAR schedule to be met, while still providing an instrument which meets science requirements. Both SSUSI and NIS use a single telescope mirror, a grating spectrometer, and a scan mirror providing a 140° field of regard. The support electronics are nearly identical in SSUSI and NIS. The detectors, bracket, in-flight calibration plaque, and DPU are new designs for NIS. SSUSI was designed by APL for the DMSP program with SSG, Inc. as the subcontractor for manufacture of the optics and housing. SSG retained this role for NIS, building the optics and housing and the support electronics.

The detectors and their electronics are built in separate housings which bolt onto the optics housing. Each detector is a linear array of 32 elements. The detectors and their associated electronics are passively cooled to around −35 °C. The detector data are digitized within the detector housings, and communicated digitally to the DPU. The optical and electronic signal flow is illustrated in the block diagram in Figure 4.

Table I lists key specifications and characteristics of the instrument. The table shows the spectral range of NIS as 0.8 to approximately 2.4 μm. The long wavelength end of the spectrum is expressed as an approximation because the maximum usable wavelength is determined by the SNR required. The instrument was designed to operate usefully up to 2.6 μm, but actually outputs data up to 2.7 μm. The SNR is falling dramatically at the longest wavelengths because the detector is transitioning into cutoff. High SNR is maintained up to about 2.4 μm (see Section 3.2).

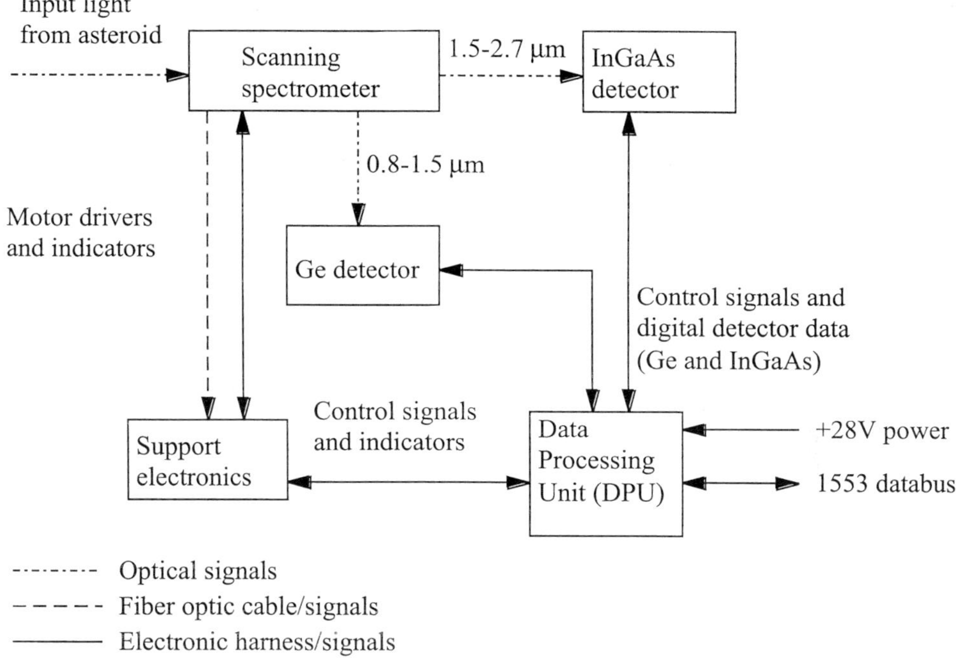

Figure 4. Block diagram of NIS.

Figure 5 is a photograph of the instrument on the NEAR spacecraft. The bracket places NIS in the right orientation to obtain the desired field of regard. The orientation of the 140° field of regard is shown in Figure 6. The wide field of regard allows NIS to operate during different phases of the mission. During orbital operations around Eros, the line of sight will be approximately out the 'side' of the spacecraft, near the line of sight of the imager and other instruments. The boresight reference for NIS, which is coaligned with the other instrument boresights, is called instrument nadir. The field of view will be pointed (using the scan mirror) in the vicinity of instrument nadir during the orbital phase. During the flyby phase, the spacecraft will pass between the asteroid and the Sun at low speed (5 m s^{-1} nominally) to measure the gravitational deflection and also to permit NIS data to be acquired at low phase angle. This requires the line of sight to be positioned approximately 90° in the anti-Sun direction from the instrument nadir. The asteroid will subtend a small field of view at flyby, so only a few steps will be required to scan across the asteroid. The maximum limit of the line of sight is 110° anti-Sun of nadir, permitting flexibility in timing of the 180° roll needed near closest approach. In the sunward direction, the line of sight can be positioned up to 30° from instrument nadir. At this position, referred to as the start position, the instrument is viewing the solar illuminated calibration plaque. Approximately the first 10° from the start position are not used because they are partially viewing the calibration plaque.

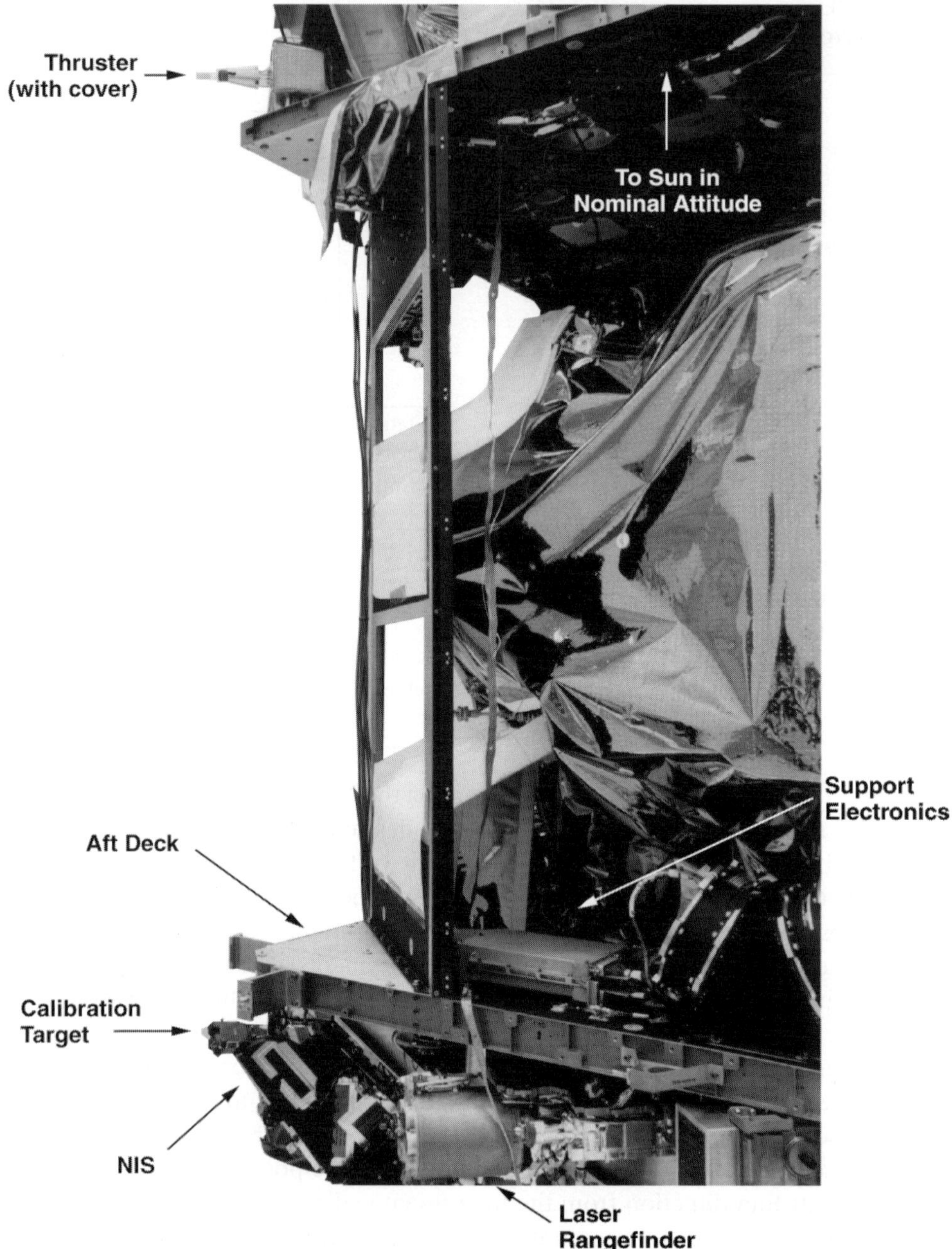

Figure 5. Photograph of the NIS on the NEAR spacecraft during assembly. The spacecraft side panels and thermal blankets are removed allowing interior components to be seen. The optical surface of the calibration target, which is gold, can be seen facing upwards (this is a spare calibration target; the flight unit was kept covered when mounted to the spacecraft). The thruster labeled on the forward deck partially shadows the calibration target at zero yaw angle, so a slight spacecraft yaw will be used for calibration observations.

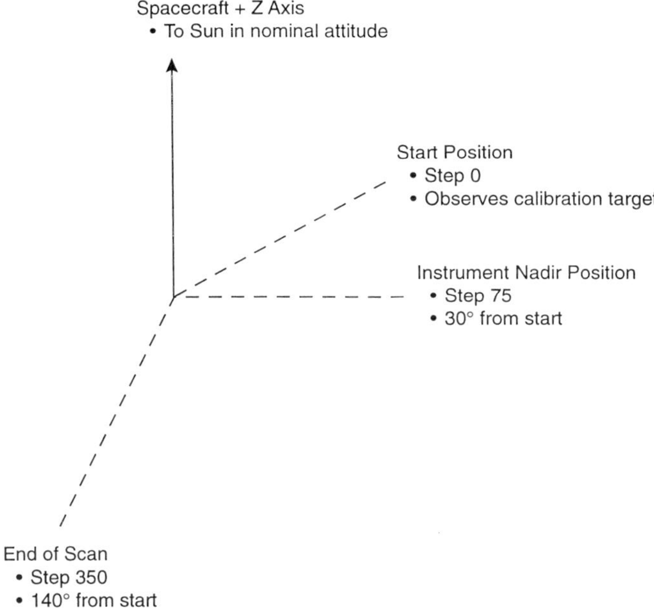

Figure 6. Field of regard of the NIS in relation to the NEAR spacecraft. The start and instrument nadir positions are indicated by a fiducial using an optical sensor. During orbital observations, the field of view usually will be oriented near the instrument nadir position, which allows co-ordinated observations with the MSI. During the flyby the field of view will be oriented approximately in the anti-Sun direction since the flyby will provide zero phase angle.

2.2. OPTICS

A spectrometer designer will find the NIS to be an unusual design for an infrared instrument. The reason for the selected design is straightforward; insufficient development time was available to produce a new instrument so it is a modification of another instrument – the ultraviolet imaging spectrograph SSUSI – designed for a very different purpose. The design task was to convert an earth-scanning ultraviolet spectrograph operating from 115 to 180 nm into one for use in the range 0.8 to 2.7 μm at an asteroid.

2.2.1. *Optics Description*

The NIS optics consists of a scan mirror, an aperture stop, a telescope mirror, and a Rowland circle spectrometer using the first and second orders from the diffraction grating. The optical layout of the NIS is shown in Figure 7.

The NIS scan mirror provides a 140° field-of-regard as described previously. Between the scan mirror and the imaging mirror is a 20 by 25 mm rectangular aperture stop. Sensitivity analysis demonstrated that these dimensions are adequate for the NEAR mission. The location of this stop between the mirrors was a compromise. If it were closer to the imaging mirror the scan mirror would have to be larger,

 JEFFERY W. WARREN ET AL.

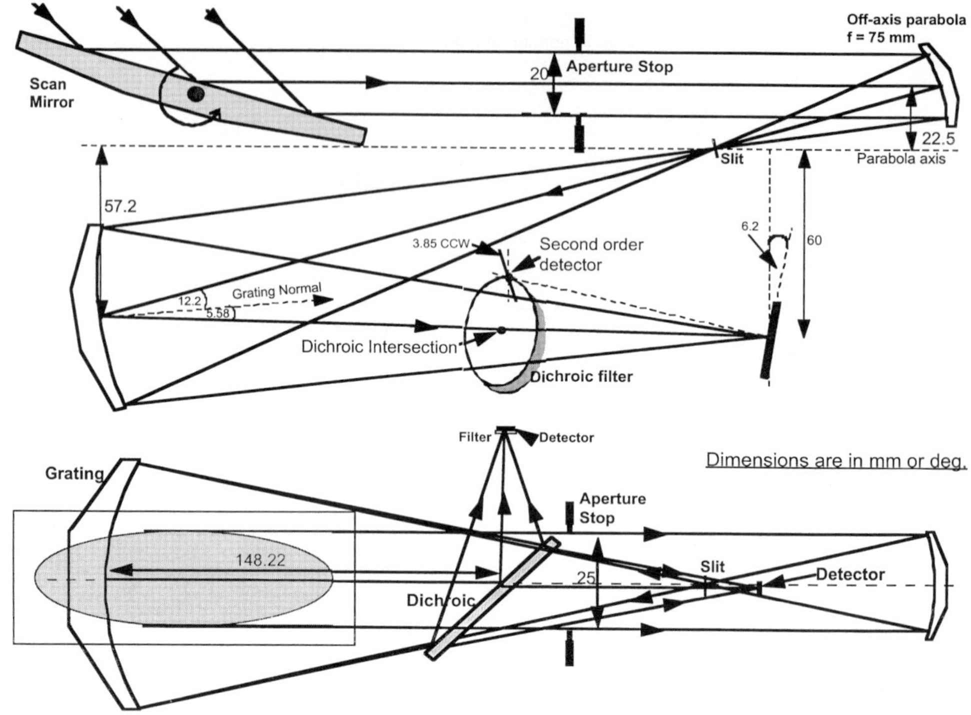

Figure 7. Optical schematic of the NIS.

and if it were closer to the scan mirror the imaging mirror would be larger. (Note that the 11.8° FOV of the SSUSI requires a wide scan mirror and a large telescope mirror which are not needed on the NIS, but remain part of the design.) The shaded area of the mirror shows the elliptical area of the mirror used when the FOV is at an extreme angle. This area varies with the FOV direction and is a minimum when the FOV is directed to the end of scan.

Imaging is performed by a single off-axis parabolic mirror which images the scene at the field stop, the entrance slit of the spectrometer. This mirror has a focal length of 75 mm and its center is 22.5 mm off-axis. It provides adequate image quality over the small FOV. It images the scene at the field stop with a scale of $0.76°$ mm^{-1}. There are two selectable slits; a movable narrow slit with dimensions of 1×0.5 mm giving a spatial FOV of $0.76° \times 0.38°$, and a fixed 1 mm square slit with an FOV of $0.76°$ square. The narrow slit is planned as the primary data taking configuration. Therefore, the slits are located so that the narrow slit is in best focus, while the wide slit has a very slight defocus. A shutter allows the field stop to be closed for recording dark signals. The imaging mirror also images the aperture stop at the diffraction grating. This ensures that the grating has the minimum size and is evenly illuminated. The diffraction grating and mirrors have reflecting surfaces of gold which provides optimum reflectance for the spectral range above 0.8 μm.

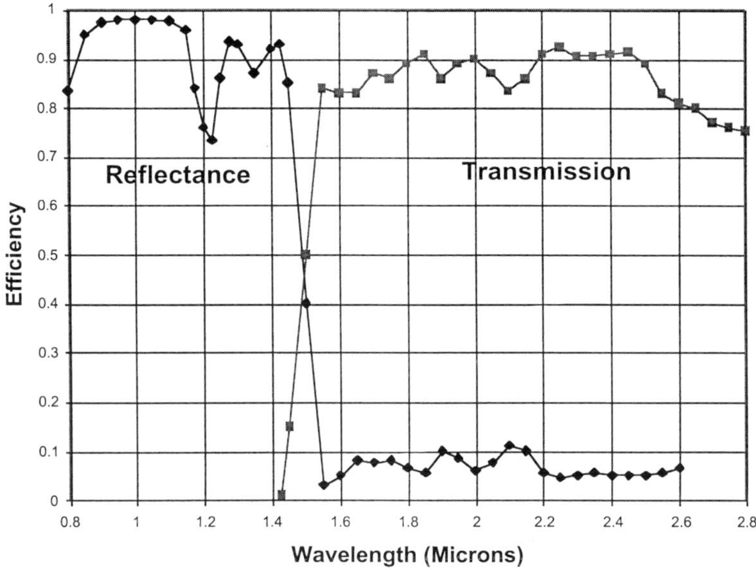

Figure 8. Reflectance and transmission characteristics of the NIS dichroic beamsplitter.

The principal difference between the NIS and the SSUSI is the spectral range. The SSUSI operates over the spectral range 115 to 180 nm but the NIS has the range 0.8 to 2.7 μm. A single grating spectrum cannot accommodate this full range with acceptable efficiency so it was broken into two ranges; 0.8–1.5 and 1.3–2.7 μm. A single grating provides both ranges by operating in the first and second orders. These orders are separated by a dichroic beamsplitter which transmits the long wavelengths and reflects the short wavelengths, as shown in Figure 8. It has high transmission above 1.55 μm and high reflectance below 1.45 μm, with 50% efficiency achieved near 1.5 μm. Since the shorter wavelength detector provides much better performance, the transition was designed to be at 1.5 μm to effectively use all the elements of this array, while the first few elements of the long wavelength array get very little light.

The diffraction grating operates in the Rowland configuration, i.e., it both disperses and reimages the energy. It has a radius of curvature of 200 mm and a ruling of 57 grooves mm^{-1}. These combine to produce a spectrum which has a dispersion of 88 nm mm^{-1} in the first order. The traditional problem with a Rowland circle spectrometer, astigmatism, is reduced by making the grating toroidal. The grating radius of curvature in the direction perpendicular to the dispersion direction is 193.7 mm. This cancels the astigmatism at the center of the spectrum although it is still present at the extreme wavelengths. This allows adequate spatial resolution in both the spectral and spatial directions. Reducing the astigmatism also concentrates the energy on the detector.

Because the two detectors are the same size, the spectral range of the second-order detector must be half that of the first-order detector. If the detectors were aligned the same relative to the optical axis, the minimum and maximum wavelengths in the second-order detector would be exactly half those for the first-order detector. A tilt of the dichroic beamsplitter is used to shift the second order spectrum to cover the desired wavelength range. The instrument is aligned so that the center of the first-order spectrum, 2.0 μm, is on the optical axis. The best image quality is at this wavelength. Therefore the second-order image quality is best at 1.0 μm, which is not in the middle of the spectral band but is near the most important absorption bands. The longest wavelengths in the second order thus experience more aberrations than the shortest wavelengths. The results of this alignment and aberrations are clearly seen in the alignment calibration data in Section 3.5.1.

2.2.2. *Spectral Crosstalk*

A major problem with any spectrometer is the elimination of higher spectral orders, or shorter wavelengths, from the desired signal. This problem becomes even more complex in a spectrometer using both first and second orders. Figure 9 shows how the first, third and higher orders can overlap the second order. In the second-order (Ge) detector, the first-order light is attenuated by the low reflectance of the dichroic but several percent is still reflected to the detector. Fortunately, the Ge detector is in cut-off (no response) at wavelengths over 1.8 μm. A small correction has been determined by pre-launch calibration for removing 1.6–1.8 μm light from signals of the 0.8–0.9 μm channels. A single filter to eliminate the third and higher orders from the Ge detector would have to remove all wavelengths below 1.0 μm. As this includes part of the desired wavelength range, more than one filter is required. The solution was to use a two zone filter; one zone attenuates light below 0.77 μm and is effective for channels at wavelengths of less than 1.15 μm (1.5 × 0.77 μm). The second zone attenuates light below 1.0 μm falling on channels between the first zone and 1.5 μm. The boundary between the two zones is centered at 1.12 μm, causing a 40% loss of light which is confined to a single channel. Calibration results show that the filter was very effective in removing the higher orders. Because this order sorting filter must be very close to the Ge detector, the filter is actually mounted to the detector housing rather than to the rest of the optics.

For the first-order detector, the primary goal is to prevent second-order light at shorter wavelengths from reaching the detector. This is readily achieved in the design of the dichroic beamsplitter which has transmission of less that 0.1% for wavelengths below 1.3 μm.

2.2.3. *Focal Plane Dimensions*

The NIS has two, one-dimensional image planes which are each 1 mm by 16 mm. Detector arrays with 32 elements are used; each element is 1 mm high by 0.5 mm wide. This detector element size is unusually large, particularly for the longer wavelength detector. The magnification from the entrance slit to the detector is

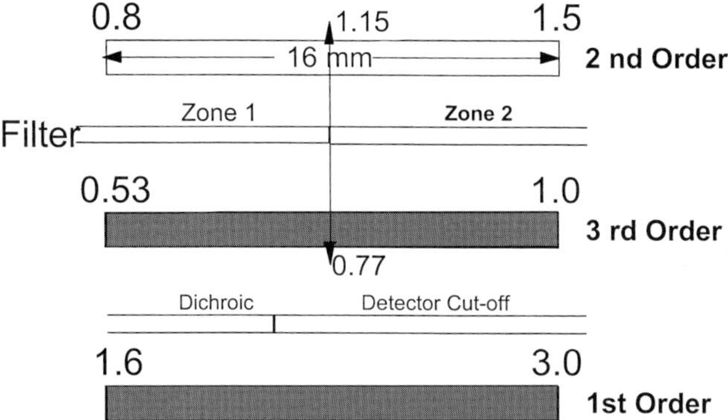

Figure 9. Spectral crosstalk of the first and third spectral orders onto the second-order (Ge) detector. The desired signal on the Ge detector is second order light from 0.8 to 1.5 μm. The detector is also illuminated with first-order light from 1.6 to 3.0 μm and third-order light from 0.53 to 1.0 μm. The first-order light is rejected by low reflectance of the dichroic beamsplitter and at the longer wavelengths by low detector responsivity. Since the dichroic beamsplitter reflectance is not zero, a correction is required for light between 1.6 and 1.8 μm. Third-order light is attenuated with a two zone filter over the detector and no correction is required.

unity, so each detector element is the same size as the narrow slit. Reducing the size of the slit was a design option, and would have permitted smaller (and improved performance) detectors to be used. The 0.5 mm dimension of the slit yields a field of view of 0.38° in the direction that the field of view is scanned. Since the step size is 0.4°, making the field of view any smaller in this direction would have resulted in unacceptable gaps in coverage between steps. The 1 mm dimension results in a field of view of 0.76°. Some flexibility was available here, but a relatively large field of view was desired to cover the MSI field of view with a few scans of NIS (a spacecraft roll maneuver is needed between each scan in some scenarios). Selecting 1 mm as the detector height allowed the use of a commercially available array for the second order, while permitting the MSI field of view to be observed with only three NIS scans. As will be discussed, this did result in an unusually large first-order detector with some impact on performance.

2.2.4. *In-flight Calibration*

For in-flight calibration a diffuse gold plaque is located at the scanner end of the instrument where it extends beyond the edge of the aft deck and intercepts direct sunlight. The viewing (emission) angle of the plaque by the NIS optics is fixed at 75°. With the spacecraft in its 'normal' orientation, with the solar arrays facing the Sun, the solar incidence angle on the plaque is 45° (Figure 5). This may be increased by 30° with spacecraft yaw, giving an incidence angle of 75° resulting in specular geometry. The plaque is made with Infragold LF (trademarked), which is a porous metal surface with a gold coating. This was selected for its acceptable spectral range, stability over time, and reasonably low specular reflection. It was chosen over Spectralon[TM] because the NIS calibration plaque cannot be protected with a cover during flight, and Spectralon has been shown to undergo wavelength-dependent optical degradation after long term exposure to solar UV radiation (Bruegge et al., 1993; Stiegman et al., 1993).

While the plaque is designed to be as diffuse as possible, a specular reflection component remains (see calibration data in Section 3.6). It is therefore desirable to operate at incidence angles as far from the specular condition as possible, which implies operation at small yaw angles. The preferred incidence angle on the plaque is about 50° (5° of spacecraft yaw) to avoid shadowing by a small thruster on the forward deck (see Figure 5). Avoidance of large yaw angles also prevents direct specular reflections of sunlight incident on the side of the spacecraft from reaching the calibration plaque, which could otherwise contaminate measurements.

The calibration plaque was designed so that its brightness would be comparable to that of Eros, and any nonlinearity of detector response that might have occurred would not strongly affect calibration. To meet launch vehicle mechanical constraints while neither sacrificing too much field of regard nor compromising the stray light design, the plaque was sized to fill only about 30% of the aperture of NIS. As a direct result the optical surfaces are not uniformly sampled. The grating in particular consists of multiple zones which have different efficiency, and not all of these are sampled by the calibration plaque observations. The in-flight data using the plaque must be related to pre-launch calibrations or preferably lunar and Earth observations (during Earth flyby) to establish the final calibration. Once this is accomplished, the calibration plaque can be used to correct for changes which might occur in the detectors or electronics.

2.2.5. *Stray Light*

Stray light from outside of the FOV was not a major issue with NIS since minimization of stray light was an integral part of the SSUSI design, and the SSUSI mission results in a much more stressing stray light environment. Both the scan and telescope mirrors had originally been designed and specified for use in the ultraviolet so they have very low scatter surfaces. Baffling along the light path prevents stray reflections from reaching the focal plane. Of more concern is light scattering from one part of the spectrum to another. This depends on the quality of the grating. The stray light levels are described in the calibration section.

2.3. DETECTOR ELECTRONICS

2.3.1. *Overview*

Figure 10 shows a simplified schematic with information applicable to both detector electronics packages. Each element of the photodiode arrays has a dedicated transimpedance preamplifier which incorporates a low pass filter. Each element acts as a separate photodiode with a common cathode for all 32 diodes in each array. For each detector and its 32 associated preamplifiers, a switch is used to multiplex the outputs to a single analog to digital (A/D) converter. This conversion

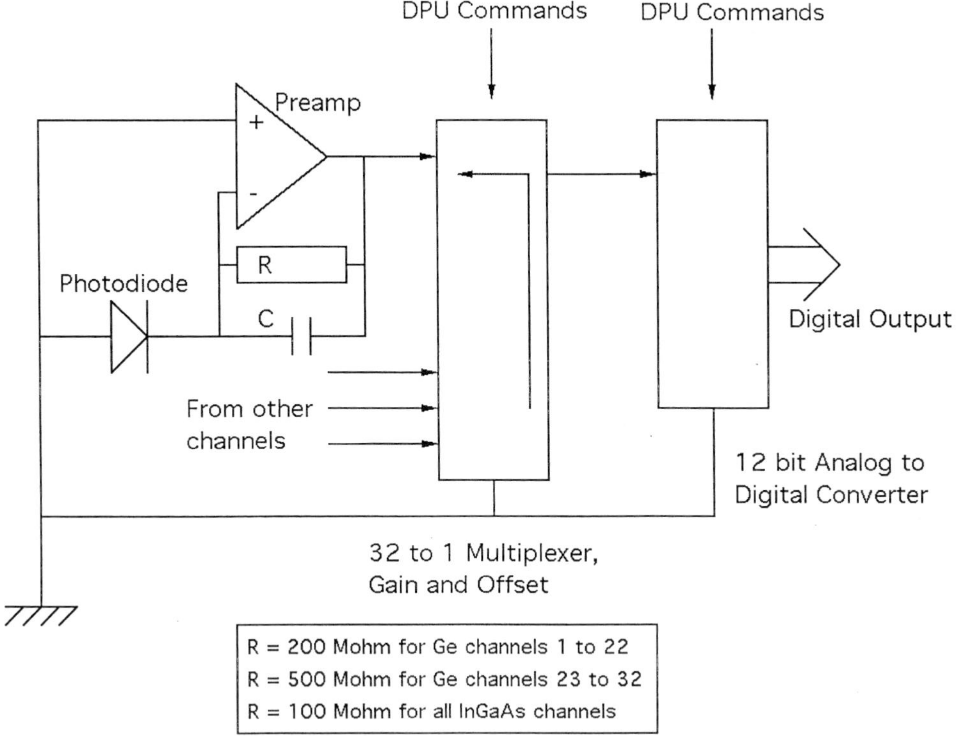

Figure 10. Block diagram of detector electronics. Each channel has a dedicated transimpedance amplifier with low pass filtering. A multiplexer sequentially connects each of the 32 preamplifier outputs to the analog to digital converter. For the Ge detector only, a selectable gain amplifier is located between the multiplexer and the digitizer, with a gain of either $1\times$ or $10\times$.

is fast in comparison with the response time and thus a very simple system can be used. For the Ge detector only, an amplifier with a selectable gain of either one or ten is located between the multiplexing switch and the A/D converter. The same converter also digitizes several temperatures in each unit. Each detector assembly functions independently and they are both controlled and powered by the DPU. Data are obtained once per second from both detectors. Design constraints included low power consumption, passive cooling, radiation tolerance, operation in vacuum, and robust compact mechanical design.

2.3.2. *Detector Arrays*

Two detector arrays are required, with sensitive wavelengths ranging from 0.8 to about 2.6 μm, sensitivity in the nW range, and linearity to better than 1%. The existing optical design defined the detector array size.

2.3.2.1. *Ge Detector.*   A 32 element Germanium (Ge) photodiode array was chosen for the second order (shorter wavelength) detector. This device is commer-

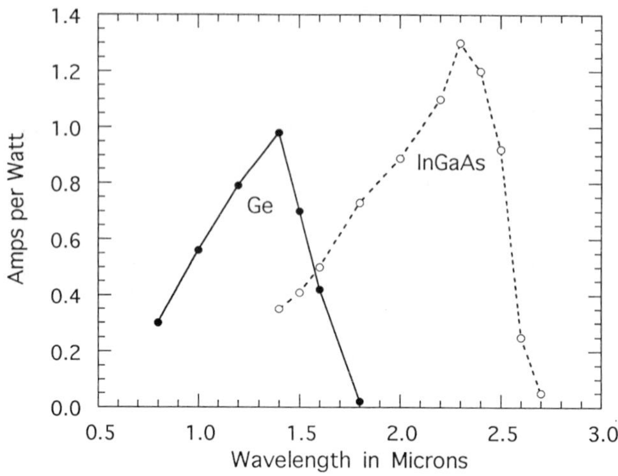

Figure 11. Spectral responsivity of Ge and InGaAs detectors.

cially available from EG&G Judson. Each element is 0.5 mm wide by 1 mm high, for a total array size of 1 by 16 mm. The spectral response curves of both detectors are shown in Figure 11. The timescale and budget of NEAR precluded a special purchase of flight qualified parts. Instead commercial parts were purchased and extra screening tests were performed in house. Initial screening showed a good degree of uniformity in shunt resistance and responsivity between different elements, and from chip to chip. The shunt resistance was within 20% for all but one of the 10 samples screened, and responsivities varied by ±3% within a detector array. This level of performance was retained after thermal cycling, and on this basis nine units were retained as potential flight devices.

2.3.2.2. *InGaAs Detector.*   For the first-order (longer wavelength) array no equivalent model to the Ge array was commercially available. Sensors Unlimited, Inc. made a smaller Indium Gallium Arsenide (InGaAs) array of a formulation which had a suitable long wavelength cutoff of 2.6 μm (Figure 11). They undertook to make a linear array of the same size as the Germanium array. This was done by paralleling groups of ten of an existing 50 μm wide photodiode array design, and increasing the height from 0.25 to 1 mm. This is a very large size for an InGaAs detector; obtaining consistently good performance is difficult.

The diode can be modeled as a current source with a parallel, or shunt, resistance, a shunt capacitance, and a series resistance which can usually be ignored. An ideal device would be a perfect current source, and the additional model elements of a real device may seriously limit performance, making the amplifier dominate the noise and stability performance of the system. For this application the most important detector specification is the shunt resistance. SNR is directly proportional to photodiode shunt resistance. Since the shunt resistance of the photodiode nom-

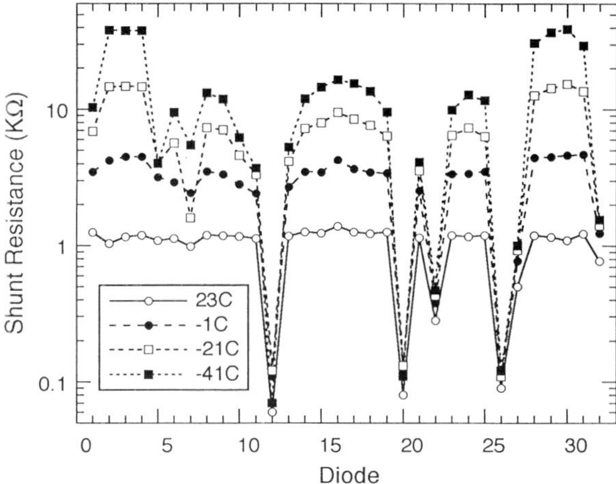

Figure 12. Shunt resistance (on a per channel basis) of an InGaAs detector array, but not the unit selected for flight. High shunt resistance is required to obtain high SNR. Shunt resistance measurements were not obtained on the flight detector for schedule reasons; the flight detector was selected on the basis of actual performance in a breadboard circuit.

inally doubles for each 8C reduction in temperature, cooling can greatly improve performance.

The requirement for the InGaAs diodes was $10\,\Omega$ shunt resistance at $-30\,°C$. The manufacturer experienced some difficulty in producing these large area diodes with the required shunt resistance. Figure 12 shows shunt resistance data for a candidate device. Some elements (12, 20, and 26) are clearly defective at all temperatures and this device would not be selected for flight in consequence. Other elements (5, 11, 21, and 32) appear reasonable in a room temperature test but fail to improve at low temperatures. The actual selection of a flight detector from among the available arrays was performed by measuring performance in a breadboard circuit, rather than inferring performance from resistance measurements. The array selected for flight has no defective elements at room temperature but a few low resistance elements when cold. This manifests itself in the assembled unit as channels with greater noise than average. The financial and especially time constraints of the program led us and the selected manufacturer to accept arrays with somewhat less than ideal performance. Nevertheless the overall system level performance objectives were achieved, albeit with a requirement for obtaining dark data more frequently than originally planned.

2.3.3. *Electronics Design*

A preamplifier circuit providing high current sensitivity but low bandwidth is required. For the InGaAs detector this must be accomplished despite the low shunt resistance of the photodiodes. A number of fairly elaborate ideas were tried before adopting the simple transimpedance or virtual ground circuit shown in Figure 10.

The preamplifier has a noise gain of one plus the ratio of feedback resistance to diode shunt resistance. The amplifier input offset voltage and the amplifier voltage noise are multiplied by the noise gain to obtain the output offset voltage and output voltage noise, respectively. The InGaAs photodiode shunt resistance is much less than the feedback resistance and as a result the noise gain is high, and voltage noise and offset voltage of the amplifier are important. Under these circumstances good amplifier voltage specifications are more important than good current specifications, and the amplifier performance dominates system performance. In principle, reverse bias on the photodiodes could be used to raise the shunt resistance, but the narrow bandgap photodiodes would produce unacceptably high leakage current. The Ge photodiode shunt resistance (when cold) is comparable to the feedback resistance, which requires a balance of good current and voltage noise specifications for the amplifier. Amplifier offsets are much less important. In both circuits quantization by the A/D converter contributes to noise, but it is the dominant noise source only for the Ge circuit when operated in the unity gain state.

The effective input resistance of the transimpedance circuit at low frequency is approximately the value of the feedback resistance divided by the operational amplifier open loop gain. Provided that this is much less than the diode shunt resistance most of the signal current will be measured. Since the diode resistance varies with temperature, finite open loop gain is a possible source of measurement inaccuracy. High open loop gain is therefore required for the preamplifier in the InGaAs circuit.

Variations in diode resistance also cause the noise gain to vary, resulting in an output voltage shift with no light on the photodiode; a zero or dark drift. Drifts in the amplifier offset with temperature or time also cause a dark drift. Some method of periodically inspecting the zero level and perhaps setting it to zero may be used, leading naturally to an input light chopper as the system with the best performance but also the most complexity. For the present system this complexity is not needed and periodic measurement of the dark signal is a satisfactory alternative.

The feedback resistors for the preamplifiers are shunted by a capacitor to give approximately 0.25 s time constant or 1 Hz noise bandwidth. Since data are obtained once per second, this would give a maximum possible mixing between time adjacent measurements of about 2% (e^{-4}). When the slit, shutter, or scan mirror are moved, any of which could give a substantially different optical input, no data are taken in the next second. This eliminates any concern about mixing with the previous measurement.

2.3.3.1. *Ge Electronics.*   The feedback resistor for the Ge transimpedance amplifier is 200 MΩ for channels 1–22, and 500 MΩ for channels 23–32. These values were chosen to give close to full scale output for the maximum anticipated signal for each channel, taking into account the nominal optical design. The shunt resistance of the Ge diodes is about 5 MΩ at room temperature and is in the hundreds of MΩ at operating temperature. Room temperature noise gain of about 100 approaches unity at operating temperature. For this detector any FET input operational amplifier can be selected on the basis of low current noise, low power supply current, and convenience. An AD 648 dual packaged operational amplifier was chosen.

The output of the A/D converter is referred to in units of DN, or data number, where one DN corresponds to one least significant bit from the A/D converter. The 12 bit A/D converter works over a five volt range, so one DN is equal to 1.22 mV at the output of the preamplifiers when the Ge electronics gain is set to one. When the switchable 10 times gain stage is selected (between the transimpedance amplifier and the analog to digital converter), one DN is equal to 0.12 mV at the output of the preamplifiers. The 10 times gain state provides higher sensitivity for use with small signals because the noise level is low.

The Ge data are quantization noise limited when operated with unity gain state, and nearly so when operated with 10 times gain state. The voltage noise predicted using values from the amplifier data sheet is about 0.2 mV typical at room temperature and less than 0.1 mV cold. The current noise term is about 0.01 mV warm and even less cold. The resistor and intrinsic diode noise are unimportant. Although these are typical rather than maximum values from the amplifier data sheets, test results are consistent with these calculations. The equivalent quantization noise of the A/D at the preamplifier output is 0.35 mV or 0.035 mV for the unity and 10 times gain states, respectively. The offset voltage is significant warm but not cold, and the bias current always has less effect than the voltage offset. Even though the noise gain for this detector is low, the voltage noise is the largest source of noise for this circuit. Although better amplifier voltage specifications are possible with a different type of amplifier, use of a FET input operational amplifier, with its excellent current specifications, provides the best overall performance.

2.3.3.2. *InGaAs Electronics.*   For the InGaAs amplifier design the very low detector shunt resistance means that the transimpedance amplifier will be operating at very high noise gain. With a 100 MΩ feedback resistor a value of 10,000 would be typical. This means that an operational amplifier with very high open loop gain and low voltage noise will be needed as well as low offset voltage and drift if a simple circuit is to be used. All of this is in addition to low current noise, bias current, and also of course low power and high radiation tolerance.

Since low offset voltage was crucial, single packaged operational amplifiers were used and after a broad selection process, the OP97A was chosen. Several otherwise excellent operational amplifiers were rejected because of poor radiation tolerance. The OP97A is a bias compensated bipolar operational amplifier with good noise specifications and has proved to be successful. The noise gain is so high, because of the low diode shunt resistance, that voltage offsets at the microvolt level are significant, and it has been found that board resistance can affect offsets. This occurs because the transimpedance amplifiers are on two boards joined by a short length of flexboard, with all even numbered diode outputs amplified on one board, and odd outputs on the other board. The resistance of the ground sets up a small potential difference between the boards, and an odd/even offset for the outputs. Ground straps were added to reduce this effect which could also have been reduced by redesign of the flexboard had there been sufficient time.

With a noise gain of about 10,000, the maximum amplifier input offset voltage (over temperature) of 60 μV gives an output offset of 600 μV, and the presence of slight potential differences across the board add to this. To enable these offsets to be digitized a zero shift of approximately 1.25 V ($\frac{1}{4}$ full scale!) was used. The bias current contribution to offset of 25 mV is quite negligible in comparison with voltage effects.

The predicted voltage noise using values from the data sheet for a 0.1 to 10 Hz bandwidth is 5 mV, which is about four DN. The current noise, which is higher than it would be with a FET input operational amplifier, is still negligible. Measured performance is better than this for many channels, but comparable for others. The noise performance is discussed in greater detail in Section 3.2. The 10 Hz upper limit on bandwidth in the specification is higher than the one Hertz noise bandwidth of the preamplifiers, so the calculation overstates the noise somewhat. Also, the amplifiers appear to be performing better than typical specifications. The most important aspect of the noise is its behavior at near DC frequencies. The dark signals change slowly over a period of many seconds. This noise behavior can be thought of as a drift in the dark signals. Each time dark data are taken, the outputs can be effectively 're-zeroed' by subtraction of the dark signal during data analysis. The lower limit of the effective noise bandwidth, and hence the noise level, is determined by how often dark data are taken. Selecting the frequency of dark data acquisition will be a key operational decision.

The noise and offset performance of the InGaAs detector electronics is driven by the low shunt resistance of the photodiodes. The channels with the most offset and noise, which are mostly at longer wavelengths, have consistently high offset and noise. Had more time been available a prolonged layout effort on the boards might well have improved the offset performance, but it would not have greatly improved the noise performance. With more time the vendor may have been able to make higher shunt resistance photodiodes, although as previously stated the large size makes that a difficult task. As will be discussed in Sections 3.2 and 4.2, the science objectives can be met despite the higher drift by obtaining dark data more frequently than had been originally planned, as well as averaging multiple observations in a manner consistent with mission constraints.

2.4. MECHANICAL DESIGN

The structure of the NIS is based very heavily on the SSUSI design, with only very minor changes. The structure is built out of aluminum in two sections. The scanner section contains the scan motor and scan mirror, and mounts the calibration plaque. The telescope section contains the primary mirror, slits and shutter, grating, dichroic beamsplitter, and detector mounting flanges. All the reflective optics are manufactured on aluminum substrates. The combination of aluminum optics and aluminum structure provides for an optical system with minimal sensitivity to bulk temperature change. Although the reflecting surfaces were changed to gold for NIS, the basic SSUSI design was maintained.

2.4.1. *Scanner, Slits, and Shutter*

The scan mirror is designed to scan rapidly from horizon to horizon on a low earth orbiting spacecraft for the SSUSI application. A total field of regard of 140° is covered about every 20 s. The scanner design is thus optimized for continuous high speed operation. Because of the very low speed of NEAR at Eros, high speed scanning is not needed. The scan mirror on NIS can therefore best be thought of as a pointer rather than a scanner. Typically the same spot will be viewed for at least several seconds before the scan mirror is pointed to the next position.

The line of sight is positioned by rotating the scan mirror with a stepper motor. A four phase stepper motor is used, which rotates 15° per step. This is coupled to the scan mirror through a 75:1 reduction gear. This provides 0.2° of rotation per step. Because the line of sight is determined by a reflection off the scan mirror, the step size of the line of sight is twice the mechanical step size, or 0.4° per step. Unless we explicitly state that we are looking at mechanical rotation, we always refer to scan angles as they relate to the line of sight.

The start position, which views the calibration plaque at the most sunward end of the field of regard, is called step 0. The instrument nadir position is at step 75 (30° away from start). The end of scan position is at step 350 (140° from start). Independent verification of mirror position can be obtained at the start and instrument nadir positions. This is provided by two discrete optical monitoring sensors. A vane is attached to the axle of the scan mirror, which rotates with the scan mirror. On one side of the vane are light emitting diodes, one each for the start and instrument nadir positions. On the other side of the vane are two fiber optic cables, aligned with the two light emitting diodes. The vane has two slots machined in it that correspond to the start and instrument nadir position. When at one of these positions, the light from the corresponding light emitting diode is transmitted through the slot in the vane and captured by the fiber optic cable. The fiber optic cable is connected to the support electronics. The light is sensed there and a digital signal is generated when the light is detected. The feedback we have is then a digital signal whenever the scan mirror is positioned at start or instrument nadir. These signals are referred to as the start and nadir fiducials.

Using one of the fiducials the position of the line of sight can be established. This is necessary every time power is turned on to the instrument, or after the scan mirror is unpowered. The software is designed to do this using the nadir fiducial, but it could be accomplished with the start fiducial. After the position has been established, commands can be issued to move the scan mirror. While ground commands can specify a specific step to position the mirror, the DPU must provide motor control signals for each step of the mirror. The DPU keeps track of the commanded signals to determine the current mirror position. Only when the mirror position is commanded to the start or nadir positions can an independent check of the mirror position be obtained. The process of counting commanded steps to determine mirror position is a very robust one. No instances of losing count have been noted in ground testing of the NIS. While we plan to check the fiducials occasionally during routine operation, they are mainly required to re-establish position after power has been off.

The limits of operation of the scan mirror are steps 0 and 350. This is the range over which the NIS can optically function properly. The mirror itself can move beyond these ranges. At launch, the mirror is stowed parallel to the cover and held securely in place by the cover. Even after the cover is opened, the instrument is not capable of operating in this configuration, since the primary mirror is viewing the end of the scan mirror instead of its surface. This position is step −49. After the cover is first opened, the mirror will be commanded forward by 49 steps to step 0, where the start fiducial should indicate on. After this initial operation, the mirror should never again move to positions below step 0.

If power to the scan motor is removed, the mirror will be moved by the anti-backlash spring to a mechanical stop beyond the end of scan (beyond step 350). This is not a calibrated position. To re-establish the position of the scan mirror, a command called 'find home' is used. This steps the mirror backwards until the nadir fiducial is activated, where the mirror is held. The scan mirror position counter is then automatically set to step 75.

The scan mirror is normally moved at a rate of 20 steps per second when moving to a specified position or 10 steps per second when searching for the nadir fiducial during the 'find home' command. These rates can be changed by ground command, but these values work well for currently planned events. During data taking, the scan mirror usually operates in a single step mode that is not really described by a scan rate. To move the scan mirror one step, one phase of the stepper motor is de-excited while the next phase is simultaneously excited. This results in a very rapid motion from one step to the next. An anti-backlash spring is used to help insure smooth scanning in the forward direction. It is this spring that moves the scan mirror past end of scan when power to the stepper motor is turned off. The mechanical oscillations following the step damp out in under 50 ms, so data could be taken in the same second as the mirror step, although for detector electronics reasons, one second is waited before taking data.

There are three identical motors on NIS to operate the scan mirror, slit, and shutter. Only one phase of a motor is excited at any time. The motors can be operated in either 20 V or 15 V mode (the nominally named 15 V mode actually provides a potential of 14.4 V). The 15 V mode, with a power dissipation of about 2 W per energized motor, is normally used to conserve power and reduce heating and the resulting temperature rise of the motors. The 20 V mode, with a power dissipation of 4 W per energized motor, provides extra torque margin to insure the motor can move the mechanical load if a degradation of the mechanism occurs in flight.

When only the scan motor is energized, the 20 V mode actually involves switching between 15 and 20 V, with the higher voltage used for just 100 ms whenever the motor is commanded to make a step, and 15 V being used to hold the mirror in position the rest of the time. Extra torque can thus be provided when needed (when the mirror moves) with only a minor increase in average power. This operational scheme also allows improved thermal control of the motor (due to lower power dissipation) which ensures lower motor temperatures with corresponding increased long term operational reliability. A copper braid thermally couples the scan motor to the NIS optics housing to further minimize the scan motor peak temperature during operation. As mentioned previously, the scan motor must be energized at all times while operating or the mirror will be moved by the anti-backlash spring beyond the end of scan.

The slit assembly allows NIS to be operated with either the wide slit, narrow slit, or shutter in place. The wide slit is fixed in position within the slit assembly. The narrow slit is mounted to a lever arm which can be rotated to move the narrow slit in and out of the optical path. After it reaches position, the slit is held in place magnetically with no further power required.

For reliability, both the scan and slit motors are required to provide a 100% or more torque margin. This means the motor can provide twice the torque required to do its job. Both motors meet this requirement even in 15 V mode (torque is directly proportional to voltage). Each stepper motor also has both primary and secondary windings, providing backup in case of failure of the motor windings.

The shutter is also a motor operated lever arm within the slit assembly, but with no opening in it. A different design philosophy was used than for the slit. Since a failure of the shutter in the closed position would result in total failure of the NIS, the shutter is equipped with a return spring to insure the shutter opens when power is removed. The motor, with both primary and secondary windings, can be used to open the shutter if need be, but normally the spring opens the shutter when it is not actively being held closed. To hold the shutter closed, therefore, the stepper motor must be continuously energized. The shutter motor is dissipating power whenever the shutter is closed. Because the return spring exerts a force that the stepper motor must overcome with added torque, the torque margin to close the shutter is just under 50%. The tradeoff is to provide less design margin for a failure mode (shutter stuck open) that has marginal impact on the science ability, in order to provide a redundant mechanism to overcome a failure mode (shutter stuck closed) that results in total failure of the instrument. Implicit in this philosophy is the idea that failure of the shutter in the open position is an acceptable degradation of the instrument. The InGaAs detector in particular needs to obtain fairly regular dark data, so this would seem to be a problem. The solution is that the scan mirror can be positioned at high speed to look into space (which is quite dark compared to our sensitivity level), providing functional redundancy.

The calibration plaque is held on a short aluminum bracket off the scan mirror end of the instrument. While this is not the best place for a calibration plaque optically because it is observed at a high phase angle, it was the only option available mechanically. The NEAR ground rules and schedule precluded use of an articulated calibration plaque that could have been moved in and out of the field of view and observed at a lower phase angle. Because of spacecraft Sun pointing limitations, a calibration plaque fixed at the other end of the field of regard could not have been solar illuminated. In the final design, the calibration plaque just reaches the keep out zone defined for the payload fairing on the rocket, at the same time that the back of the InGaAs detector just reaches the keep out zone for the clamp band that holds the spacecraft to the third stage (and then separates them). All available space was used to obtain as favorable a viewing geometry as possible. This configuration also results in a higher angle of incidence on the scan mirror during calibration observations than during asteroid observations, but this effect can be compensated through calibration.

As discussed earlier, the available $140°$ field of regard is well matched to the mission requirements. To match the orientation of the SSUSI field of regard to the desired orientation for NEAR, the instrument must be mounted at a $45°$ angle to the aft deck of the spacecraft. Because the instrument itself is fairly large, the bracket is also quite large. To minimize weight, the bracket is built of magnesium, which keeps its mass down to 1.3 kg. Because the bracket is a critical factor in determining launch loads, all vibration testing of the NIS was performed on the flight bracket. Vibration testing verified that the line of sight of the instrument was stable before and after vibration. The bracket and its associated mounting hardware are designed to ensure that the NIS is thermally isolated from the spacecraft aft deck. The bracket itself has a nominal amount of inherent thermal resistance based on its geometry and material composition; however, the majority of the thermal isolation is provided by the 10 G–10 spacers which are bonded to the bracket at the aft deck mounting interface. By specifying these spacers at the maximum thickness and minimum cross section permitted by structural and vibration considerations, and ensuring that the thermal path through the low conductivity titanium mounting hardware was further reduced by implementing G-10 shoulder washers between the bolt heads and the bracket, the heat flow path from the spacecraft aft deck to the NIS was minimized to the greatest practical extent.

2.4.2. *Cover*

The cover is designed to be opened one time only in space. The cover has multiple functions. It helps to maintain cleanliness of the optics prior to and during launch, it serves to hold the scan mirror in place during launch, and it adds to the structural strength of the housing during launch by providing structure across the large opening for the mirror.

The cover is held closed at launch with a pin, within a pinpuller assembly, mounted to the cover. The exposed end of the pin is held in a latch mounted to the housing. The pin is pulled out of the latch by a bellows pyro that pushes on a tab on the pin. Once the pin has been pulled out of the latch, the cover is pushed open by a spring mounted to the hinge. A microswitch indicates whether or not the cover is closed, and another microswitch indicates whether or not the cover is fully open.

The cover to housing interface was modified for NIS (compared to SSUSI) to account for the long term exposure to cold vacuum conditions prior to opening of the cover. The principle concern is that the surfaces which must separate after being in contact for long periods can stick together if not selected properly. For SUSSI, the parting surfaces were metal against elastomer. While metal to metal parting surfaces were preferred, this was not practical for the existing mechanical design. After evaluation of numerous options, the parting surfaces in the final design are Kapton against metal, which provide good long term stability in the expected conditions and did not compromise the mechanical integrity of the housing.

The pinpuller assembly contains redundant pyros and tabs, located on opposite sides of the pin. Only one pyro will nominally be fired to open the cover in flight. If for some reason this pyro fails, the second pyro can be fired to open the cover. The pyros for the NIS share the same relays with the laser rangefinder. Therefore, the NIS and laser rangefinder covers must be opened at the same time.

The NIS cover does not have a window in it. Therefore, failure of the cover to open would lead to total failure of the instrument. Also, no optical data can be obtained during cruise phase (from the calibration plaque) until the cover is opened. While a portion of the cover could have been made of a transparent material, this would not have helped since the mirror is stowed during launch and only the end of the scan mirror is imaged by the primary mirror.

2.5. THERMAL DESIGN

The NEAR mission configuration ensures that the solar arrays and high gain antenna always point within 30° of the Sun except for brief isolated periods (during which NIS will not be operated). The instruments mounted on the outside of the spacecraft aft deck are therefore always in the shadow of the spacecraft, except for components right at the edge of the deck which may or may not be illuminated depending on the spacecraft rotational angle selected for instrument pointing. By limiting the solar illumination of the aft deck instruments, the constrained spacecraft Sun angle pointing simplifies the design process for these components by reducing the variation in their thermal environments. For the NIS in particular, this favorable and relatively stable thermal environment provides an excellent opportunity for an entirely passive thermal control system augmented by selected heater circuits. This simple, reliable thermal management system relies only on radiative heat rejection to space of the dissipated NIS heat loads in order to cool the optics housing and detectors to their required temperatures. The heater control circuits ensure that the NIS is maintained above minimum survival temperatures during dormant periods and within the specified temperature range for optimum opto-mechanical performance during operational sequences. The primary objective of the NIS thermal design is to cool the detectors to between $-30\,°C$ and $-45\,°C$

during operational sequences. This detector temperature range is critical for obtaining good sensor performance. Also of importance is the temperature of the optics housing. In order to minimize IR radiation within the cavity, the optics and housing must be maintained below 0 °C during data acquisition periods. In addition, the temperature gradients within the optics housing structure must be less than 10 °C in order to prevent optics misalignment induced by thermal distortion effects. The primary design challenge dictated by the specified NIS thermal requirements was the thermal accommodation of the spacecraft mounting interface that can vary from −30 °C during cold cruise mode to +40 °C during hot asteroid mapping mode. The varying parasitic heat leak to the cold spacecraft bus or heat soakback from the hot spacecraft bus needed to be minimized in order to provide a workable NIS design envelope within the instrument mass and power budget constraints. The most stressing case is when the aft deck is at maximum temperature. The design philosophy is to induce progressively decreasing temperatures from the hot aft deck interface to the cold detectors by including high thermal impedance isolation provisions at each mechanical joint. The largest temperature drop is across the thermal isolators and mounting hardware that are used to attach the bracket to the aft deck. The bracket itself has a relatively minor thermal resistance based on its geometry and magnesium composition, therefore the temperature gradient across the bracket structure is minimal.

The NIS is bolted to the bracket through two titanium flexures which are commonly referred to as the NIS 'feet'. The relatively small cross-sectional area of the feet in conjunction with the low thermal conductivity of the titanium material provides a substantial thermal resistance across the feet and hence the potential for a large temperature gradient. However, if this gradient exceeds 30 °C there is the danger of inducing stresses in the optics housing that could result in the misalignment of the line of sight of the two detectors. Therefore, an operational heater circuit maintains the NIS optics between the thermostat control points of −16 °C and −10 °C. A redundant operational heater circuit is included to ensure proper thermal performance in the event of a primary heater circuit failure. Each circuit is sized (17 W capacity at 33.5 V) to account for the worst case heat loss through the open NIS cover aperture and thermal blankets during extreme mission cold conditions, yet the thermal mass of the instrument precludes excessive temperature cycling during the hottest mission scenarios. The heat rejected through the open cover aperture is greater than the sum of the NIS motors and optics housing electronics heat loads even in the mission extreme hot condition, therefore the NIS housing is under positive heater control during all operational phases which ensures fine-tuned thermal control. Extensive thermal balance testing of the fully integrated NIS instrument verified that the temperature gradient across the feet would not exceed 15 °C in either hot or cold mission extreme conditions, therefore misalignment induced by thermal stresses is not an issue for flight. The two detector assemblies, which include the actual sensors as well as their associated processing electronics, are bolted to the optics housing. Ideally the detectors and optics housing would be designed to operate at the same temperature which would simplify the detector interface and packaging design. However, because the NIS housing must be controlled to an average temperature of about -13 °C in order to minimize the temperature gradients across the titanium mounting feet, and the detectors are required to run between -30 °C and -45 °C, the design of the detector assemblies was not trivial. Each detector assembly consists of an external aluminum chassis, the electronics processing boards and their mounting hardware, the detector chip and mounting bracket, and the interface spacer and attachment hardware. In order to minimize parasitic heat soakback from the relatively warm NIS to the cold detector assemblies during operational sequences, considerable effort was dedicated to designing a thermal interface with the smallest possible heat conduction path. ULTEM 1000 Series resin was selected as the spacer material due to its extremely low thermal conductivity ($k = 0.0056$ W/in-°C) which is 33% lower than the conductivity of the G-10 material typically used for this application. The thermal spacer was designed with the minimum cross-sectional area that would guarantee a closed optical path and also satisfy structural and vibration requirements. The spacer was also designed at the maximum thickness permitted by the instrument space envelope, and titanium attachment hardware incorporating ULTEM 1000 shoulder washers under each bolt head was specified in order to further reduce the conduction path.

The effective thermal isolation at each detector assembly-to-NIS housing interface allows each detector to be designed and analyzed as a standalone unit. Selected external surfaces on assembled detector housings have unobstructed views to space and also receive either minimal off-angle solar illumination or no illumination even during mission hot extreme conditions. Therefore these surfaces can be used as radiators (with a 5 mil Silverized Teflon finish) to reject the internally dissipated heat to space. The detectors and processing electronics are conductively coupled to the radiator surfaces through aluminum mounting brackets and mounting hardware in order to minimize the temperature gradient between these components and the cold radiator. The remaining externally exposed surfaces on each detector assembly are covered with thermal blankets.

The effectiveness of the detector assembly thermal control scheme was critical to the overall performance of the NIS, therefore it was deemed necessary to verify the thermal efficiency of this design early in the design process in order to reduce risk. An engineering-level thermal vacuum test on a detector assembly prototype verified that acceptable detector temperatures (less than -30 °C) could be achieved for a NIS housing interface temperature as high as 0 °C. The detectors would therefore have ample temperature margin when attached to the NIS housing which is not expected to exceed -10 °C during the mission. The detector assembly thermal design was subsequently re-verified during the instrument-level NIS thermal balance test in which the fully integrated instrument was in the final flight configuration.

The thermally isolated detector assembly housings with integral radiators provide optimized thermal control during operational periods; however, during NIS dormant periods additional heater power is required to maintain the assembly components above minimum survival temperatures. This additional power is provided by redundant thermostatically-controlled survival heater circuits (3.0 W per detector for each circuit) with the heater elements and bi-metallic thermostats bonded to non-radiator surfaces of each detector housing. The thermostat set points for the detector assemblies secondary survival heater circuits (on $-44\,°C$/off $-41\,°C$) are offset below the thermostat set points for the detector assemblies primary survival heater circuits (on $-42\,°C$/ off $-38\,°C$) in order to prevent simultaneous heater operation and the resulting double peak power draw. It would be advantageous to select thermostats with lower closure (on) points in order to allow larger radiator sizes with corresponding colder operational detector temperatures; however, the selected thermostats are already at the minimum available range offered by the vendor. Software control of the detector survival heaters would have allowed lower heater set points; however, this option was not supported by the heater wiring scheme selected for the NEAR spacecraft.

An interesting aspect of the NEAR mission is that the spacecraft's power margin is lowest at first aphelion (before the Earth flyby), even though this is during cruise with all the instruments turned off. The key power specification for the NIS instrument is thus the survival heater power, not the operational power. This has a significant impact on when the cover should be opened. The primary survival heaters have enough capacity to keep the instrument warm without the secondary survival heaters turning on as long as the cover is closed. Since the external surface of the cover is insulated with thermal blankets, opening the cover and exposing the interior of the instrument to space results in a substantial additional heat leak. This additional heat leak will cause the secondary survival heaters to operate in addition to the primary survival heaters, which adds 4.75 W to the peak survival heater load. There is not sufficient power available at aphelion to supply this extra power and still maintain the desired power margins. The cover was therefore opened after aphelion, on September 24, 1997.

As mentioned earlier, one important factor in the design was the desire to minimize IR radiation within the optics housing due to blackbody radiation. The spectral irradiance of the blackbody radiation is determined by the temperature of the housing. At the low operational temperatures expected during the mission, the radiation is predominately long wavelength IR, with only a weak tail at shorter wavelengths. Only the InGaAs detector responds appreciably to this radiation. The blackbody radiation induces a DC signal in the InGaAs detector channels, which creates a small offset. Since there is a large offset present already, which is corrected with dark measurements, this added offset is of no consequence as long as it is relatively constant. The temperature of the optics is not constant at a single temperature, but rather cycles up and down as the thermostatically controlled heaters cycle on and off. A temperature range of less than 10 °C is expected in flight, with the variation (corresponding to the heater duty cycle) occurring on a time scale of about an hour. The result is variations in the blackbody radiation-induced signal of at most 2–3 DN as the housing temperature varies. This does not cause a problem since frequent dark readings will be taken to compensate for drifts in the InGaAs electronics. While this effect is not an issue in the performance of the NIS, it illustrates why the optics were designed to operate at a relatively low temperature. Had the optics been operated at room temperature, this effect would have been severe.

2.6. DATA PROCESSING UNIT

The NIS is controlled by a Data Processing Unit, or DPU, one of three on the spacecraft. The DPU controlling NIS also controls the magnetometer (MAG), and hence is known as the NIS/MAG DPU. The DPU interfaces to the spacecraft Command and Telemetry Processor (CTP) over the redundant MIL-STD-1553 databus, provides power conditioning for both instruments, and provides digital control and data readout for both instruments. A more extensive description of the DPU hardware and software, much of which is common for all NEAR instruments, is contained in Hawkins et al. in this issue.

Power is supplied to the DPU from the main spacecraft power bus, with a relay to turn power to the DPU on or off. The DPU automatically boots up when it receives power, and sequentially turns on the magnetometer and NIS. The NIS operational heaters are wired in parallel with the DPU, so when the DPU is turned on the NIS operational heaters also receive power, under thermostatic control. While the NIS or magnetometer can be individually turned off or on with the DPU running, the NIS operational heaters cannot. The DPU also contains the power conditioning electronics that convert the spacecraft bus voltage to regulated supply voltages for the instrument electronics.

The DPU software handles uplinked commands, collects NIS data, and sends telemetry data to the CTP. There are commands for configuring the NIS mechanisms; these include positioning of the mirror, shutter, and slit, selecting primary or backup motor windings, and selecting 15 or 20 V motor operation. All software and commanding is a function of the DPU; the support electronics provide the motor driver and fiducial monitoring electronics. There are also commands for defining and running NIS data acquisition sequences.

NIS data acquisition sequences integrate a specified number of spectra, telemeters the result, and moves the mirror a specified amount. The process is repeated a number of times as specified in the sequence. The sequence can also be setup to periodically shorten the normal integration and insert an integration with the shutter in. All of the accumulated spectra, dark and light, are telemetered. After the specified number of observations and mirror movements are completed, the mirror is returned to its original position. The entire process can be run up to 65,535 times (or forever if zero repetitions are specified). The DPU is always running a sequence. When a sequence completes, a default sequence is run. Up to 16 sequences can be defined, which can then be executed as needed. One of the 16 sequences is the default sequence, which is programmed to telemeter no spectra.

Spectra sent to the CTP are usually placed on a solid state recorder for later downlink. Depending on available downlink bandwidth, NIS data can also be sent directly to the ground. The DPU software periodically collects housekeeping data and sends it to the CTP. The housekeeping data include instrument temperatures and currents, mechanism status, and current sequence information. Housekeeping data are also included with each set of spectra telemetered, which summarize the instrument state at the time the spectra were taken.

The DPU software can be uploaded in flight, where it is stored in an electrically erasable programmable read only memory (EEPROM). Should there be a need to adjust operation of the NIS which can be accomplished by software, then these changes can be made during the mission.

NIS does not perform any data compression before sending data to the ground, but it can sum multiple frames of data onboard (up to 16 frames without possibility of rolling over bits). While the data itself is well suited for lossless compression, the total telemetry bandwidth needed by the NIS is a small fraction of the available bandwidth. Each spectrum is contained in a science record of 1216 bits, including science housekeeping data. Reducing this amount of data further by data compression has not been deemed necessary to accomplish the mission. If detailed operational planning showed sufficient benefit to the mission from implementing data compression, this would require uploading new software to the DPU in flight.

The NEAR instruments operate in one second frames, under control of a one pulse per second (1 PPS) clock signal from the spacecraft. The NIS detector data are digitized at the very end of the one second frame. As discussed in Section 3.3.3, the detector data are not uniformly weighted over the one second frame, but is low pass filtered such that the optical signals at the end of the one second frame are weighted most heavily. The time tag stored by the NIS instrument is taken at the next 1 PPS signal. The time tag thus corresponds to a time a few milliseconds after the data are digitized, and most of the optical energy in the recorded data was received in the one quarter second prior to the time tag. While the time tag is after the time the data are acquired, it is the time closest to when the data was acquired. Because of details in the interaction of the DPU and CTP, the command to take data must be issued in the one second frame before the frame in which the data are taken. The mission elapsed time (MET) of the spacecraft command is therefore two seconds earlier than the MET of the time tag in the data.

3. Pre-Flight Calibration and Characterization

3.1. Overview and Facilities

Calibration of the NIS can be divided into three sequential phases, each of which builds on the previous phase: (1) characterization of individual components, (2) calibration of the integrated instrument before launch, and (3) in-flight calibration. The majority of the pre-launch calibration results reported here were performed at the instrument level. The in-flight calibration activities, which will build on the pre-flight calibration, are described in Section 4.1.

Instrument level calibration tests were performed in the optical calibration facility (OCF) of the Applied Physics Laboratory. In this facility the instrument is mounted on a motion stage inside a vacuum chamber with an internal diameter of 1.3 m and a length of 2 m. The motion stage can rotate the instrument in azimuth and elevation through over 180° (limited by NIS cables). The internal volume of the vacuum chamber is surrounded by a cooling shroud. An external refrigerator circulates a cooling fluid through coils on the shroud to reduce the internal temperature environment to as low as $-40\ °C$. A diagram of the OCF is contained in Hawkins et al. (this volume).

There are two methods of illumination for the chamber. In one orientation the instrument can face a chamber window which has a diameter of 200 mm. Various sources can be mounted in air outside the window. In the other orientation, a rotation of 180° allows the instrument to look along a vacuum beam tube into an off-axis parabolic collimating mirror. Sources that can be located at the focus of the collimator include a vacuum monochromator, or, using a window, a point source, a radiance source, or rock and mineral samples. The monochromator can be illuminated by an incandescent lamp with quartz optics for spectral calibration. At the input slit is a set of neutral density filters to attenuate the light by known amounts. At the exit slit a selectable long-pass filter is used to remove higher orders. The wavelength can be changed manually or can be set and scanned under computer control. The exit slit of the vacuum monochromator is located at the focus of the collimator which has a focal length of 1.43 m and operates at $f/7$. The collimated beam is folded into the beam path to the instrument by a plane folding mirror.

3.2. DARK SIGNALS AND NOISE

Extensive pre-launch testing was conducted in the OCF thermal vacuum chamber under simulated mission conditions to characterize the dark signals from the detectors, and the noise in the measurements. Initial in-flight dark data are now available, so the data shown for dark signals and for SNR predictions in this paper are based on the in-flight data. The pre-launch data were very similar to the in-flight data obtained to date.

Figure 13 is an example of some of the NIS dark spectra obtained in flight. All channels of the Ge data have an offset of about 80 DN, with very little variation between channels for unity gain, and only a ± 2 DN variation for gain 10. The 80 DN offset is deliberately introduced in the electronics to insure that any electronics drift or offset changes that occur during four years in flight will not result in dark signals of under zero volts, which could not be measured by the digitizer.

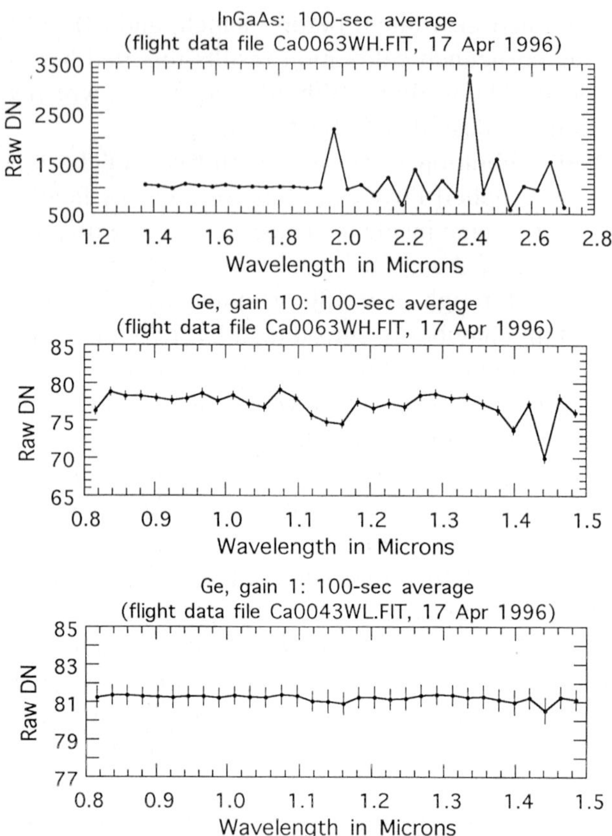

Figure 13. Representative NIS dark spectra obtained from in-flight measurements on 17 April, 1996 (60 days after launch). Dark spectrum for germanium detector at gain = 1. Dark spectrum for germanium detector gain = 10. Dark spectrum for indium-gallium-arsenide detector. The spectra are averages of 100 one-second dark measurements, with one-sigma error bars indicated.

The actual preamplifier offsets are small, since the noise gain is small. The offset for the Ge channels is approximately the same for both the unity and 10 times gain states, since the voltage offset is added into the signal after the selectable gain amplifier. The InGaAs channels have a 'baseline' offset of about 1000 DN, which is deliberately introduced for the same reason as for the Ge detector. Many channels differ significantly from this baseline offset due to preamplifier offsets. The preamplifier offsets can be positive or negative and are added to the intentional offset. The significant pre-amplifier offsets are a direct result of the low shunt resistance of the InGaAs photodiodes. The odd/even offset effect due to finite ground resistance is quite evident at the longer wavelengths.

Figure 14 shows the expected signal levels and SNR for the initial flyby of Eros. The signal levels in the InGaAs band are low enough that despite the high offsets

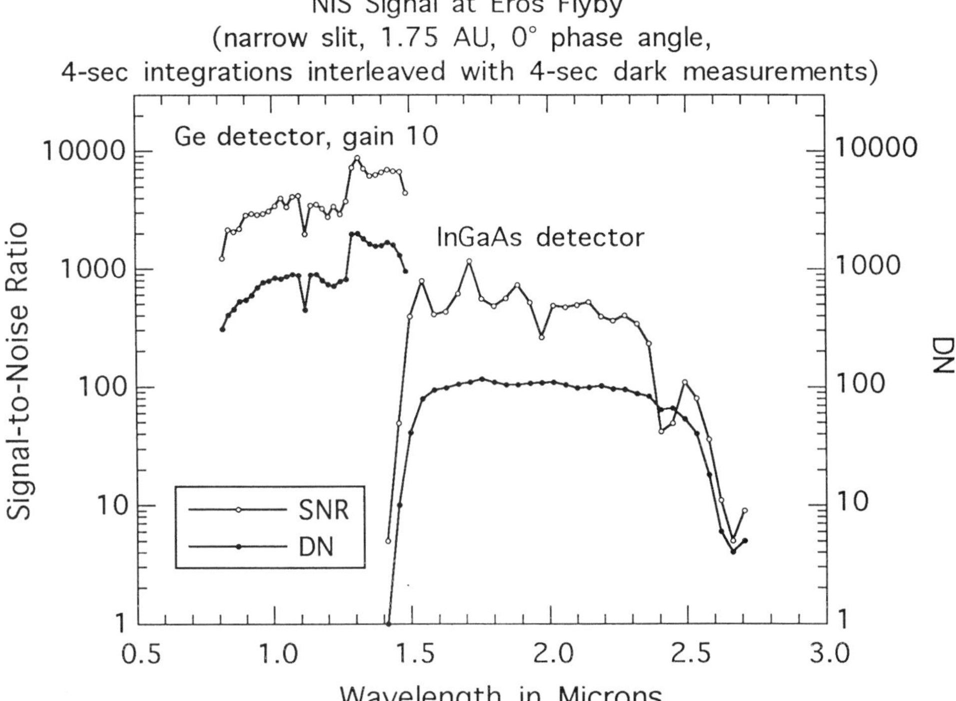

Figure 14. Predicted NIS DN levels and signal-to-noise ratio (SNR) for the low phase angle Eros flyby observations. Signal levels were estimated using nominal values for solar flux, the reflectance of Eros, heliocentric distance ($D = 1.75$ AU), the phase angle at the time of the flyby ($\alpha = 0$ deg), and radiometric calibration constants determined from pre-flight calibrations. The noise was estimated using in-flight dark signal measurements obtained on 21 May, 1996. The assumed spectrum acquisition sequence uses the narrow slit over a 10-s time interval (4 s of Eros data, 2 s for shutter changes, 4 sec dark data).

there should be no saturated channels. The large change in predicted DN for the last 10 Ge channels is due to the larger feedback resistors used for those channels.

The SNR shown in Figure 14 is calculated assuming that 10 s are used for each observation. While much more time is possible for each observation during the flyby, and thus the flyby SNR could be further improved, this is consistent with the amount of time available during low altitude orbital operations if contiguous mapping is desired on a single orbit. For this calculation the 10 s for each observation is assumed to be split between four seconds of dark data and 4 s of target data, with one second unused between each. The SNR for the Ge band is seen to be very high. The ultimate accuracy of this data will be limited by calibration rather than SNR, since the SNR is so high. The InGaAs SNR is very good out to about 2.4 μm, after which it falls dramatically. The region of good SNR encompasses the pyroxene absorption feature between 1.9 and 2.3 μm. There is one channel at 1.95 μm with somewhat lower SNR; this same channel shows a high offset in Figure 13.

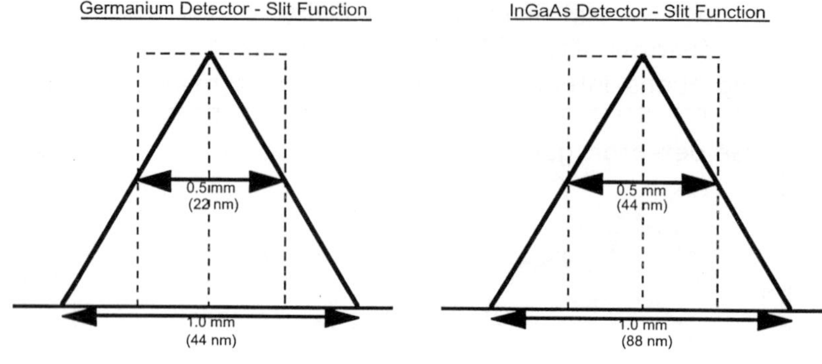

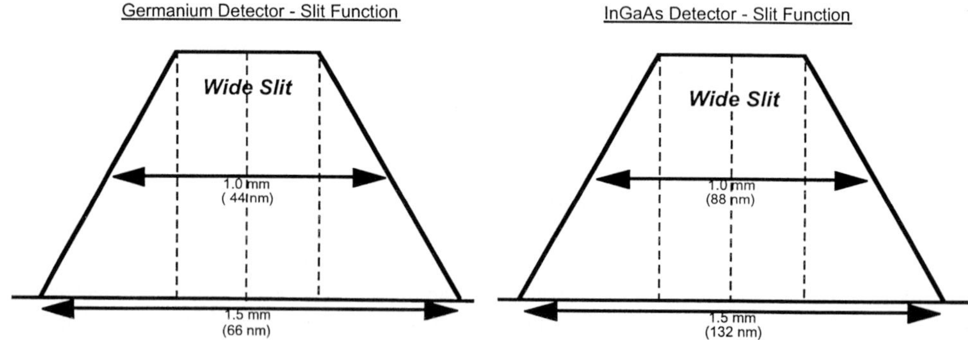

Figure 15. Ideal slit functions for the narrow and wide slits.

3.3. WAVELENGTH AND RADIOMETRIC CALIBRATION

3.3.1. *Spectral Calibration*

The spectral response of any single detector element is the 'slit function' of that element. Both the width of a detector element and the width of the slit affect the wavelength range of light falling on the element. The spectral resolution for a field filling source is the convolution of the slit image with a detector element. This is shown schematically in Figure 15. These idealized theoretical functions do not include the roll off created by aberrations.

Spectral calibration was performed by imaging the filament of a stable (but uncalibrated) incandescent source onto the entrance slit of the monochromator whose exit slit was at the focus of the collimator. The image of the monochromator exit slit subtends an angle of 0.01° by 0.2°. This is fully contained within the instrument field stop, which has angular dimensions of 0.76° × 0.38°. The long dimension of the monochromator slit coincided with the short dimension of the

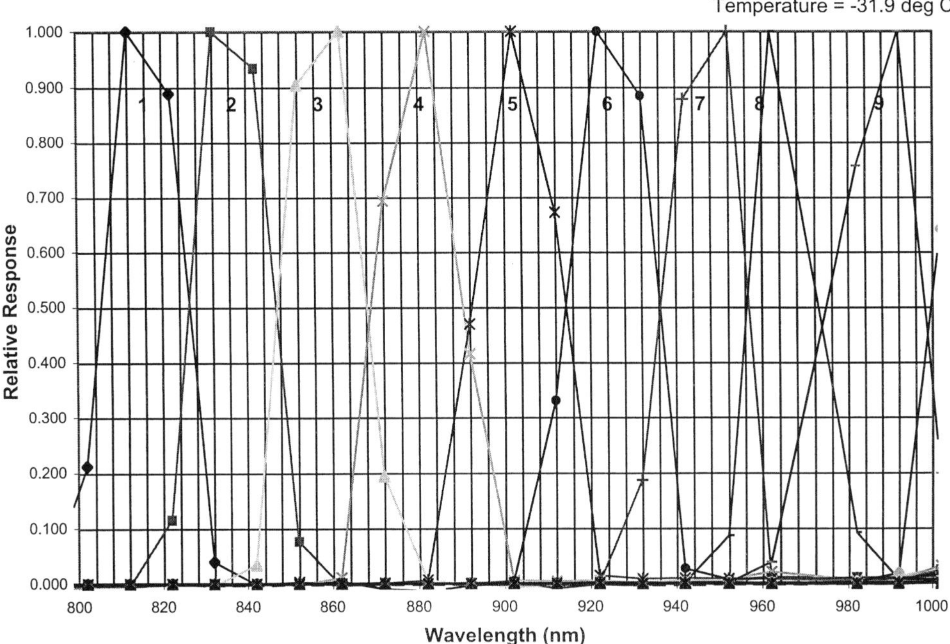

Figure 16. Spectral response of several of the Ge channels used in determining spectral calibration.

instrument slit. Long pass filters were not used in the monochromator so that both its first and second orders could be used for spectral calibration. (It is important to realize that this second order is a product of the monochromator and not a crosstalk problem in the spectrometer.) The signals from the detector arrays were recorded as the monochromator wavelength was stepped in 5 nm increments over the second-order wavelength range (Ge detector) and 10 nm increments over the first-order wavelength range (InGaAs detector). For each detector element we plotted the relative response as a function of the wavelength. One example of the results is shown in Figure 16.

The numbers identify the detector element and each profile shows the relative response as the monochromator is scanned across it. The width, defined by the separation between crossing points of adjacent elements, is very regular. The curves cross close to the 50% point and the half widths are close to the expected value of 22 nm. Occasionally when a data sample was not taken, such as at 970-nm in this case, the profiles are distorted. The spectral calibration consisted of reading the wavelength at which two profiles crossed, or using the 50% response point in the cases where there is no crossing. For each element the spectral calibration is defined as this response point.

This detailed spectral calibration was performed at three temperatures, $-7\,^{\circ}\mathrm{C}$, $-17\,^{\circ}\mathrm{C}$, and $-23\,^{\circ}\mathrm{C}$. The average results for three data sets are shown in Figure 17. This figure shows the wavelength at the starting edge of each of the Ge and

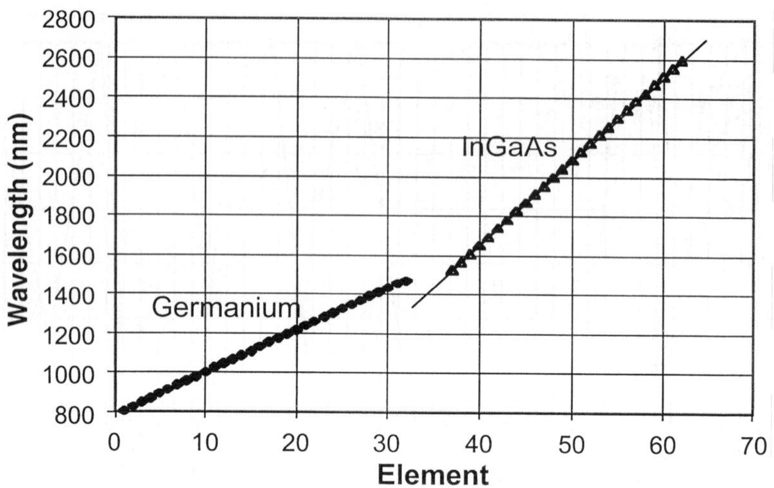

Figure 17. Spectral calibration data and linear regression.

InGaAs detector elements. Converting from the starting wavelength to the central wavelength of each element, the Ge spectral calibration is given by:

$$\lambda \, (\mathrm{nm}) = 794.6 + 21.61n \, ,$$

in which n is the element number between 1 and 32. The accuracy of this equation for the three data sets is approximately ± 0.5 nm.

For the InGaAs elements the center wavelength is given by:

$$\lambda \, (\mathrm{nm}) = 43.11n - 50.8 \, ,$$

in which n is the element number between 33 and 64. The accuracy of this equation for the three tests is approximately ± 3.5 nm. The error is primarily a systematic error between tests performed at different temperatures.

The full range of the Ge array is 805.4 to 1497 nm and the full range of the InGaAs array is 1307 to 2730 nm. These ranges are further limited, however, by the falling responses of the InGaAs detector at the long wavelengths and the efficiency range of the dichroic filter. (There are several extra elements at each end of the InGaAs spectrum with unusable signals.) The Science Team requested that the Ge detector array be aligned such that the fifth element was centered at approximately 900 nm, matching the center wavelength of the 900 nm MSI filter. Element five is centered at about 903 nm according to the above equation.

Spectral calibrations of the wide slit show that the average offset of the wide slit spectrum relative to the narrow slit is -3.3 nm for the Ge detector, and twice this for the InGaAs detector. This corresponds to a linear offset of 0.075 mm between the slit centers.

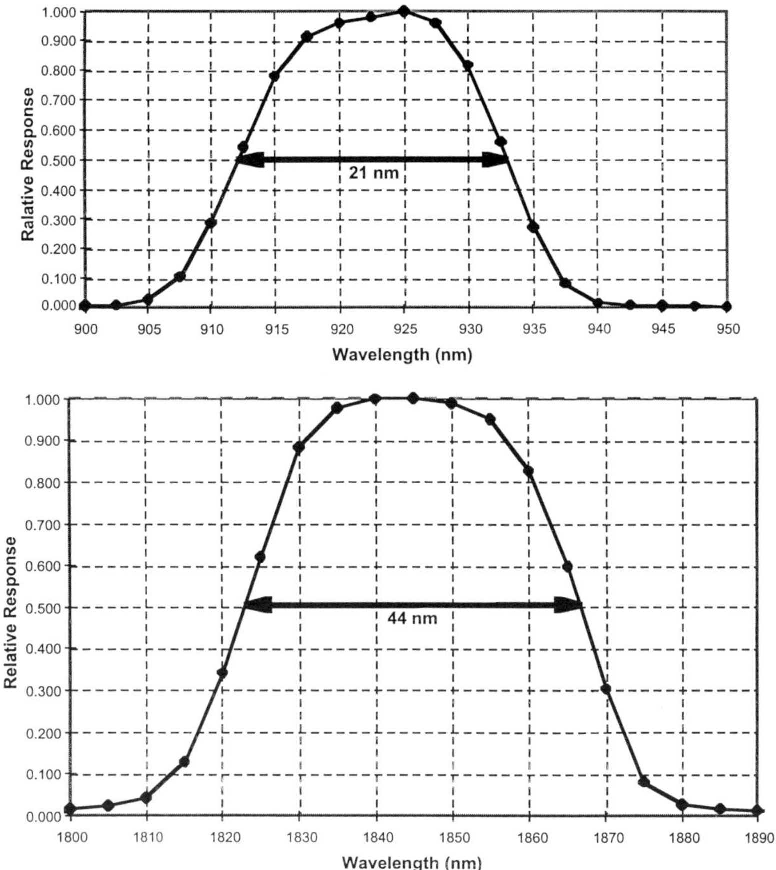

Figure 18. Measured slit functions with the wide and narrow slits.

The spectral resolution is determined by a detailed spectral scan across a single detector element. The spectral range 900 to 950 nm was scanned taking samples at 2.5 nm intervals. This covers Ge element No. 6 which has the detailed profile shown in Figure 18. Figure 18 also shows the data for the InGaAs detector element No. 44. The profile is normalized to the peak signal after background subtraction. It shows that the slit function in Figure 15 is a good simulation of the actual data except for the rounding produced by the finite spectral range and aberrations.

Figure 19 shows the detailed transmission profile of element No. 15 which was performed with the same setup as for the spectral resolution test. This appears strange as it has two peaks with a dip in the middle. This is caused by the opaque boundary between the two zones of the long-pass filter used to eliminate higher orders from the grating. This zone has a width of 0.2 mm, or 40% of the width of a detector element. As the filter is about a millimeter above the detector surface the defocus also affects the area obscured. These data are not normalized to the peak

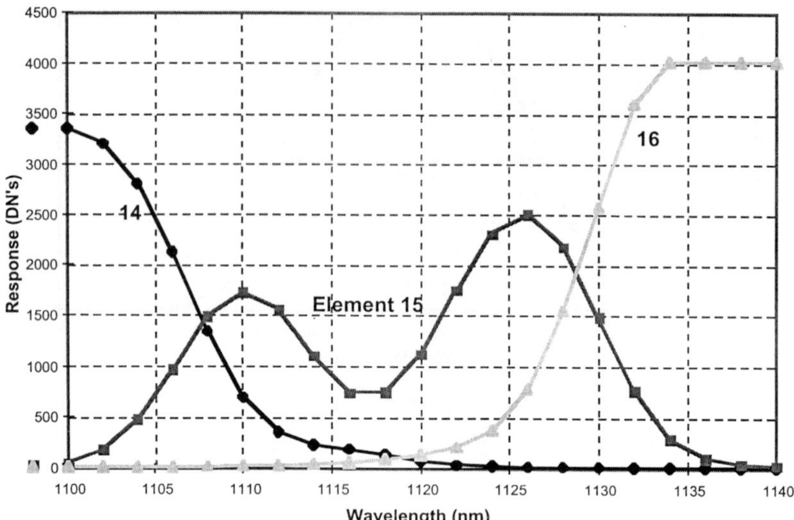

Figure 19. Spectral response across the boundary of the two zone-order sorting filter. The output of channels 14–16 is shown as wavelength is scanned. Channel 15 is 40% obscured by the opaque boundary between the two zones of the order sorting filter mounted over the Ge detector. The dip in the element 15 output occurs when light is blocked by the opaque region.

so the intensity is representative of the signal strength as compared with elements 14 and 15. (In this data channel 16 reaches saturation.)

The location of the filter boundary was designed to be within 50 nm of 1100 nm. The Science Team requested that the opaque bondline be centered over one element, so that only that element would have its signal reduced, and element 15 was selected as the location of the boundary. The measured location is at 1117 nm, meeting the optical requirement and confining the obscuration to one element. Although the signal for this element is reduced, the SNR is still high and the effect is well calibrated so there should be little or no impact on its scientific utility.

3.3.2. *Spectral Crosstalk*

All wavelengths which satisfy the condition, $m\lambda = K$, in which K is a constant, are diffracted in the same direction by a diffraction grating. The values of m are the spectral orders. Thus at any point in the focal plane several wavelengths are present. The challenging part of spectrometer design is to eliminate all the unwanted wavelengths. In the NIS this is done using a combination of dichroic reflection and transmission, long pass filters, and the detector response limits. During the design phase an analysis was performed to estimate the spectral crosstalk produced by the design. Actual component efficiencies and estimated detector responses were used to calculate the source of the signal produced by the Ge detector. The InGaAs detector has no problems with spectral crosstalk because the dichroic filter

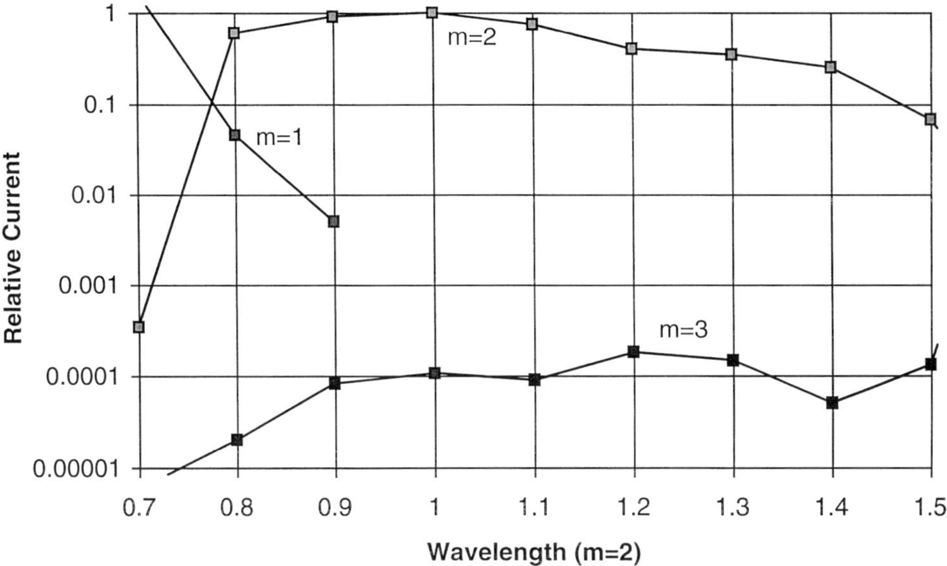

Figure 20. Prediction of relative signal for different spectral orders from the Ge detector.

has effectively zero transmission for the short wavelengths which could interfere. Figure 20 summarizes the results of the analysis.

The lower wavelength axis is the second-order wavelength falling on the Ge detector. The plot is normalized to a response of unity at a wavelength of 1 μm for $m = 2$ and the values are based on the expected signals from the asteroid. The other curves show the relative signal for the first and third orders. The third order is attenuated by at least three orders of magnitude so it can be ignored. The first order is attenuated at long wavelengths by the low detector response but there is still some response to first-order wavelengths below 1.8 μm. The results predict a relative detector signal from the first order of 7.6% at 0.8 μm and 0.55% at 0.9 μm. Thus, the first few Ge elements will have detectable spectral crosstalk.

Several tests were performed to measure the spectral crosstalk and to determine a method of correction. The most direct method is to use an extended spectral source and record the signal with and without the wavelengths below 1 μm filtered out. With the filter in the beam any signal from the first few Ge elements must be from the longer wavelengths rather than the second order. Another method is to record the signal from both arrays as the wavelength is stepped through the range of the first order.

When all the data collected for wavelengths between 1630 and 1900 nm (815 to 950 nm in the Ge range) are summed, the signal distribution is shown in Figure 21. The crosstalk signal falls off rapidly as expected because of the rapidly decreasing sensitivity of the detector to longer wavelengths. There is an increase in signal from element No. 1 to element No. 2 because the reflectivity of the dichroic beamsplitter

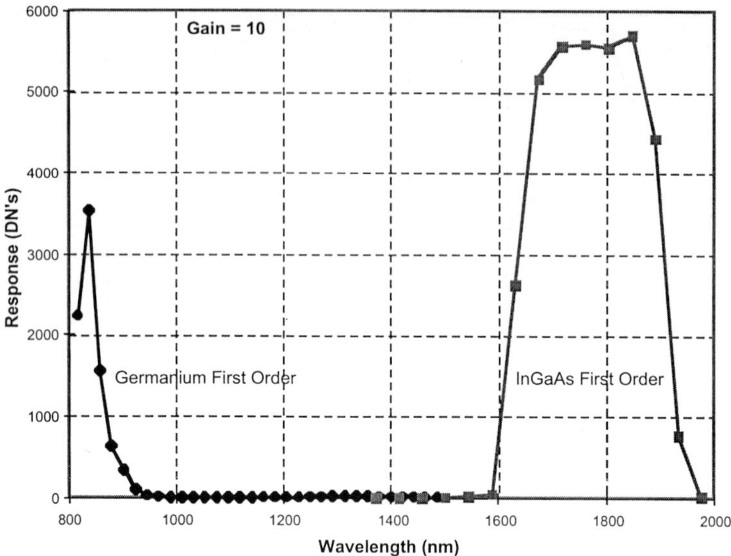

Figure 21. Measured Ge response to first-order wavelengths, and corresponding InGaAs data. Note that while the crosstalk to the Ge detector is a large fraction of the InGaAs signal, the Ge signal will be an order of magnitude larger than the InGaAs signal in operation, so the required correction is a small fraction of the Ge signal.

is increasing between 1.6 and 1.65 μm (see Figure 8). When the Ge signal is divided by the corresponding signal from the InGaAs detector, the resulting value is the correction required to the Ge channel based on the InGaAs. The anticipated fractional correction is shown in Figure 22.

From the scatter on the fractional corrections, the probable error is about 10% of the value. As the asteroid signal in the Ge channel is expected to be about ten times the InGaAs signal in this wavelength range, the correction factor does not exceed 10% of the signal so the error of the correction is only 1% in the worst case for the first two Ge elements.

3.3.3. *Radiometric Calibration*

The radiometric response of the NIS was determined in a number of tests by recording the response while viewing a calibrated field and aperture filling integrating sphere. The radiometric response of the NIS to a radiance source is shown in Figure 23. The Ge data show the loss of signal at 1125 nm caused by the boundary between the two zones of the filter, and higher response above 1290 nm because of higher amplifier gain. The effect of spectral crosstalk has been corrected using the method described in section 3.3.2 before deriving the radiometric calibration coefficients.

The two slits have widths corresponding to 0.38° and 0.76°. Thus the nominal ratio between their signals is 2. Whenever a calibration was performed data were collected with each slit. The results of ratioing the signals from the two slits has

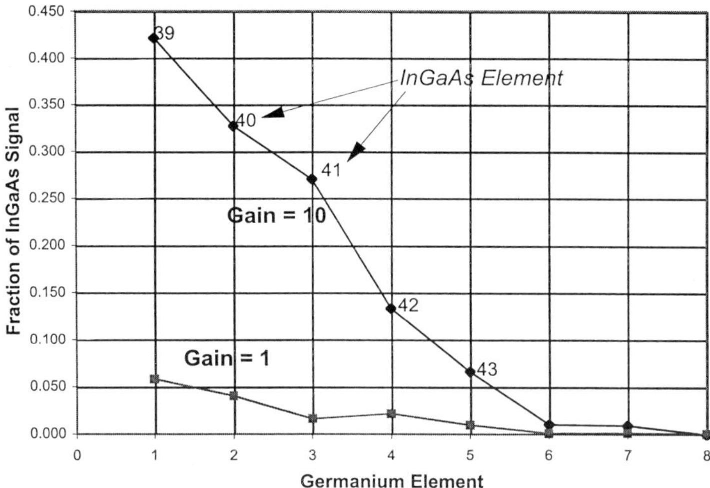

Figure 22. Correction factors for first-order spectral crosstalk onto the Ge detector.

the unusual form shown in Figure 24. Rather than the expected smooth curve with little dependence on wavelength, the ratio is systematically channel-dependent. The average is about 2.1 with variations of as much as 10%. Eight data sets are plotted in the figure so the small amount of scatter at each point shows that the variations are real.

The reflectivity of the scan mirror, and therefore the radiometric calibration, is expected to change with scan angle since the angle of incidence is changing. This effect was the subject of several tests performed when the NIS was pointed at the spheres through the rear window of the OCF. The scan mirror was pointed at the center of the extended source outside the vacuum chamber and readings were taken. The scan mirror was stepped in 10° increments (corresponding to 25 mirror steps), and the motion stage was moved by the same amount to maintain the same view of the source. This procedure was repeated over the full angular range of the scan mirror. The background signal was subtracted and the values expressed as a fraction of the signal at a mirror step of 188. Figure 25 shows the average variation (for most detector elements) of relative signal with mirror step. These curves also show that the variation is similar for either slit. It should be noted that most observations will be made at scan mirror positions of step 70 or higher, and thus the most significant variation in responsivity versus scan mirror position will be avoided in the asteroid data. However, the calibration plaque must be viewed at step 0. This emphasizes the importance of obtaining data during the Earth flyby to compare the calibration plaque data to known references.

The angle of incidence at the mirror varies linearly between 77.5° at step 0 and 7.5° at step 350. The theoretical reflectance of gold at 1.0 μm over this range is shown in Figure 26. At 2 μm the theoretical reflectance has a similar

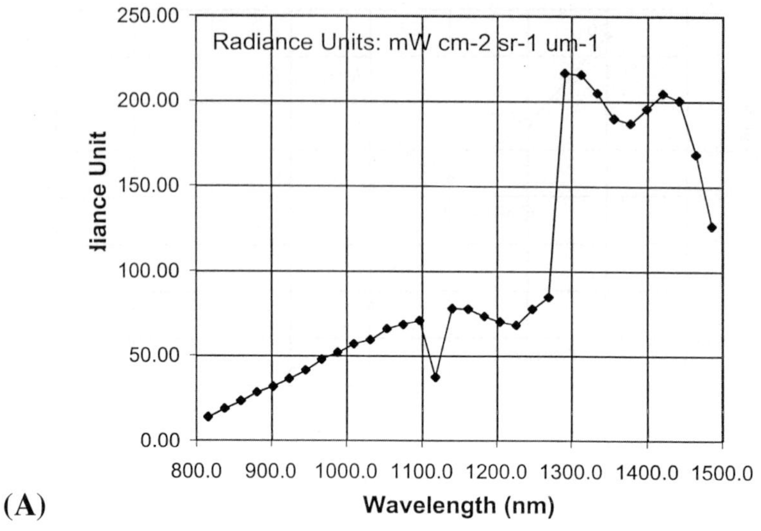

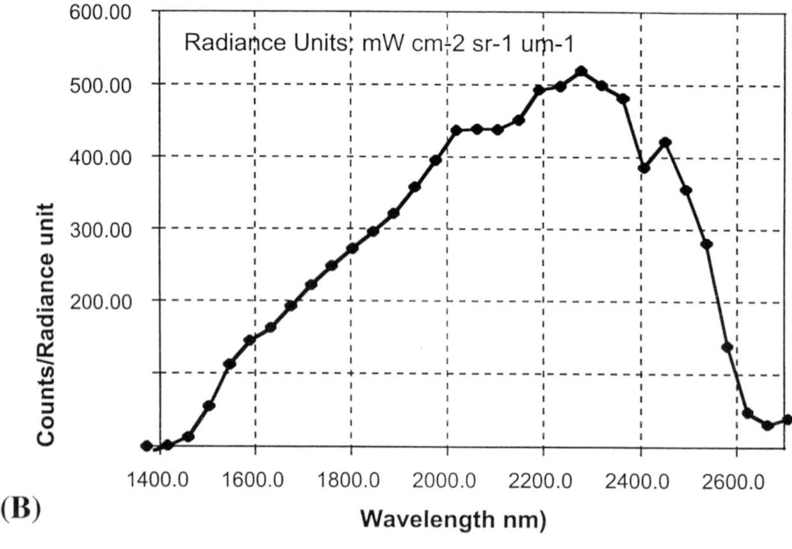

Figure 23. Radiometric calibration of the NIS for both Ge and InGaAs detectors. The units of radiance are mW/cm^2 sr μm.

distribution except the minimum total reflectance falls to 0.955. In these curves the total reflectance change is only 1% at 1.0 μm and 2% at 2.0 μm. This is considerably less than the measured 8% change in Figure 25.

3.3.4. *Polarization Sensitivity*

The percentage polarization of light is defined by the equation:

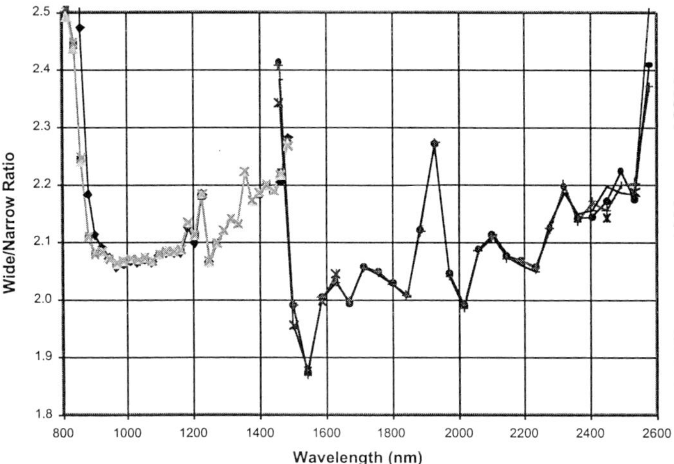

Figure 24. Response ratio between the wide and narrow slits.

$$P(\%) = ((I_H - I_V)/(I_H + I_V)) \times 100\% ,$$

in which I_H and I_V are the signals for horizontally and vertically polarized light. Horizontal polarization is parallel to the long dimension of the slit.

This equation gives the ratio between I_H and I_V as:

$$I_H/I_V = (100 + P)/(100 - P) .$$

As an example, for the Moon the value of P varies between -1.5% at a $15°$ phase angle and about 8% at a phase angle of $100°$ (Markov, 1962). For a polarization of 8% the ratio of I_H/I_V is 1.17. If the instrument is polarization sensitive this can cause a change in the recorded signal as a function of the degree of polarization of the light reflected from the asteroid. The NIS is expected to be polarization sensitive because the scan mirror, the grating, and the dichroic are all polarization sensitive components. The sensitivity of each component is not known individually so the effect of the total instrument can only be determined by measurement. It is also expected that it will depend on the angle of reflection from the scan mirror.

To measure the polarization sensitivity a polarized source, consisting of an integrating sphere covered with an infrared polarizer, was viewed directly with no intervening optics other than the chamber window. After the scan mirror was set to the desired angle and the motion stage was oriented to view the sphere a data set of five measurements was collected. These were: dark, no polarizer, polarizer transmission axis vertical, axis horizontal, and axis at $45°$. The percentage polarization was calculated for each data set. The measurements were repeated for the range of scan mirror angles.

Figure 27 shows the percentage polarization of the Ge and InGaAs channels as a function of wavelength for one mirror position. The mirror is at scan position

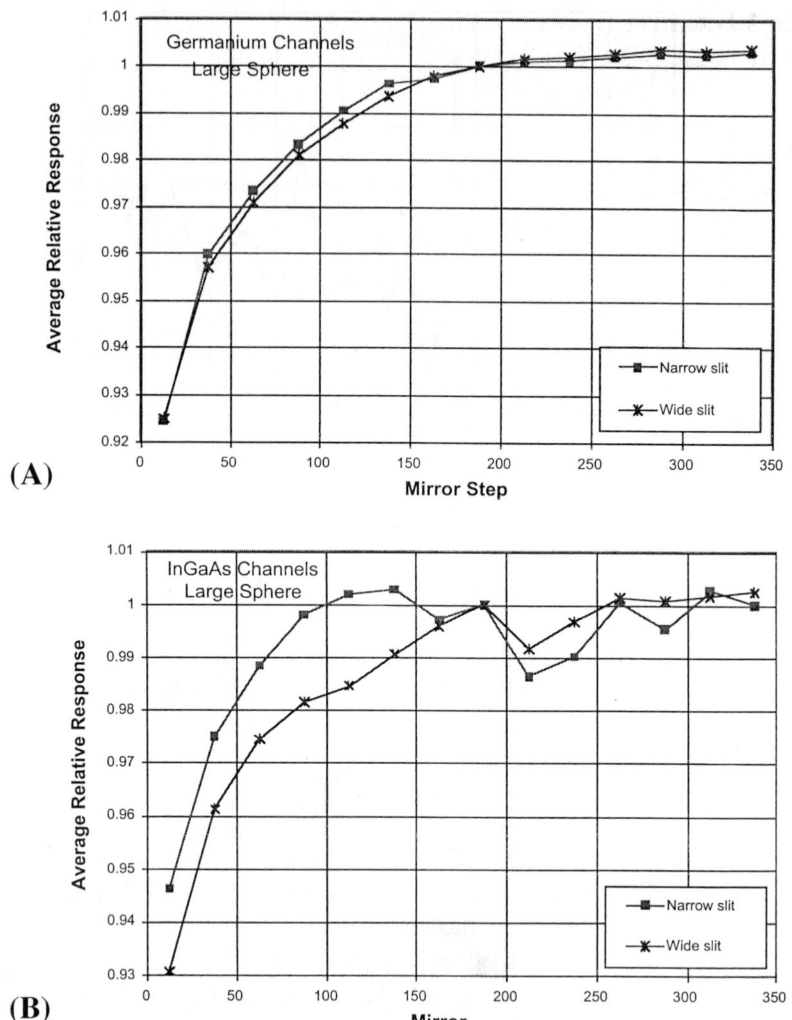

Figure 25. Average signal variation with scan mirror angle for the (a) Ge and (b) InGaAs detectors.

188. There are some significant variations in the polarization versus wavelength, particularly for the Ge channels. The first five Ge elements show a rapidly changing polarization starting at −20% at 800 nm and reaching zero at 880 nm. The polarization falls rapidly for the longest wavelength Ge channels and rises rapidly in the initial channels of the InGaAs array. Between 1.1 and 1.3 μm there is first a dip and then a rise in polarization, followed by a return to the nominal level. At each of these wavelengths the dichroic beamsplitter is undergoing a significant change in efficiency (Figure 8). Interference filters are inherently polarization sensitive devices, and in any wavelength region where there is a significant change in efficiency there will be a substantial polarization dependence.

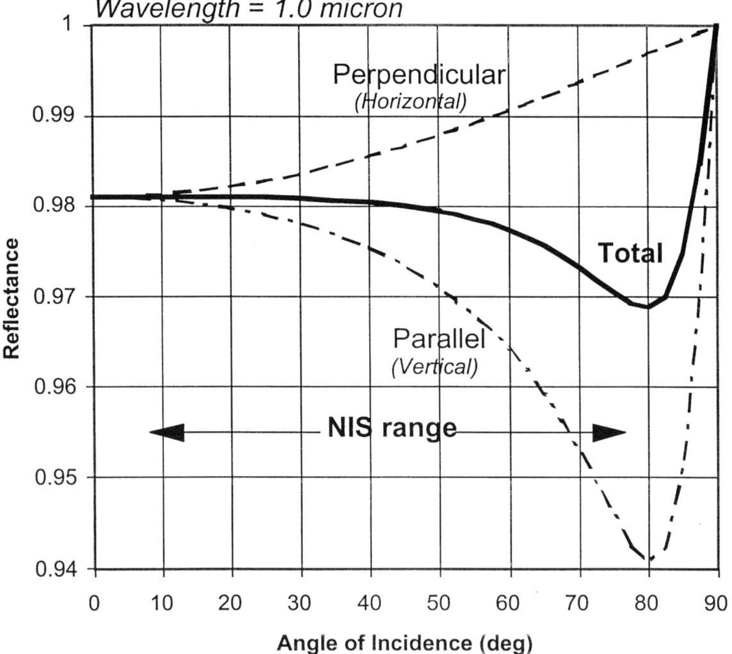

Figure 26. Theoretical reflectance of gold at a wavelength of 1.0 μm.

Any oblique reflection off a dichroic or a metal will produce polarization which is a function of the incidence angle. The variation of the percent polarization with mirror angle is plotted in Figure 28 for three of the Ge wavelengths and three of the InGaAs wavelengths. Two of the wavelengths selected for Ge, at 903 and 1313 nm are in the well-behaved regions of the spectral curve, and the third, at 1227 nm, is at the peak of the anomalous region. All three curves have the same trend with angle and the two from the well-behaved region are close together. In each case the percent polarization decreases with increasing mirror step. The curve from the anomalous region remains high over the full angular range. The curves for the InGaAs detector have a similar variation; they decrease steadily as the angle increases. The relationship between the angle of incidence on the mirror and the mirror step is given by: $I = 77.5 - 0.2 \times$ Step. Thus the angle of incidence varies from 77.5° at step 0 to 7.5° at step 350. Each mirror step changes the angle of incidence by 0.2°. For large step values the angle of incidence is small so the polarization introduced by the scan mirror is small. Figure 26, showing the theoretical reflectivity of gold versus incidence angle, demonstrates the effect of incidence angle on polarization. In this figure the perpendicular and parallel are the orientations of the electric vector of the polarized light relative to the plane of incidence. These correspond to the horizontal and vertical orientations of the polarizer, respectively, in the NIS tests. The percent polarization of the instrument

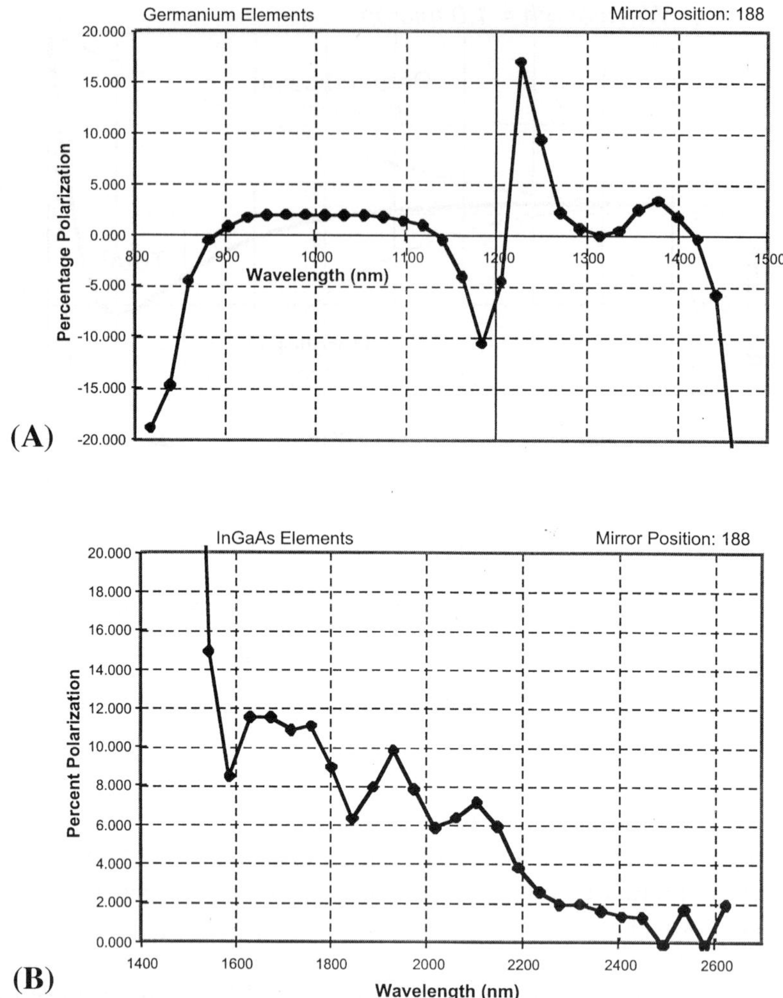

Figure 27. Polarization versus wavelength for the (a) Ge and (b) InGaAs channels.

is thus expected to increase with increasing incidence angle on the scan mirror (decreased step number), which is observed in Figure 28.

The results clearly show the mixed effects of the scan mirror, the grating, and the dichroic filter. The practical effects at Eros will depend on the polarization of reflected light from Eros at different phase angles, and the response difference of NIS for the two components at the scan mirror position used to observe those phase angles. The two limiting cases are low phase angles (mirror position $\sim$325), to be observed by NIS only at limited times, and phase angles near 90° (mirror position $\sim$75) which will normally be observed from Eros orbit. The polarization properties of Eros have been described by Zellner and Gradie (1976) over the phase angle range 9°–53°. At low phase angles polarization is approximately 0.8%. For

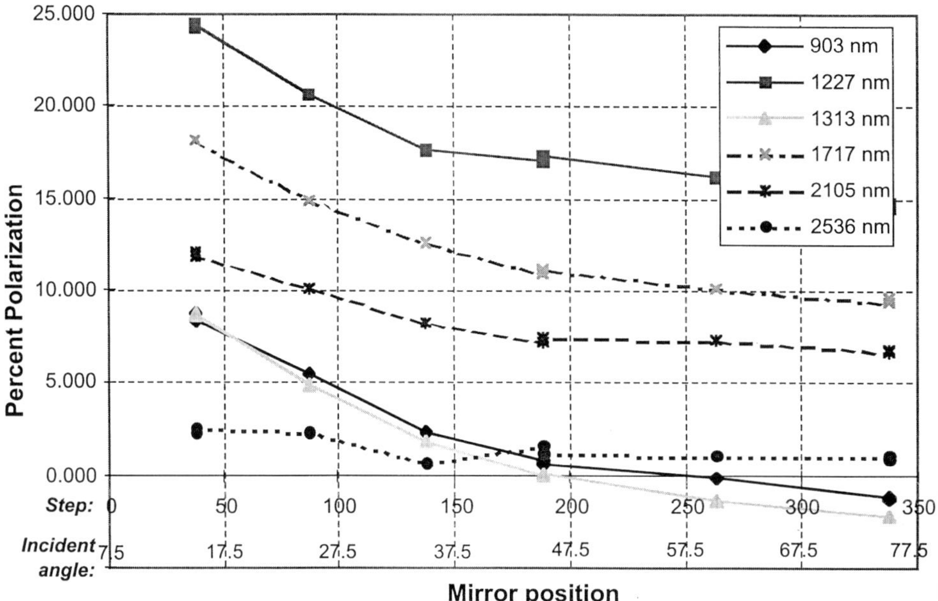

Figure 28. Percentage polarization for selected channels versus the scan mirror angle.

a source with this linear polarization in the worst orientation, the equation defining polarization above shows that the maximum and minimum components are in the ratio 1.016. For two signals with intensities of 1.016 and 1.0 and an instrument with no polarization sensitivity the total signal would be 2.016. The instrumental response difference for the two components in the worst channels is $\sim$15% at this mirror position, so the measured signal will be $(1.016 \times 1.075 + 1.0 \times 0.925)$ for a total signal of 2.017, an increase of $\sim$1/1000 which is near the noise level at low phase angles (Section 4.2). At moderate phase angles polarization is expected to be of order 6%, based on extrapolation of Zellner and Gradie's results to higher phase angles. Repeating the above exercise for mirror position 75, the polarization properties of NIS would lead to an increase of signal of $\sim$1/160 in the worst channels, which is again near the noise level at 90° phase angle. Thus, although polarization is an interesting property of the NIS, it should have an insignificant effect on the data accuracy.

3.4. OFF-AXIS RESPONSE

Measurements of the stray light within the NIS scattered from outside the FOV show it to be reduced by 4 or 5 orders of magnitude from a source about 2° from the slit.

The off-axis response of the NIS was measured by using the motion stage to scan a small source across the FOV. (The source was the exit slit of the monochromator located at the focus of the collimator. With the monochromator set at the zero

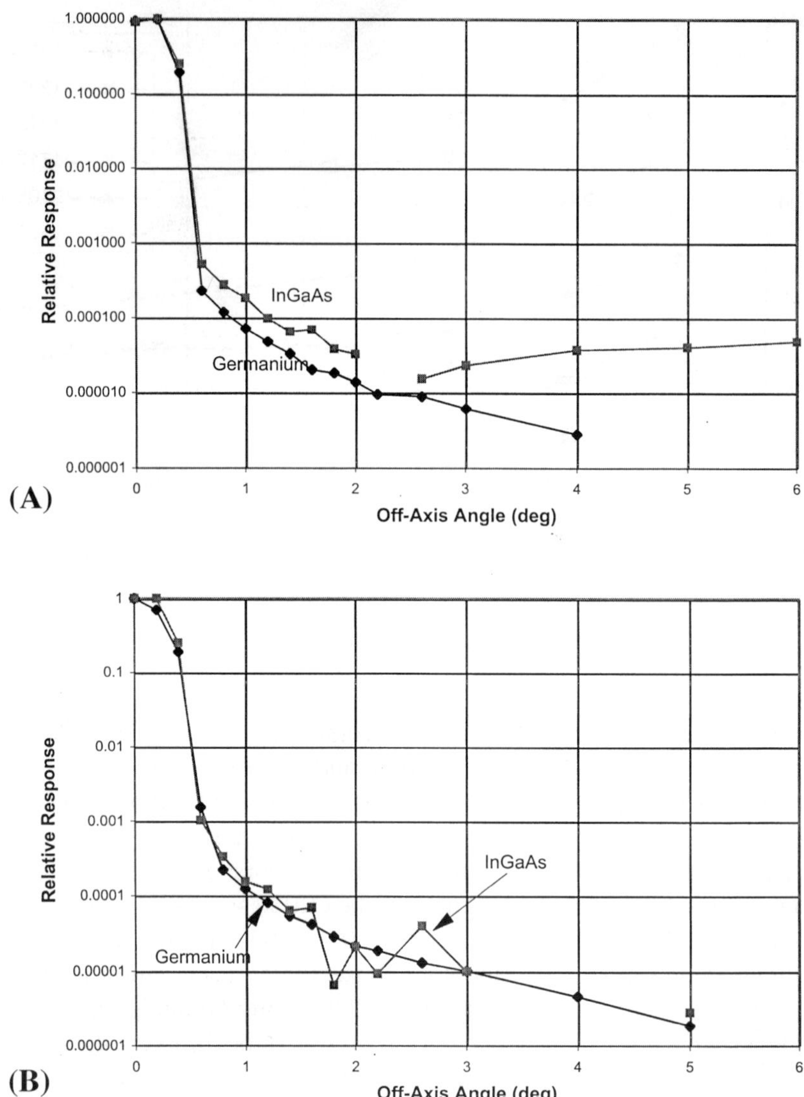

Figure 29. Off-axis response of the InGaAs and Ge detectors for a point source scanned in (a) the spectral direction and (b) the spatial direction. Because the scattered light signal is very weak, a few elements with lower SNR on each end of the arrays were not used for this calculation.

order and illuminated with white light, all wavelengths entered the instrument.) The neutral density filters in the monochromator, the variable exit slit width, and the two gains of the Ge channels were used to extend the dynamic range so the stray light could be traced to low signal levels. The FOV was scanned across the source in both spatial and spectral directions.

First the background signal was subtracted and then the data were normalized to the axial signal. The average results for the Ge elements 5 through 25 are shown in Figure 29. The top plot is the relative response as the source is scanned in the spectral coordinate of the slit. A better name for the relative response is the Point Source Rejection Ratio (PSRR) of the NIS. At $0.3°$ from the edge of the slit, the signal is only 0.02% of the peak signal. These data have been tracked to the limits that the accuracy will allow. The measurement accuracy is at best 1 in 4095, or 0.025%, however, this has been reduced by averaging ten samples for each data point. The lower plot in Figure 29 shows the corresponding results in the spatial direction. The same degree of off-axis rejection is achieved demonstrating that the level of off-axis light being scattered into the FOV of the NIS is very low. We can trace it down by about five orders of magnitude, the limit of our measurement accuracy.

To place this into an operational context, this data has been used to calculate the expected signal from outside the nominal field of view when the instrument is observing an infinite, uniformly illuminated plane. Assuming average behavior over the spectral range of NIS and the wide slit, the estimated fraction of the total observed signal that comes from outside the field of view is 10^{-3}. Note that the light from outside the field of view will fall in the correct spectral channels. If the surfaces outside the field of view have spectral properties similar to the observed region, then the effect of the stray light signal is further minimized. Of course, in some cases the region outside the nominal field of view may be brighter than the observed region due to lighting conditions in the observed region. In this case the effect of stray light is increased in proportion to the relative brightness of the two regions.

3.5. ALIGNMENT

3.5.1. *Detector Array Co-Alignment*

The dichroic beamsplitter divides the energy between the two detector arrays. Each array has been aligned independently for focus, position, and angular orientation. For efficient energy collection the entrance slit must be imaged squarely onto each array and for each array to see the same target area the two arrays must be co-aligned. Tests were performed to determine the accuracy of the alignment. The source for the NIS was the monochromator exit slit located at the focus of the collimator. The wavelength was set to zero so all wavelengths in the zero order were transmitted. Test data were recorded as the OCF motion stage was rotated in azimuth (the spatial direction in the NIS) in steps of $0.02°$ over a range of $1°$. The data were processed by subtracting the dark signal and normalizing the angular variation to the peak response. When Ge elements 10–20 and InGaAs elements 40–50 (near the center of each array) are averaged, the resulting distributions are as shown in Figure 30.

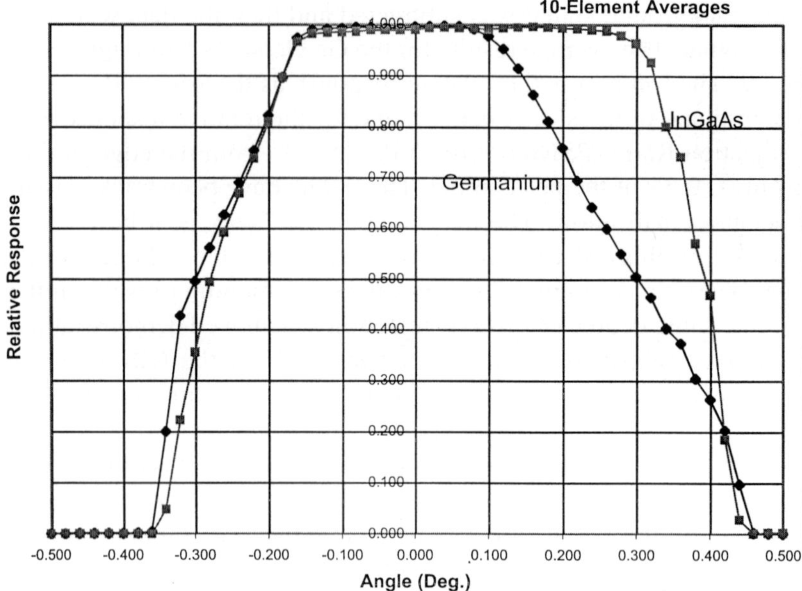

Figure 30. Spatial alignment of the InGaAs and Ge detector arrays, with 10 elements averaged in each array.

The Ge array shows the effect of aberrations, notably astigmatism which causes the signal to fall off. The InGaAs energy follows a different optical path after it passes through the dichroic beamsplitter. It is also closer to being on axis at the center of the range, whereas the Ge array has to be offset to intercept the desired wavelength range. The edges of each array are the same, however, so they are spatially aligned to an acceptable accuracy of about $0.01°$ or a position of 0.014 mm. There is a small difference in the integrated area of the target sampled because of the different aberrations in the two optical paths.

The motion stage was also scanned in the spectral direction to move the point source across the narrow dimension of the slit. The resulting distribution for all elements is shown in Figure 31. Perfect alignment of the two arrays is, of course, expected as the profile is controlled by the entrance slit and not by the width of the detector arrays. Some variation along the Ge array is evident, however. This is shown in Figure 32 which shows the response distribution for all the Ge elements as the source is scanned across the width of the slit. At the 50% points the width of this distribution is $0.35° \pm 0.02°$.

The spatial profile of a number of elements is shown in Figure 33 for the narrow and wide slits. These curves were used to determine the effective widths of each slit, the width ratio of the two slits, and to measure the spectral offset between their centers.

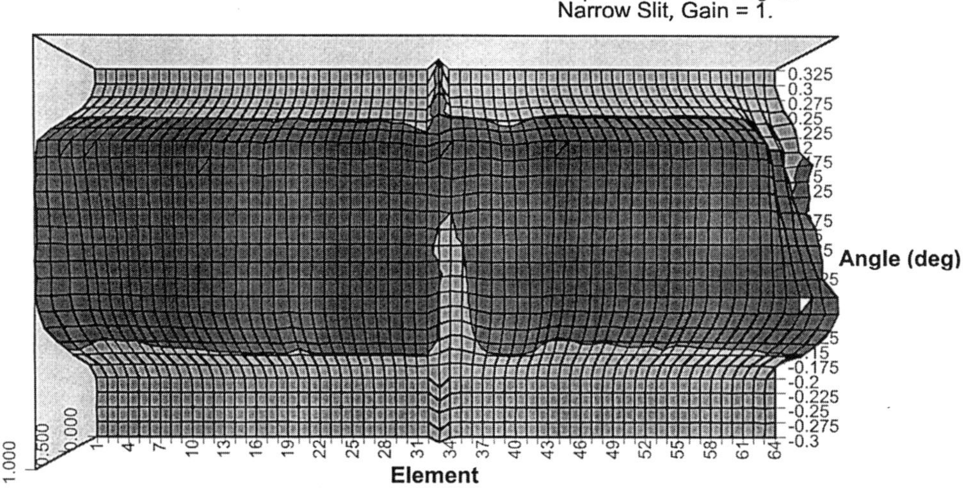

Figure 31. Response for all detector elements as a point source is scanned across the slit in the spectral direction.

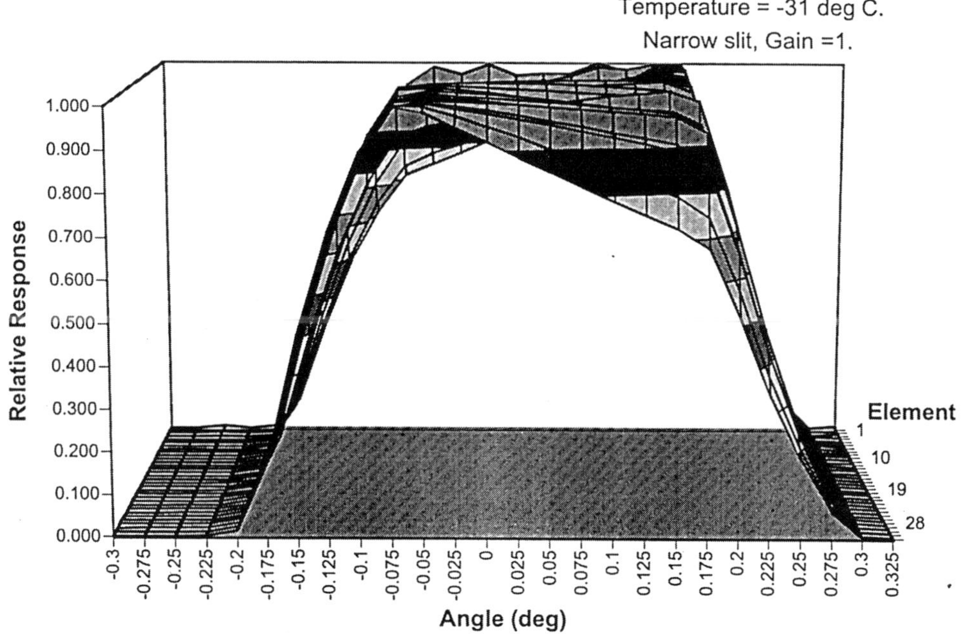

Figure 32. Ge detector response profiles in the spectral direction.

3.5.2. *Co-Alignment with Other Instruments*

Each optical instrument on NEAR had its boresight measured relative to a reference optical cube mounted on the instrument. For the NIS, this was accomplished

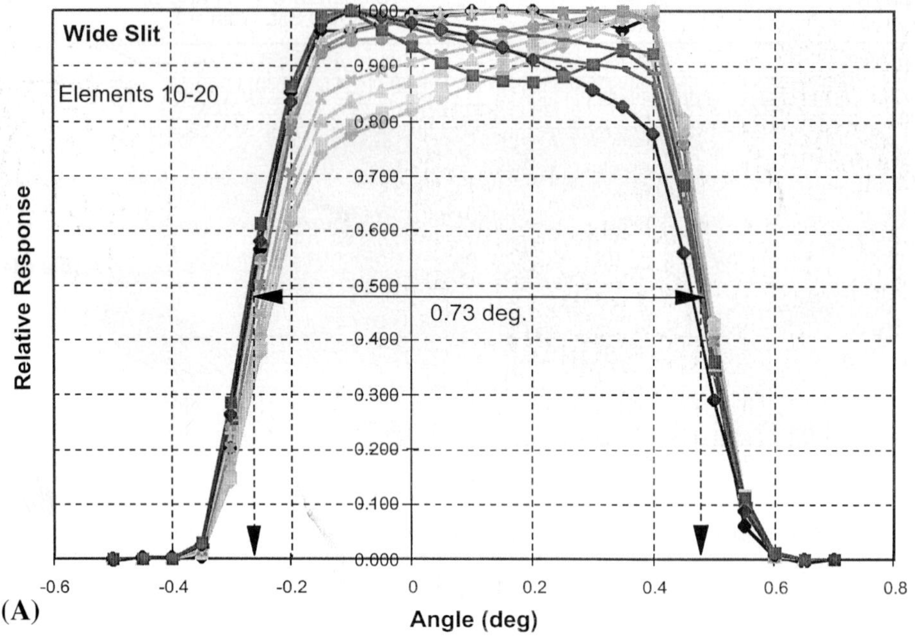

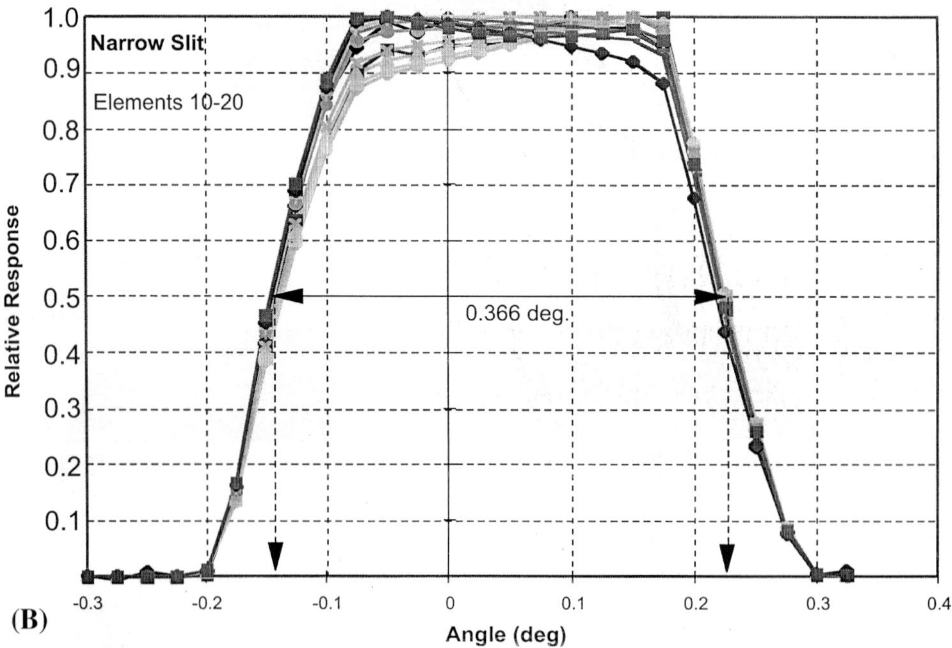

Figure 33. Variation in response as a point source is moved across the slit in the spatial direction for (a) the wide slit, and (b) the narrow slit.

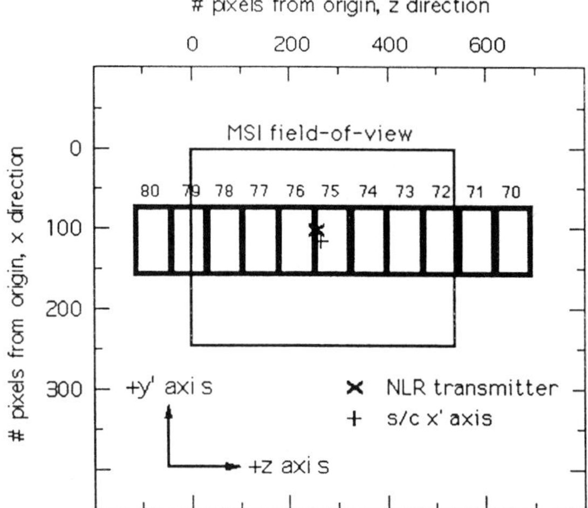

Figure 34. Coalignment and orientation of NEAR optical instruments from preflight measurements. The x' axis, the spot formed by the NLR transmitter, and the NIS field-of-view through the narrow slit for different mirror positions are shown in MSI pixel coordinates. (The corresponding step number is written beside each NIS field of view that is shown).

using a theodolite to measure the boresight looking into the entrance slit, and mapping this to the reference cube. Alignment of the boresights of MSI, NIS, and the NEAR Laser Rangefinder (NLR) relative to each other and to the nominal common boresight (the spacecraft x' axis) were measured on ground, after spacecraft integration, using the reference optical cubes on each instrument. For NIS, the boresight is at mirror step 75, or instrument nadir, which views an angle 90° from the Sun in the nominal spacecraft orientation. For MSI the boresight is the center of the field-of-view, and for NLR it is the direction of the beam emitted from the transmitter. From this information, the locations of NLR and NIS footprints and the x' axis within an MSI image were derived (Figure 34).

In addition to instrument coalignment, the design peculiarities of MSI, NIS, and the spacecraft control the spatial pattern of instrument overlaps from the viewpoint of a data user. The z direction on the spacecraft, which points toward the Sun with the spacecraft in its nominal orientation, is within the scan plane of the NIS mirror. This z axis also lies parallel to the long dimension of the 2.9° × 2.25° MSI field-of-view. The ideal common instrument boresight is called the x' axis, and it also by definition lies within the mirror's scan plane; the y' axis is orthogonal to the x' and z axes. An MSI image is displayed in 'landscape' mode on most video monitors, with the first pixel in the upper left of the screen. 'Up', from the viewpoint of an observer sitting on the spacecraft, is therefore at the right side of an image. Consequently, scanning the NIS mirror moves its footprint horizontally across an MSI image as it is displayed on a video monitor.

Figure 34 shows the instrument footprints in MSI pixel coordinates, as displayed on a video monitor, with the first pixel in the upper left of the screen. Uncertainty is of the order of 1 MSI pixel. At the instrument nadir position of NIS (mirror step 75), the spacecraft x' axis, the spot formed by the NLR transmitted beam, and the center of the MSI frame all fall within the NIS narrow slit field-of-view. The footprints for 7 NIS mirror steps through the narrow slit (72–78) all fall entirely or almost entirely within the MSI field-of-view; the footprint for step 79 straddles the boundary of the MSI frame in the $-z$ direction.

As these measurements were done under conditions of 1 g and room temperature, conditions experienced by NEAR in space will distort the coalignment between instruments slightly. During the Earth flyby, coalignment between the instruments will be re-measured as described in Section 4. Because shifts in coalignment are to be expected, the inflight coalignment results should be used for processing of flight data.

3.5.3. *Scan Mirror Calibration*

The preceding discussion relates the field of view of NIS with MSI and NLR when NIS is at the instrument nadir position. For other scan positions, the line of sight moves by nominally 0.4° per step. Because the stepper motor and reduction gears are not perfect, however, this step size is not exactly 0.4°. The difference between the actual and nominal line of sight has been measured as part of ground calibration. Figure 35 shows the result. The data were obtained at 10 step intervals. The maximum difference is about three quarters of a step. This builds up over many steps, so there is no place where there is a big gap between adjacent steps. The various periodicities seen in the data are a result of the various number of steps required to complete a full rotation of components in the motor and reduction gear. These are repeatable effects, and the figure shows two independent tests of the pointing, which overlay nicely. Data were also obtained in one step increments at two points in the field of regard to ensure that there were no high frequency variations that would be aliased in the 10 step data.

3.6. IN-FLIGHT CALIBRATION PLAQUE

The calibration plaque will serve as the main source of in-flight radiometric calibration, so the stability of its properties and knowledge of its spectral reflectance are critical to making full use of the high SNR data to be obtained by NIS. The need to quantify the reflectance properties of the calibration plaque that might be encountered under these flight conditions led to a rigorous program of ground measurements, which were conducted at two of the top U.S. laboratories in reflectance spectroscopy, the U.S. Geological Survey in Denver, and the RELAB facility at Brown University. USGS conducted directional-hemispherical measurements relative to Halon (in which reflection from an incident beam is collected by an integrating sphere), and RELAB conducted bidirectional measurements at a ref-

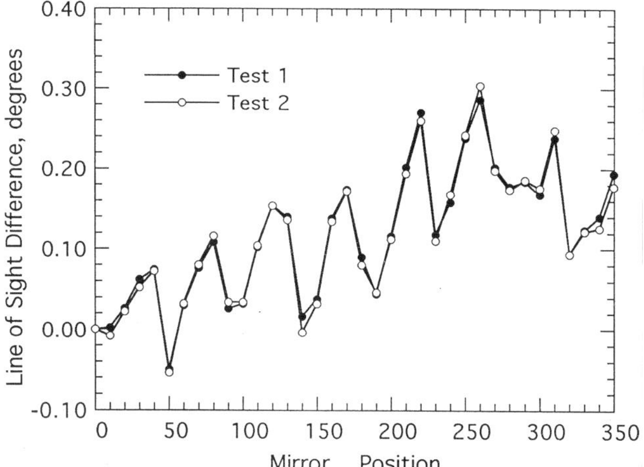

Figure 35. Pointing calibration of the NIS scanner. The x axis is the step number, and the y axis is the difference between the measured line of sight and the ideal line of sight. A positive number indicates that the actual line of sight is further away from start than the ideal line of sight. Data were taken at 10 step increments. The periodic variations in scanner pointing calibration are due to mechanical tolerance in gears and are repeatable effects.

erence geometry ($I = 30°$, $e = 0°$, where I is the incidence angle, and e is the emission angle) and at representative geometries to be encountered in-flight ($I = 45°$, $55°$, $65°$, $75°$; $e = 75°$). The bidirectional measurements presented somewhat of a problem, because the partially specular nature of plaque reflectance necessitated use of a calcite depolarizer. Calcite has its own absorption features at >2200 nm, rendering data at those wavelengths subject to spectral artifacts. The directional-hemispherical measurements thus provide superior characterization of the spectral properties of the plaque, yet not on its absolute reflectance at a given bidirectional geometry; bidirectional measurements in turn provide the dependence of plaque reflectance on illumination geometry.

Figure 36 shows the spectral properties of the calibration plaque from both sets of measurements. In this presentation, the directional-hemispherical reflectance has been scaled to bidirectional reflectance at <1300 nm, measured at $I = 30°$, $e = 0°$. At this bidirectional geometry, spectral artifacts of the calcite depolarizer are minimal. Both data sets agree closely on the shape of the spectrum, which is relatively flat above 1000 nm with a small falloff at lesser NIR wavelengths. Figure 37 shows reflectance of the plaque relative to that at the reference geometry of $I = 30°$, $e = 0°$. The behavior is only partly diffuse, with a residual specular component. At an incidence angle of 45° (expected at the nominal spacecraft orientation, with no yaw), it is brightened by over 25% relative to the reference geometry. At higher incidence angles and spacecraft yaws, brightening relative to the reference geometry approaches 200%. In-flight, incidence angles of about 50° and

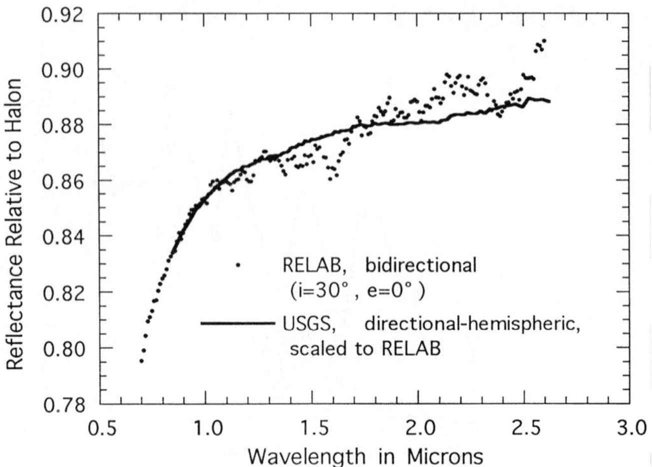

Figure 36. Reflectance spectra of the calibration plaque, based on measurements conducted at the U.S.G.S. (Denver) and at RELAB (Brown University).

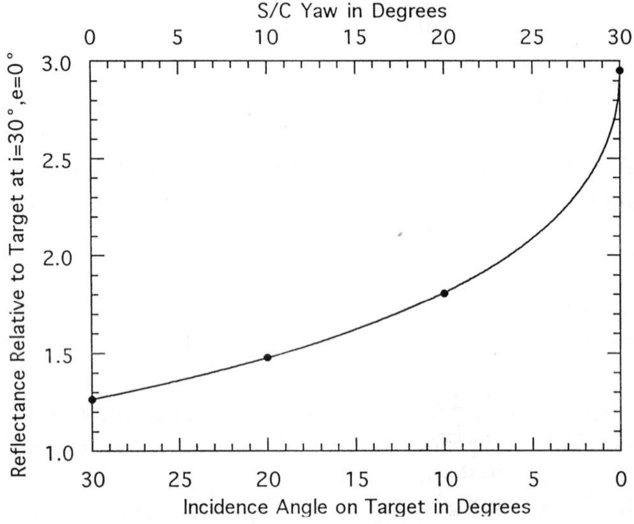

Figure 37. Variations of plaque brightness with illumination geometry, based on measurements conducted at RELAB (Brown University).

spacecraft yaws of about 5° are expected, to maintain the similarity of calibration plaque brightness to Eros and avoid operation at near specular conditions.

3.7. MEASUREMENTS OF GEOLOGIC MATERIALS

As part of the pre-flight calibration and characterization of NIS, spectra of more than 30 rock, mineral powder, meteorite, and other samples were measured while the instrument was under vacuum at nominal operating temperatures. This was accomplished by mounting a 45° fold mirror near the output port to the optical calibration facility collimator window, and then placing small sample containers on a stage below the 45° mirror, at the focus of the collimator. The samples were illuminated with a quartz halogen lamp and measured using all combinations of slit width and gain states.

Among the samples that were measured by NIS were many of the 'classical' mineral powder samples originally measured by Hunt, Salisbury, and colleagues (e.g., Hunt and Salisbury, 1970; Hunt et al., 1971) and more recently re-measured and re-analyzed by Clark et al. (1993). Additional measurements of standard calibration samples and the NIS calibration plaque were also obtained. Examples of some representative NIS spectra of these materials, calibrated relative to a halon standard also observed by NIS, are presented in Figure 38. The spectra are completely consistent with previous spectra of these materials, and demonstrate the spectral resolution and fidelity that can be achieved by NIS. Some of these samples were also measured using the multispectral filters on MSI during its pre-flight calibration period, and Figure 38 indicates that the agreement between NIS and MSI spectra for the region of spectral overlap (900 to 1050 nm) is excellent. The sample spectra also serve as a useful database for development and refinement of data reduction and analysis algorithms prior to the collection of actual Eros spectra in 1999.

4. Preliminary Mission Planning

4.1. IN-FLIGHT CALIBRATION

In-flight calibrations of the NIS are designed to monitor and measure: (1) the instrumental dark current and its variations with time and temperature; (2) the radiometric calibration coefficients needed to convert instrument DN into flux units; (3) the effects of stray and scattered light in the instrument; and (4) the spatial and spectral alignment of the Ge and InGaAs detectors. The baseline NIS mission calibration plan is outlined in Table II.

Dark current measurements are being taken periodically early in cruise (with the NIS cover closed) in order to verify the health of the instrument, to measure the level of instrument noise as a function of temperature and spacecraft activity level, and to characterize the thermal and power performance of NIS for comparison with thermal vacuum chamber measurements and model predictions. Pre-flight tests indicated that several of the InGaAs channels exhibited large (10–20 DN) drifts in dark current on timescales of a few minutes. The effect of these relatively slow drifts can be completely accounted for in the data as long as dark spectra are frequently interspersed with actual asteroid spectra. Once at Eros, such dark measurements will be obtained by using the NIS internal shutter to block out incoming light and/or by pointing the scan mirror towards a dark region of space.

                JEFFERY W. WARREN ET AL.

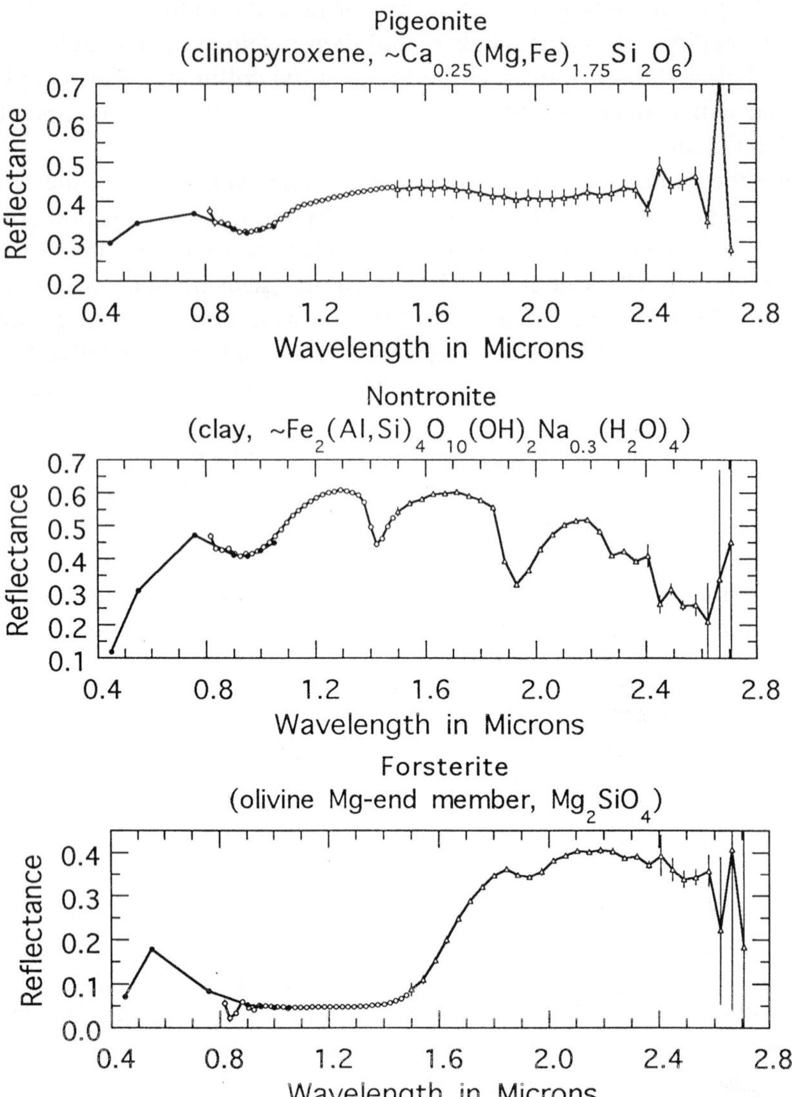

Figure 38. Spectra of typical mineral and rock samples measured by NIS during pre-flight calibrations at APL. Data from the Ge detector are shown as open circles; data from InGaAs are open triangles. Shown for comparison are the 7-color spectra of these same samples as measured by the MSI instrument. Both the NIS measurements and the MSI measurements were calibrated through observations of the same halon reflectance standard. This plot thus shows how well the two instruments agree in the region of spectral overlap. Error bars in MSI spectra include variance due both to noise in the data and to shadows in the images of the samples. Top: Spectrum of pigeonite powder HS199. The diagnostic 1000 nm and 2000 nm pyroxene absorption bands are clearly detected. Middle: Spectrum of nontronite powder SWa-1.a. Narrow spectral features near 1400 nm and 1900 nm are caused by adsorbed and structural water within this clay mineral. Bottom: Spectrum of forsterite fragments GDS70.c. The strong olivine absorption feature centered near 1200 nm dominates this spectrum. All samples kindly provided by Roger Clark of the U.S.G.S.

Table II
NIS inflight calibration plan (with major events noted)

Time	Target	Calibration objectives	Procedure
		Launch	
Repeatedly during early cruise	Cover closed	– Dark current level and variability	Acquire long-period dark measurements
		Cover opened	
After cover opened, just before earth gravity assist	Calibration target	– Radiometric response – Stray light from calibration target	Multiple spectra of cal. target, at spacecraft attitude where reflected sunlight is $\sim$20° off specular geometry. Also scan through mirror positions near calibration target to quantify stray light in different positions
		Earth gravity assist	
Close apporach to Earth	Ocean, Antarctica	– Wavelength calibration	Spectra of water and ice to verify wavelength calibration using absorptions.
Close approach to Moon	Moon	– Radiometric response	Multiple spectra of Moon
Distant Moon observations	Moon	– Coalignment of NIS and MSI – Detector alignment	Take data with both instruments while slowly scanning across Moon in 2 orthogonal directions.
Several times during ate cruise	Calibration target	– Radiometric response	Multi spectra of cal. target, at sppacecraft attitude where reflected sunlight is $\sim$20° off specular geometry.
		Approach to Eros	
Frequently	Calibration target	– Stray light from calibration target	Scan through mirror positions near calibration target to quantify stray light in different positions; use same spacecraft geometry as for calibration measurements.
Eros approach	Eros	– Coalignment of NIS and MSI – Detector alignment	Take data with both instruments while slowly scanning across Eros in 2 orthogonal directions.
		Eros orbit insertion	
All orbital altitudes at Eros	Eros limb	– Quantify magnitude of stray light	Scan swath across Eros limb.
Several times during Eros orbit	Shutter closed	– Dark current behavior	Acquire one hour of continuous dark measurements
Frequently	Calibration target	– Radiometric response	Multiple spectral of cal. target, at spacecraft attitude where reflected sunlight is $\sim$20° off specular geometry.
Interleaved with spectra of Eros and cal. target	Shutter closed	– Dark current level	Acquire dark measurements close in time to Eros and cal. target spectra.

Radiometric calibration of NIS spectra will be achieved primarily from observations of the onboard diffuse gold calibration plaque. The calibration plaque will be measured frequently in flight once the NIS cover is opened, by orienting the spacecraft to a geometry that allows sunlight to be reflected off the calibration plaque and into the instrument. Using the reflectance properties and the photometric model of the plaque determined in pre-flight calibrations (see Section 3.6), knowledge of the spacecraft geometry and heliocentric distance, and knowledge of the NIS temperature, it will be possible to derive updated radiometric calibration coefficients from a model of the calibration plaque data. During the Earth–Moon flyby in January 1998, additional calibration measurements of the Earth and Moon are planned to verify instrument health and wavelength calibration (by observing terrestrial H_2O absorption features) as well as to provide an independent assessment of the NIS radiometric calibration through observations of well-characterized lunar surface regions. Pre-flight radiometric calibrations of NIS achieved using integrating spheres and well-calibrated light sources (Section 3.3) serve as an additional backup and crosscheck on the NIS in-flight radiometric calibration.

Other characteristics of NIS will also be measured during cruise and at Eros in order to perform the best possible calibration of the data. During the Earth–Moon flyby, the coalignment with MSI will be measured by scanning swaths of measurements from both instruments across the disk of the Moon along both axes of the imager field-of-view. Correspondence in the signal levels from NIS with the disk's position in the MSI frame will show the relative position of the two instruments' fields-of-view. For NIS, this measurement will also confirm that spatial alignment of the Ge and InGaAs detectors has been maintained. The magnitude of out-of-field signal will be quantified by acquiring swaths of spectra across the limb of Eros, during different orbits as the source of stray light (Eros) changes in its size and illumination. These results will also be augmented by the results of the more controlled investigations of stray and scattered light conducted during pre-flight calibrations (Section 3.3).

Figure 39 shows a schematic representation of the NIS data calibration stream. The determination of radiometric calibration coefficients from the calibration plaque observations proceeds separately at frequent intervals during the primary orbital mission, thus allowing for the characterization of any drifts in the NIS radiometric calibration with time. The calibration of both the calibration plaque and Eros spectra rely upon accurate measurements of the dark current bracketing the Eros observations close in time, as well as upon the application of the other critical instrumental corrections determined in pre-flight calibrations.

4.2. PLANS AT EROS

NIS observations of Eros will be highly coordinated with MSI imaging observations in order to fully realize the synergy between these two instruments. Observation sequences will usually be designed so that full NIS spectral coverage is obtained

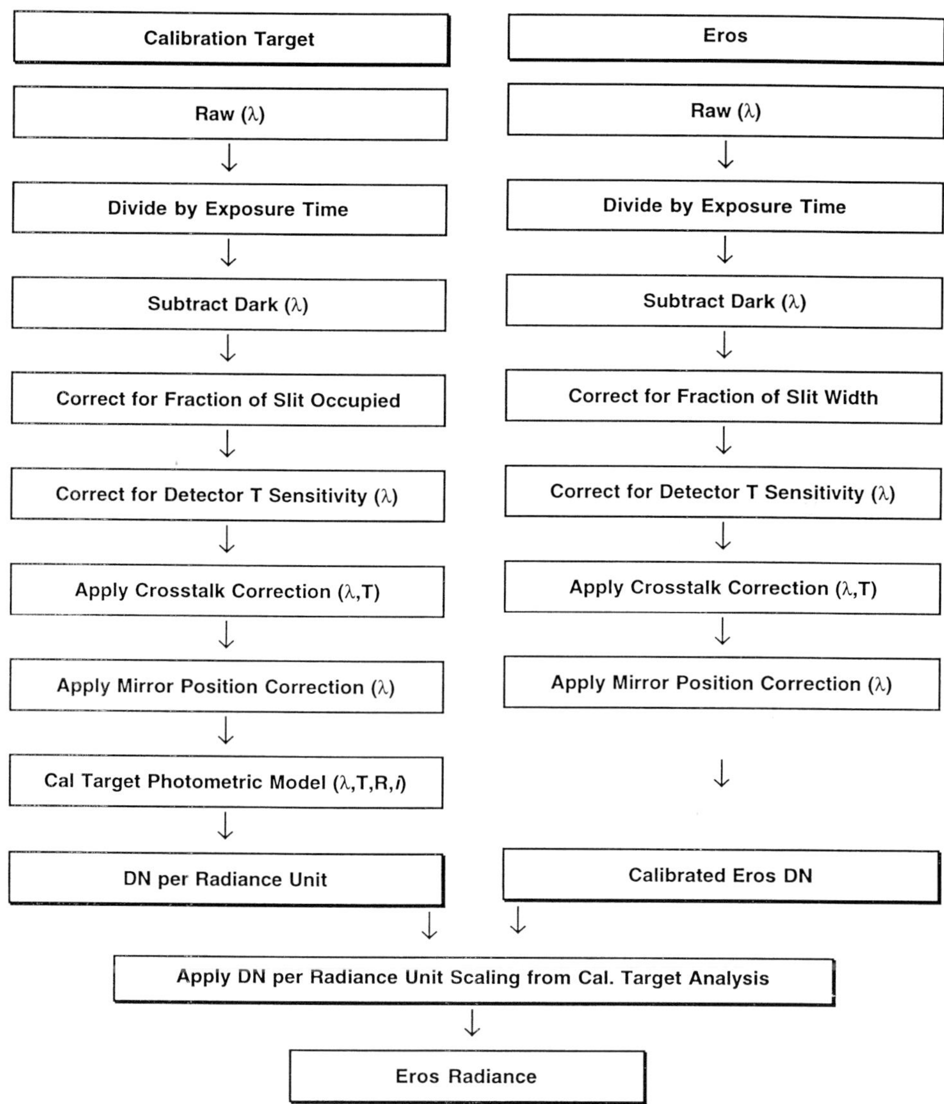

Figure 39. Schematic of NIS data calibration pathway. There are two parallel paths involved: one for the NIS onboard calibration target, and another for the Eros spectra. Both pathways rely on a combination of calibration results from in-flight and pre-flight instrument and calibration target tests.

for each MSI field of view. This corresponds to approximately 24 NIS spectra (a 3 × 8 grid) for each MSI image (Veverka et al., 1996; Hawkins et al., 1997). NIS coverage along the long axis of the MSI images is achieved by moving the internal scan mirror, while coverage along the short MSI axis will be achieved either by spacecraft orbital motion or by rolling the spacecraft slightly off the nadir direction.

Each NIS spectrum will be tagged with pertinent instrument and spacecraft housekeeping data as well as observation geometry information (latitude, longitude, incidence angle, emission angle, etc.). This information will allow each NIS spectrum to be mapped onto the surface of Eros in near-real-time so that the Science Team can keep track of which regions have been covered. The fact that the spacecraft will be in orbit around Eros means that there will be ample opportunity to go back and measure regions not covered, or to measure the same region at a range of incidence and emission angles in order to derive surface photometric properties. The end result will be a globally-registered and map-projected 3-D image cube (spatial × spatial × spectral dimensions) that will allow minerals to be detected and spatially mapped across the surface of Eros.

As mentioned above, the NIS Eros observations will need to be bracketed by frequent dark spectra in order to most accurately calibrate all of the spectral channels. To increase the SNR, it is also possible to add up to 16 NIS spectra at a time before transmitting the data to Earth. Using the pre-flight calibration data to estimate the expected noise and instrumental radiometric coefficients, and using the telescopic spectrum of Eros and reasonable assumptions for heliocentric distance and phase angles, we can estimate the expected performance of NIS at Eros. The results are shown in Figure 40 for typical observational scenarios during the early portion of the orbital phase. Conditions for this calculation are a heliocentric distance of 1.75 AU, 90° phase angle, and 10 s total integration time (including dark signal acquisition). Typical SNR in the 800 to 1500 nm range is in excess of 200, while typical SNR in the 1300 to 2700 nm range is approximately 50. These SNR values will allow very good characterization of the positions, widths, and strengths of absorption features from many types of minerals expected to occur on asteroid surfaces. Later in the mission the asteroid is approaching perihelion at 1.13 AU with a consequent improvement of SNR. Also, much longer integration times can be used where required to further improve the SNR.

The expected signal levels are low enough that the 10 times gain state of the Ge detector should be used for most data. During the Earth flyby and late in the mission the heliocentric distance is reduced and use of the unity gain state may be required to avoid saturation. The wide slit can be used to improve SNR at the expense of spectral and spatial resolution. Since predictions show good SNR should be obtained with the narrow slit in most circumstances (through adding of multiple spectra), the narrow slit is expected to be used for most observations. Dark data will be obtained frequently. The tentative plan for coordinated MSI-NIS observations is summarized by Veverka et al. (1997).

The calibrated NIS image cube data from Eros will be archived and distributed to the NASA Planetary Data System. Much of the data will be rapidly available on-line for downloading by anyone via the Internet and World-Wide Web (the NEAR home page can be found at http://hurlbut.jhuapl.edu:80/near.html). The data will be used by the Science Team as well as asteroid scientists worldwide to investigate fundamental questions about the nature of asteroid surfaces as well as the link

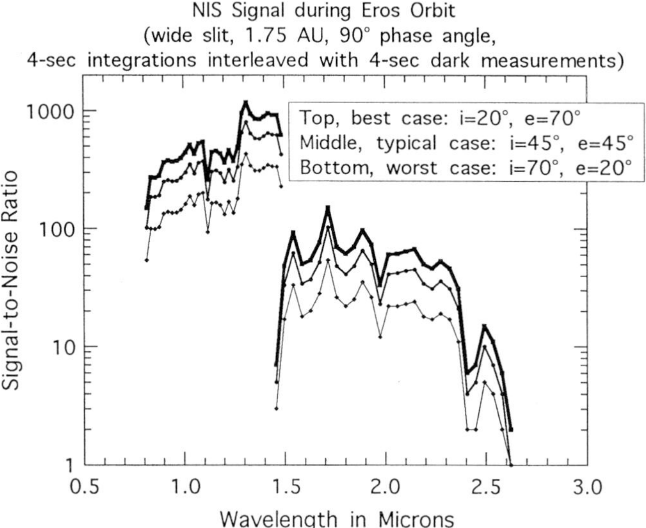

Figure 40. Predicted NIS signal to noise ratios (SNR) for the high phase angle Eros orbital obser-vations. Signal levels were estimated using nominal values for solar flux, the reflectance of Eros, heliocentric distance ($D = 1.75$ AU), the phase angle for orbital operations ($\alpha = 90$ deg), and radiometric calibration constants determined from pre-flight calibrations. The noise was estimated using in-flight dark signal measurements obtained on 21 May, 1996. The assumed spectrum acquis-ition sequence uses the wide slit over a 10-s time interval (4 s of Eros data, 2 s for shutter changes, 4 s dark data). Three cases were examined: 'Best case' (triangles) corresponding to low incidence angle, high emission angle measurements; 'Typical case' (circles) corresponding to medium incid-ence and emission angle values; and 'Worst case' (squares) corresponding to high incidence angle, low emission angle measurements. The asteroid's flux decrease with incidence and emission angle was modeled using the empirical asteroidal phase variation curves of Gehrels and Tedesco (1979).

between asteroids and meteorites. Specific science issues that can be addressed include: studying surface mineralogic variations associated with impact craters as a way to probe the interior of Eros; examining the spectra of fresh craters and steep walls or slopes as a way to assess the role of putative 'space weathering' processes in the alteration of asteroidal surfaces; providing clues to the origin and geophysical structure of Eros by searching for evidence of large-scale mineralogic variability and/or coherent spectral units; and, constraining models of the origin and evolution of ordinary chondrites and other meteorites by comparing the detailed mineralogy and chemical composition of a typical S asteroid to that of the major meteorite families. There are obviously many more issues that will be addressed by the NIS data, and there will certainly be many surprises from this first-ever close-up examination of a primitive body. By providing high-quality global mineralogic information about the surface of an asteroid, the NIS investigation will play a major role in the continuing exploration of our solar system.

Acknowledgements

The authors would like to express their thanks to the entire NIS team, whose efforts led to the success of this instrument on a very tight schedule. From concept through launch support, and mission operations after launch, the entire team has given an exceptional effort to make this instrument and mission a reality.

We are indebted to Roger Clark and colleagues at the U.S. Geological Survey Denver Spectroscopy Laboratory and Carle Pieters and colleagues at the Brown University RELAB facility for their careful measurements of the NIS flight calibration plaque under a variety of challenging viewing geometries. We also sincerely thank Roger Clark, Dick Morris, Dale Cruikshank, Ted Bunch, and the Smithsonian Institution and Johnson Space Center for the loan of rock, mineral powder, and meteorite samples that were measured by NIS.

References

Adams, J.: 1974, *J. Geophys. Res.* **79**, 4829–4836.

Allen, C., Morris, R., and McKay, D.: 1994, *J. Geophys. Res.* **99**, 23173–23185.

Bell, J. F., Davis, D. R., Hartmann, W. K., and Gaffey, M. J.: 1989, in R. P. Binzel, T. Gehrels, and M. S. Matthews (eds.), 'Asteroids: The Big Picture', *Asteroids II*, University of Arizona Press, Tucson, pp. 921–948.

Bruegge, C. J., Stiegman, A. E., Rainen, R. A., and Springsteen, A. W.: 1993, *Opt. Eng.* **32**, 805–814.

Chapman, C. R. and Gaffey, M. J.:1979, in T. Gehrels (ed.), 'Reflectance Spectra for 277 Asteroids', *Asteroids*, University of Arizona Press, Tucson, pp. 655–687.

Chapman, C., Morrison, D., and Zellner, B.: 1975, *Icarus* **25**, 104-130.

Cheng et al.: 1997a, 'Near-Earth Asteroid Rendezvous: Mission Overview,' *JGR Planets*, in press.

Cheng et al.: 1997b, *Space Sci. Rev.*, this volume.

Clark, B. E., Fanale, F., and Salisbury, J. W.: 1992, *Icarus* **97**, 288–297.

Clark, R. N., Swayze, G. A., Gallagher, A. J., King, T. V. V., and Calvin, W. M.: 1993, The U.S.G.S. Digital Spectral Library: Version 1: 0.2 to 3.0 um. U.S. Geol. Surv. Open File Report 93–592.

Cloutis, E. and Gaffey, M.: 1991, *J. Geophys. Res.* **96**, 22809–22826.

Cloutis, E., Gaffey, M., Jackowski, T., and Reed, K.: 1986, *J. Geophys. Res.* **91**, 11641–11653.

Fanale, F., Clark, B. E., and Bell, J. B.: 1992, *J. Geophys. Res.* **97**, 20863–20874.

Farquhar, R. W., Dunham, D. W., and McAdams, J. V.: 1995, *J. Astron. Sci.* **43**, 353–372.

Fischer, E. and Pieters, C.: 1994, *Icarus* **111**, 475–488.

Gaffey, M. J., Bell, J. F., and Cruikshank, D. P.: 1989, in R. P. Binzel et al. (eds.), 'Reflectance Spectroscopy and Asteroid Surface Mineralogy', *Asteroids II*, University Arizona Press, Tucson, pp. 98–127.

Gaffey, S. J., McFadden, L. A., Pieters, C. M., and Nash, D. B.: 1993a, in C. Pieters and P. Englert (eds.), 'Ultraviolet, Visible, and Near-Infrared Reflectance Spectroscopy: Laboratory Spectra of Geologic Materials', *Remote Geochemical Analysis: Elemental and Mineralogical Composition*, Cambridge University Press, Cambridge, pp. 43–71.

Gaffey, M., Bell, J. F., Brown, R. H., Burbine, T. H., Piatek, J. L., Reed, K. L., and Chaky, D. A.: 1993b, *Icarus* **106**, 573–602.

Gaffey, M. J., Burbine, T. H., and Binzel, R. P.: 1993c, *Meteoritics* **28**, 168–187.

Gehrels, T. and Tedesco, E. F.: 1979, 'Minor Planets and Related Objects. XXVIII. Asteroid Magnitudes and Phase Relations', *Astron. J.* **84**, 1079–1087.

Hawkins, S. E. et al.: 1997, *Space Sci. Rev.*, this volume.

Hunt, G. R. and Salisbury, J. W.: 1970, *Mod. Geol.* **1**, 283–300.

Hunt, G. R., Salisbury, J. W., and Lenhoff, C. J.: 1971, *Mod. Geol.* **2**, 195–205.

Markov, A. M. (ed.): 1962, *The Moon: A Russian View*, University of Chicago.

McCord, T. B., Adams, J. B., and Johnson, T. V.: 1970, *Science* **168**, 1445–1447.

Morris, R.: 1977, 'Origin and Evolution of the Grain-Size Dependence of the Concentration of Fine-Grained Metal in Lunar Soils: the Maturation of Lunar Soils to a Steady-State Stage', in *Proc. Lunar Planet. Sci. Conf. 8th*, pp. 3719–3747.

Murchie, S. L. and Pieters, C. M.: 1996, *J. Geophys. Res.* **101**, 2201–2214.

Pieters, C. M. and Englert, P. A. J.: 1993a, *Remote Geochemical Analysis: Elemental and Mineralogical Composition*, Cambridge University Press, Cambridge.

Pieters, C., Fischer, E., Rode, O., and Basu, A.: 1993b, *J. Geophys. Res.* **98**, 20817–20824.

Pieters, C. M. and McFadden, L. A.: 1994, *Ann. Rev. Earth Planetary Sci.* **22**, 457–497.

Santo, A. G., Lee, S. C., and Gold, R. E.: 1995, *J. Astron. Sci.* **43**, 373–398.

Stiegman, A. E., Bruegge, C. J., and Springsteen, A. W.: 1993, *Opt. Eng.* **32**, 799–804.

Tholen, D. J. and Barucci, M. A.: 1989, in R. P. Binzel, T. Gehrels, and M. S. Matthews (eds.), 'Asteroid Taxonomy', *Asteroids II*, University of Arizona Press, Tucson, pp. 1139–1150.

Veverka et al.: 1997. 'An Overview of the NEAR Multispectral Imager (MSI)–Near Infrared Spectrometer (NIS) Investigation', *JGR Planets*, in press.

Wendlandt, W. W. and Hecht, H. G.: 1966, *Reflectance Spectroscopy*, Wiley, New York, 298 pp.

Wetherill, G. and Chapman, C.: 1988, in J. Kerridge and M. Matthews (eds.), 'Asteroids and Meteorites', *Meteorites and the Early Solar System*, University of Arizona Press, Tucson, pp. 35–67.

Zellner, B. and Gradie, J.: 1976, *Icarus* **28**, 117–123.

Zellner, B., Tholen, D., and Tedesco, E.:1985, *Icarus* **61**, 355–416.

THE X-RAY/GAMMA-RAY SPECTROMETER ON THE NEAR EARTH ASTEROID RENDEZVOUS MISSION

J. O. GOLDSTEN, R. L. MCNUTT, JR., R. E. GOLD, S. A. GARY, E. FIORE,
S. E. SCHNEIDER and J. R. HAYES
The Johns Hopkins University, Applied Physics Laboratory, Laurel, MD 20723, U.S.A.

J. I. TROMBKA and S. R. FLOYD
*NASA Goddard Space Flight Center, Laboratory for Extraterrestrial Physics, Greenbelt,
MD 20771, U.S.A.*

W. V. BOYNTON and S. BAILEY
Lunar and Planetary Laboratory, University of Arizona, Tucson, AZ 85721, U.S.A.

J. BRÜCKNER
Max-Planck-Institut für Chemie, Mainz, Germany

S. W. SQUYRES
Cornell University, Astronomy Department, Ithaca, NY 14853, U.S.A.

L. G. EVANS
Computer Sciences Corporation, Science Programs, Lanham-Seabrook, MD 20706, U.S.A.

P. E. CLARK and R. STARR
The Catholic University of America, Washington, D.C. 20064, U.S.A.

(Received 24 January, 1997)

Abstract. An X-ray/gamma-ray spectrometer has been developed as part of a rendezvous mission with the near-Earth asteroid, 433 Eros, in an effort to answer fundamental questions about the nature and origin of asteroids and comets. During about 10 months of orbital operations commencing in early 1999, the X-ray/Gamma-ray Spectrometer will develop global maps of the elemental composition of the surface of Eros. The instrument remotely senses characteristic X-ray and gamma-ray emissions to determine composition. Solar excited X-ray fluorescence in the 1 to 10 keV range will be used to measure the surface abundances of Mg, Al, Si, Ca, Ti, and Fe with spatial resolutions down to 2 km. Gamma-ray emissions in the 0.1 to 10 MeV range will be used to measure cosmic-ray excited elements O, Si, Fe, H and naturally radioactive elements K, Th, U to surface depths on the order of 10 cm. The X-ray spectrometer consists of three gas-filled proportional counters with a collimated field of view of 5° and an energy resolution of 850 eV @ 5.9 keV. Two sunward looking X-ray detectors monitor the incident solar flux, one of which is the first flight of a new, miniature solid-state detector which achieves 600 eV resolution @ 5.9 keV. The gamma-ray spectrometer consists of a NaI(Tl) scintillator situated within a Bismuth Germanate (BGO) cup, which provides both active and passive shielding to confine the field of view and eliminate the need for a massive and costly boom. New coincidence techniques enable recovery of single and double escape events in the central detector. The NaI(Tl) and BGO detectors achieve energy resolutions of 8.7% and 14%, respectively @ 0.662 MeV. A data processing unit based on an RTX2010 microprocessor provides the spacecraft interface and produces 256-channel spectra for X-ray detectors and 1024-channel spectra for the raw, coincident, and anti-coincident gamma-ray modes. This paper presents a detailed overview of the X-ray/Gamma-ray Spectrometer and describes the science objectives, measurement objectives, instrument design, and shows some results from early in-flight data.

Space Science Reviews **82**: 169–216, 1997.
© 1997 *Kluwer Academic Publishers. Printed in Belgium.*

1. Introduction

The Near Earth Asteroid Rendezvous mission, or NEAR, was successfully launched on February 17, 1996. NEAR is the first launch under NASA's Discovery Program, an initiative for small, low-cost planetary missions. As the first spacecraft to orbit an asteroid, the NEAR mission promises to answer fundamental questions about the nature and origin of near-Earth asteroids and comets.

Launched from a Delta II rocket, the 805 kg NEAR spacecraft will follow a 2 year ΔVEGA trajectory (Farquhar et al., 1995; Santo et al., 1995). This trajectory provided a flyby opportunity of main-belt asteroid 253 Mathilde in June 1997 before marking the start of NEAR's primary mission to orbit one of the largest of the near-Earth asteroids, 433 Eros, in February 1999. During the planned 10 months at Eros, a suite of remote sensing instruments will make detailed science observations of the gross physical properties, surface composition, and morphology of the asteroid. Four months of the rendezvous are scheduled with orbits as low as 35 km, which correspond to altitudes as low as 15 km above the asteroid surface.

In 1975 Eros approached within 0.15 AU of Earth and extensive telescopic and radar observations were carried out. From these and other observations, it has been inferred that Eros is an S-type (silicate rock) object, the most common type of near earth asteroid. It is not known whether S-type asteroids come from differentiated or undifferentiated parent bodies. One of the primary objectives of the NEAR mission is to obtain global elemental composition maps of the asteroid with sufficient accuracy to enable comparison with major meteorite types and to assess the compositional heterogeneity of the asteroid.

The science payload on NEAR includes five major facility instruments: a Multi-Spectral Imager, a Near-Infrared Spectrograph, an X-ray/Gamma-ray Spectrometer, a Magnetometer, and a Laser Rangefinder. The Multi-Spectral Imager (Hawkins et al., this volume) will image the surface morphology of Eros with spatial resolutions down to 5 m. The Near-Infrared Spectrograph (Warren et al., this volume) will be the primary instrument to measure mineral abundances with a spatial resolution on the order of 300 m. The X-ray/Gamma-ray Spectrometer will measure elemental abundances at the very surface of Eros with spatial resolutions on the order of 2 km and, with much coarser spatial resolution, will measure sub-surface composition to depths on the order of 10 cm. The combined observations of the X-ray/Gamma-ray Spectrometer, the Multi-Spectral Imager, and the Near-Infrared Spectrograph should provide full coverage of Eros enabling the development of global maps of Eros's surface composition.

This paper describes the design and operation of the X-ray/Gamma-ray Spectrometer, gives an overview of the science and measurement objectives, and includes early in-flight data. A more detailed review of the science investigation using the X-ray/Gamma-ray Spectrometer may be found in Trombka et al. (1996).

2. Science Objectives

Modern models of solar system formation suggest that the planets grew largely from the accumulation of planetesimals: small solid bodies that aggregated from materials condensed from the solar nebula. Most planetesimals are now lost, swept up by the major planets. Some, however, both largely intact and highly fragmented, survive in the form of asteroids. They are remnants of the building blocks from which the planets formed. While asteroids are often thought of as 'primitive' bodies, they actually vary quite widely in the degree to which they have evolved from unaltered nebular condensate. Nearly all meteorites, and by inference asteroids, date from the earliest epoch of solar system history.

Main belt asteroids are not readily accessible for study by spacecraft, but some asteroids lie in more accessible Earth-approaching or Earth-crossing orbits. A common supposition is that these near-Earth asteroids (NEA's) provide the link between main belt asteroids and meteorites, and so can provide geological and geophysical context for the meteoritic 'hand samples' that have been intensively studied in the laboratory. The target of the NEAR mission is 433 Eros, one of the largest NEAs. The spectrum of Eros (Pieters, 1976; McFadden et al., 1984) shows evidence for olivine, pyroxene, and a 'reddening agent' such as metallic particles which make it an S asteroid. S-type objects are the second most abundant class of asteroids, and dominate the inner part of the main belt, yet their relationship to meteorites is unclear. One of the main scientific objectives of the NEAR mission is to establish the link between asteroids and meteorites, if indeed it exists.

If Eros is part of a differentiated parent body, it could show some large-scale compositional heterogeneity related to the differentiation process. A fully differentiated parent body with chondritic original composition would have basaltic rock near its surface, an olivine-rich mantle, and a metallic core. Some S asteroids may be portions of the mantles of such objects. The mean composition of the mantle would be indicative of the composition of the original material from which the parent body formed. Details of the differentiation process can affect the compositional structure of the parent body's interior. If, for example, the interior were to melt completely then substantial compositional layering within the mantle could result from the subsequent fractional crystallization, with olivine that is more magnesian at the base and less magnesian at the top.

Rather than coming from differentiated bodies, it has been suggested that S objects are portions of essentially undifferentiated bodies, and are the source of ordinary chondrites (Wetherill, 1985; Wetherill and Chapman, 1988). This possibility is particularly interesting, since ordinary chondrites comprise more than 75% of meteorite falls, yet their source has not been clearly identified. Some S asteroids might be considered reasonable candidates as the sources of ordinary chondrites, but S asteroid spectra generally are redder than ordinary chondrites, have weaker mafic silicate features, and show more olivine (Gaffey et al., 1989, 1993).

The composition of Eros is unclear. Is it related to a known meteorite type or types? Is it chemically primitive or differentiated? What were the important geologic processes in its evolution? Is it is a fragment of a larger body or composed of aggregated fragments of one or more parent bodies? All of these questions will be addressed using the NEAR X-ray/Gamma-ray Spectrometer (XGRS).

3. Measurement Objectives

One of the primary objectives of the NEAR mission is to obtain global elemental composition maps of the asteroid with sufficient accuracy to enable comparison with major meteorite types and to assess the compositional heterogeneity of the asteroid. The XGRS will provide global maps of the asteroid's elemental composition. Some of the specific measurement objectives that were considered during the design of the NEAR XGRS are given below.

3.1. DETERMINATION OF METEORITE ANALOG FOR S-CLASS ASTEROIDS

The NEAR XGRS experiment, by providing elemental compositional data of a prime example of an S-type asteroid, will help to determine whether S-type asteroids could be chondrite source material. Figure 1 gives one example of how classes and possibly sub-classes of meteorite types could be distinguishable using the XGRS data. GRS-derived results that can be expected to have the indicated uncertainties based on measurements for one month of operation are shown. These results along with results from the Multi-Spectral Imager and Near-Infrared Spectrometer should resolve the controversy on the nature of Class S asteroids.

3.2. COMPOSITION MAPPING AND THE DEGREE OF DIFFERENTIATION

The extent of surface heterogeneity on a scale of km to 10s of km, and by implication the degree of differentiation and nature of processes in the early solar system, can be determined by the XRS experiment. Anticipated spectra (Figure 2), and sensitivities (Table I) from Eros during the NEAR mission have already been calculated. Geochemical terrain maps derived from the combined results of several of the NEAR instruments should give insight into the nature of surface heterogeneity and the degree of differentiation.

3.3. DETERMINATION OF THE RELATIONSHIP BETWEEN ELEMENTAL ABUNDANCES AND SPECTRAL REFLECTIVITY MEASUREMENTS

NEAR will provide the first opportunity for a direct comparison between compositional measurements of an asteroid from X-ray and gamma-ray instruments with Earth-based spectral observations. Along with Near-Infrared Spectrometer data, a

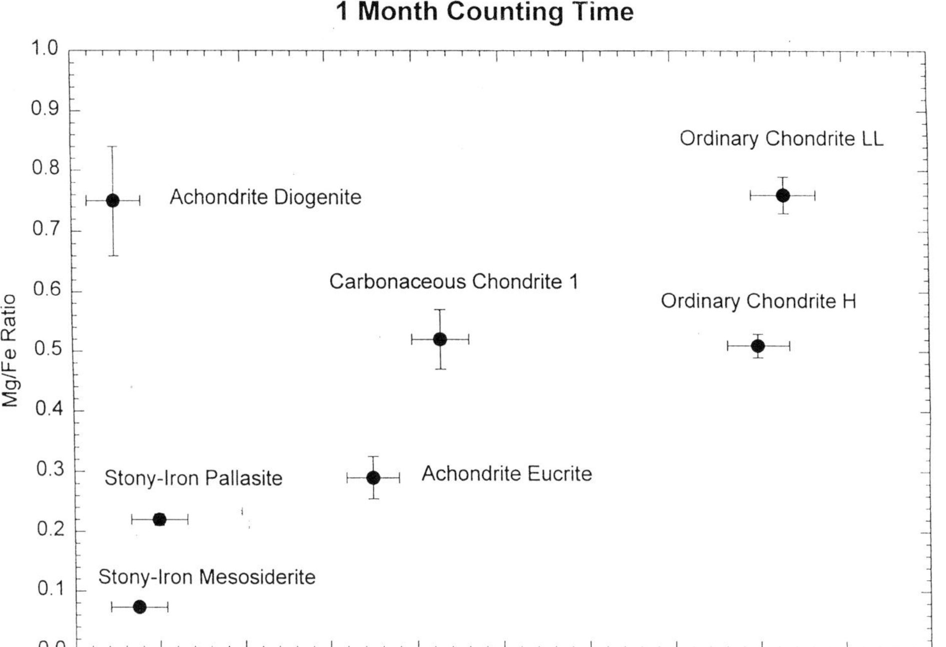

Figure 1. GRS results used to distinguish meteorite classes based on one month accumulation time. The average value of Mg/Fe versus K concentration for various meteorite classes are shown along with the expected uncertainties in the measurements.

direct geochemical basis will be established for the asteroid classification schemes developed by astronomers.

3.4. THE NATURE OF THE REGOLITH ON AN ASTEROID

The depth and extent of dust cover on asteroids has been controversial. Differences between Fe data, derived from XRF and GRS experiments should be due to differences in near-surface stratigraphy, and could put constraints on the size of a surface dust layer. Fluorescent X-rays are generated at the very surface (in the top < 1 mm), whereas unscattered gamma-rays associated with Fe are generated in the upper 10 to 20 cm. When spatial resolutions for the two results are made comparable, any differences in Fe concentration should be the result of differences in composition of a top layer, presumably a dust layer, and provide limits on the depth of such a layer.

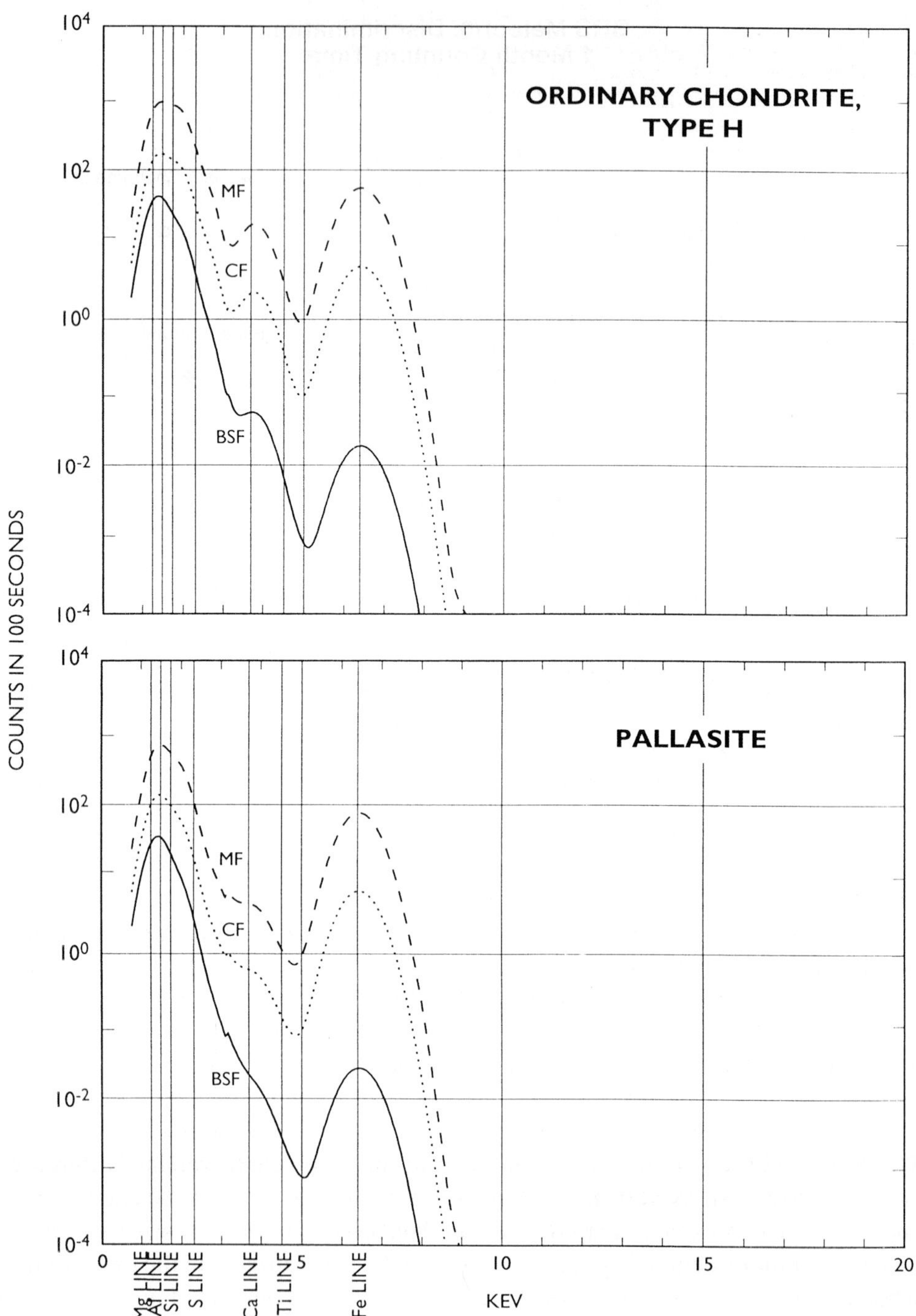

Figure 2. Expected X-ray pulse height spectrum measured during NEAR rendezvous with the three nadir pointing X-ray proportional counters. Three levels of solar activity are shown. Represented, from highest to lowest average solar output and defined in terms of GOES solar satellite X-ray instrument (1–8 Å) measured intensity, are (1) M1 Flare (10^{-5} W m^{-2}; 10–50 per year), (2) C1 Flare (10^{-6} W m^{-2}; 50–500 per year), and (3) B1 Subflare (10^{-7} W m^{-2}; baseline near solar maximum). An integration time of 100 s is desired. Two compositions are assumed: (a) ordinary chondrite, Type H; (b) pallasite composition.

Table I

Integration times required to achieve either 10% or 30% uncertainty for various chemical elements with the X-ray spectrometer during the NEAR rendezvous mission. Calculations were carried out for ordinary chondrite type H and stony iron pallasites

Element	Sensitivity	Ordinary chondrite type H		Stony iron pallasite	
		Typical	(Low flare)	Typical	(Low flare)
Fe[*]	10%	2.3 hr	(2.0 min)	1.8 hr	(1.5 min)
	30%	0.3 hr	(0.2 min)	0.2 hr	(0.1 min)
Ti[*]	10%	6.4 hr	(5.0 hr)	69.0 d	(2.2 d)
	30%	0.7 hr	(0.6 hr)	7.7 d	(4.8 hr)
Ca[*]	10%	1.8 hr	(6.0 min)	2.6 d	(1.0 hr)
	30%	0.2 hr	(0.7 min)	6.9 hr	(6.7 min)
S[*]	10%	0.2 hr	(2.7 min)		–
	30%	1.3 min	(0.3 min)		–
Si[*]	10%	1.1 hr	(0.2 min)	7.1 min	(0.7 min)
	30%	7.3 s	(1.3 s)	0.8 min	(4.7 s)
Al[*]	10%	5.3 hr	(0.8 hr)	9.0 d	(23.0 hr)
	30%	0.6 hr	(5.3 min)	1.0 d	(2.5 hr)
Mg[*]	10%	3.3 min	(0.7 min)	7.4 min	(1.4 min)
	30%	0.4 min	(4.7 s)	0.8 min	(9.3 s)

[*]Smaller integration time for areas with low-level flare activity – about 50% of area of coverage. Signal for three detectors summed.
Note: Assumption mid-cyle averages for solar activity, 5 deg FOV, 25 cm^2 detector window, and sub-flare activity during encounter, maximum solar incidence angle.

3.5. HYDROGEN DETECTION

The detection of hydrogen depends on measurement of the prompt gamma-ray (2.223 MeV) line. But this is not the only method available. If a measurement of the fast to thermal neutron flux is made, the variation in hydrogen content can also be determined and sub-surface (down to 50–100 cm) hydrogen can be detected. The presence of small amounts of hydrogen even down to depths where the 2.223 MeV line cannot penetrate will cause significant changes in the fast to thermal ratio. This ratio can be determined by looking at the prompt gamma-ray lines in Si and Fe. In both cases certain lines are produced by thermal neutron interactions (thermal capture) and others by fast neutron interactions (inelastic scatter). Thus the ratio of the inelastic line intensity to the capture line intensity for the same element is related to the fast-to-thermal neutron ratio (Evans and Squyres, 1987).

4. Operations Plan

In general, the goal of sequence design for compositional mapping is to provide coverage across the asteroid that maximizes both uniformity of coverage and spatial resolution. For the gamma-ray experiment, this is straightforward. The GRS has a wide field of view, so most reasonable pointing scenarios will fill a good part of this field of view most of the time. Also, since the gamma rays arise primarily from natural radioactivity and from galactic cosmic-ray interactions, solar illumination is not a factor. Given the lowest achievable orbit for compositional mapping (35 km circular orbit) the shortest useful integration period for the GRS will be about 20 min.

Design of X-ray observation sequences is much more complex. X-ray signal strength depends strongly on solar activity, which is inherently unpredictable. The X-ray signal is also roughly proportional to the cosine of the solar incidence angle. Due to spacecraft constraints the solar phase angle will be close to $90°$ through most of the mission. Simple nadir pointing under these conditions would always view close to the terminator, where the solar incidence angle is unacceptably low. For this reason, the preferred observational strategy is to always view at least several tens of degrees sunward of the nadir direction on better-illuminated portions of the asteroid. The irregular shape of Eros further complicates the issue. Spacecraft orbits for the compositional mapping phase of the mission are presently being modeled by the science team in order to design pointing scenarios that provide the best possible X-ray resolution and coverage distribution. The shortest useful integration period for the XRS will be about 2 min.

5. Instrument Overview

The philosophy of NEAR was to conceive a comprehensive science mission that could answer fundamental questions about the nature and origin of near-Earth asteroids, while meeting the schedule and cost constraints of the Discovery Initiative. To that end, the spacecraft has the necessary sophistication to achieve the mission goals, such as a three-axis stabilized platform, yet also has the attributes of a simple, robust design with its fixed high-gain antenna and solar panels. This philosophy translated to the science instruments as well. All of the instruments are fixed to the spacecraft and rely on a passive thermal design. For the Gamma-ray Spectrometer, this excluded the possibility of using an elaborate cryo-cooled detector mounted on a boom in the manner of Mars Observer (Boynton et al., 1992). For the X-ray Spectrometer, high cost and development time excluded the possibility of using detectors made of solid beryllium like those used on Apollo (Adler et al., 1972a, b). Despite these constraints, the X-ray/Gamma-ray Spectrometer (XGRS) contains a suitable complement of detectors which possess the reliability to survive a long-duration mission and provide sufficient sensitivity and resolution to measure and map the surface composition of Eros.

The XGRS consists of a large group of detectors, electronics assemblies, and a cabling harness, which are distributed over three deck surfaces of the NEAR spacecraft. The asteroid-pointing X-ray and Gamma-ray detector assemblies are

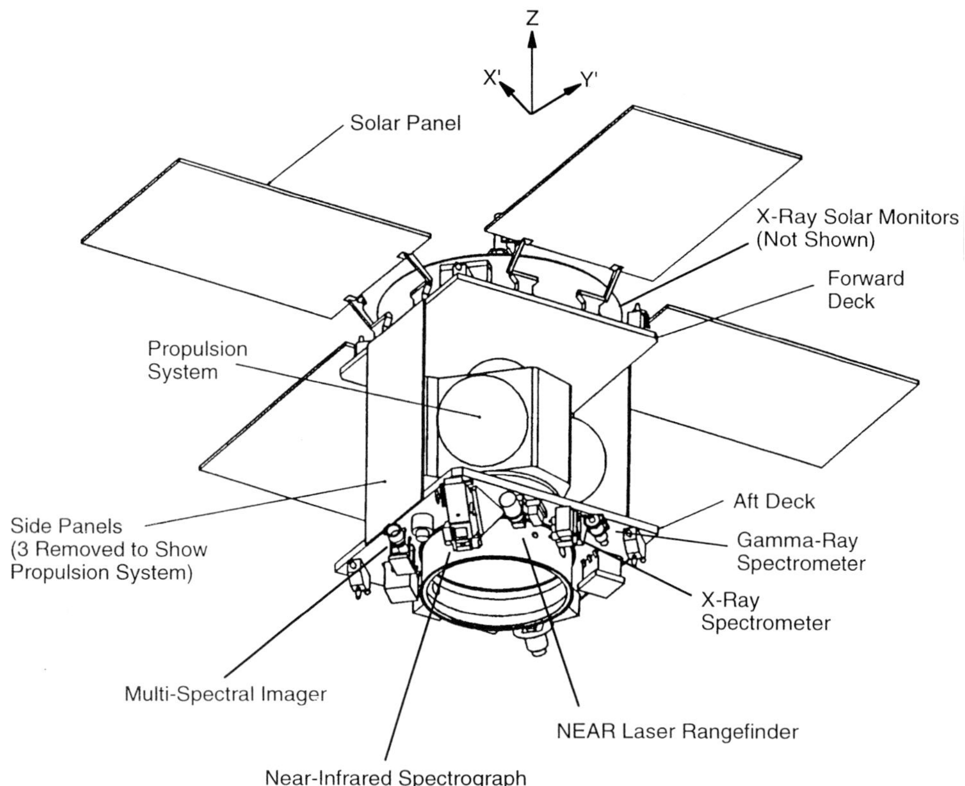

Figure 3. View of the Near Earth Asteroid Rendezvous spacecraft. The aft deck contains the instrument palette, and is almost completely in shadow. The spacecraft Z-axis is directed through the high gain antenna, nominally pointed toward Earth. The instruments are oriented so that their boresights are nearly coaligned with the spacecraft's X'-axis.

mounted next to each other on the aft deck, with their fields-of-view parallel to the X'-axis of the spacecraft (Figure 3). A shared preamp module mounted on the back of the X-ray detector assembly contains preamps for both the X-ray and Gamma-ray detectors. Nearby is the Detector Electronics which contains all of the analog signal processing functions. Behind the Detector Electronics is a dedicated Data Processing Unit (DPU) which executes all instrument software, provides the digital interface to the spacecraft, and supplies instrument power and control.

On the forward deck, two additional X-ray detectors are situated to monitor the solar X-ray flux. Their fields-of-view are parallel to the Z-axis of the spacecraft (normal to the solar panels) and are located about 10 cm above one corner of the deck so they are not obscured by the high-gain antenna. The spacecraft is generally oriented with its high-gain antenna directed toward Earth. With this orientation, the Sun is always within 30 deg of the Z-axis.

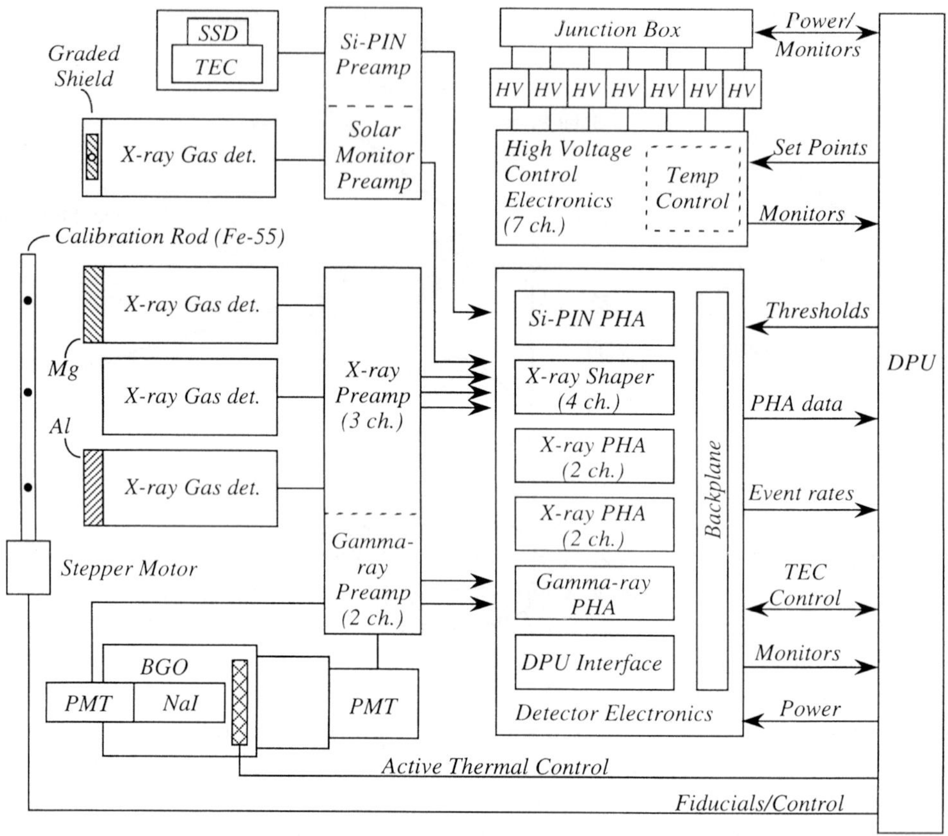

Figure 4. Functional block diagram of the XGRS. Depicted are the three nadir pointing X-ray proportional counters and gamma-ray detector with shared preamp module and the Si-PIN and gas proportional counter solar monitors with their associated preamps all feeding the central detector electronics box which contains the pulse-height analyzers. Spectra are formed in the data processing unit (DPU) which provides the spacecraft interface.

For reliability, each of the seven XGRS detectors has its own high voltage power supply. The supplies are divided into three stacks located on the inside surface of the aft deck. A separate electronics module to control the high voltage supplies is mounted on top of the Detector Electronics package.

Figure 4 shows a block diagram of the instrument, identifying all major system components. The instrument has a total mass, including the DPU, of 27 kg and dissipates 30–34 W depending on thermal conditions. The power system provides limited power switching of the high voltage subsystem, thermoelectric cooler, and operational heaters.

The NEAR spacecraft solid state recorders and downlink are limited resources and their use and allocation will regularly be negotiated throughout the mission amongst the science teams. The telemetry data rates for the X-ray and gamma-ray

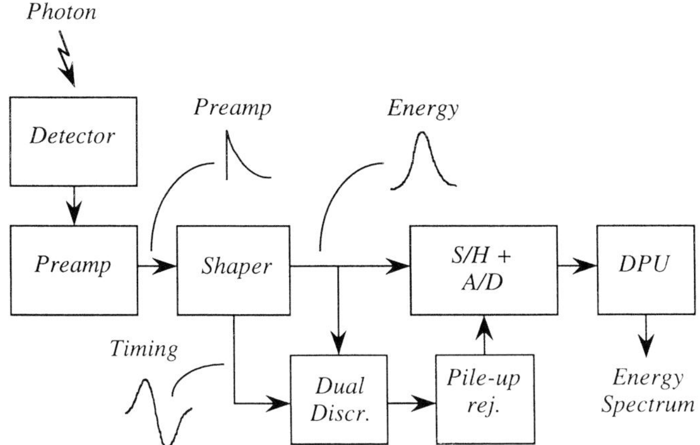

Figure 5. Generalized detection scheme for XGRS spectrometers.

experiments are independently adjustable and may be raised or lowered by commanding shorter or longer integration periods from 60 to 60,000 s. In this way, the XGRS can easily respond to changes in telemetry allocation. For example, operations planners may request that XGRS reduce its telemetry rate to accommodate a particularly long imaging sequence.

At Eros, the expected telemetry rates for the X-ray and gamma-ray experiments will nominally be 180 bits s^{-1} and 70 bits s^{-1}, respectively. During solar flare conditions, when X-ray fluorescence is more intense, the integration period for X-rays could be shortened to improve spatial resolution. Conversely, both X-ray and gamma-ray integration periods might be lengthened when merely collecting background or calibration spectra. An additional summary science mode provides a reduced set of highly compressed spectra from both experiments and combines them into a single data stream with a telemetry rate of only a few bits per second. This mode is particularly useful for checking the health and safety of the instrument or for verifying its operating configuration.

6. Instrument Design

The XGRS spectrometers are energy-dispersive spectroscopic systems rather than wavelength-dispersive systems. Figure 5 shows the generalized detection scheme. For this type of spectrometer, an incoming photon is absorbed by the detector material (solid or gaseous) and produces an output signal proportional to the energy absorbed. The detector signal is amplified, filtered, and its peak value is measured using an analog-to-digital converter. A data processing unit collects the measurements and bins them according to 'pulse height' or energy absorbed. After an appropriate accumulation time, in which many photons are processed, a complete pulse-height distribution or energy spectrum develops and is telemetered to

Table II

X-ray/Gamma-ray Spectrometer specifications

Instrument mass (including DPU)	27 kg
Instrument power	31 W
Gamma-Ray Spectrometer	
Prime detector	NaI(Tl) scintillator 2.54 × 7.62 cm
Shield detector	BGO scintillator cup 8.9 × 14 cm
Energy range	0.1 to 10 MeV, 1024-channel spectra
NaI(Tl) energy resolution	8.7% FWHM @ 662 keV
BGO energy resolution	14% FWHM @ 662 keV
FOV	≈50° FWHM @ 145 keV
Anti-coincidence background rejection	>500:1 above 5 MeV
Counting rate	10 kHz max (1 kHz nominal)
Mass	6 kg
X-Ray Spectrometer	
X-ray detectors (3)	Single-wire gas-filled proportional counters
Window	25 μm Beryllium with 2 mm grid support
Active aperture area	25 cm^2 per detector
Energy range	0.5 to 10 keV, 256-channel spectra
Energy resolution	850 eV FWHM @ 5.9 keV
FOV	5° FWHM @ 5.9 keV
Rise time rejection of background	>50% above 2 keV; >70% @ 6 keV
Counting rate	10 kHz max (1 kHz nominal)
In-flight calibration sources	Fe-55; rotated into FOV by command
X-Ray Solar Monitor	
Energy range	1 to 10 keV, 256-channel spectra
FOV	60°
Counting rate	10 kHz (level M1 flare)
Redundant detectors:	
Gas-filled proportional counter	Graded shield design; 1 mm^2 equivalent area
Solid state detector	Si-PIN photodiode 1.1 × 1.1 mm mounted on miniature thermoelectric cooler
Energy resolution	Gas: 850 eV FWHM @ 5.9 keV PIN: 600 eV FWHM @ 5.9 keV

Earth. From an analysis of the telemetered spectrum, elemental composition can be inferred.

This generalized detection scheme forms the basic approach to the design of the XGRS. Optimization maximizes the reduction of unwanted background signals, sensitivity enhancement, long-term stability, and includes in-flight calibration. Summary characteristics and specifications for the XGRS are given in Table II.

7. Gamma-Ray Spectrometer

Restrictions on mass, cost, and mission duration precluded the use of the type of gamma-ray detector used on Mars Observer, a cryogenically cooled Ge detector mounted on a boom to remove the detector from the local background radiation of the spacecraft. Instead, the NEAR GRS employs a NaI(Tl) scintillator situated within a thick cup shield fabricated from a single crystal of BGO (Bismuth Germanate). The dense BGO cup is itself an active scintillator and thus works to reduce the Compton and pair-production contributions to the unwanted background signal as well as providing direct, passive shielding from the local gamma environment. This design approach eliminated the need for a boom and allowed the detector to be body mounted to the spacecraft. Although, the energy resolution of a NaI(Tl) scintillator is not nearly as good as a cooled Ge detector (Evans et al., 1996), the NEAR GRS detector operates at room temperature, is not subject to any serious radiation damage (important for long-duration missions such as NEAR), and given the integration times planned for the mission, meets the measurement requirements for the elements cited previously.

The shield acts as an active collimator. Particles or gamma-rays entering through the shield will produce output signals in the shield detector which are coincident with signals produced in the central detector. An anti-coincidence system detects and rejects these events. The cup design confines the field-of-view to about 45 deg at lower energies. At higher energies, the field-of-view is somewhat larger. A more detailed discussion of the GRS field-of-view and BGO rejection efficiency is given in the Instrument Calibration section and in Trombka et al. (1996).

With this detector configuration, a substantial part of the composition signal is lost at higher energies due to pair production. It is likely that one or both of the gamma-rays produced by annihilation will escape the small central crystal and be absorbed in the shield. This energy loss appears in the raw NaI spectrum as secondary peaks shifted down in energy from the photopeak by either 0.511 MeV or 1.022 MeV, named the first and second escape peaks respectively. In the anti-coincidence spectrum, these escape peaks will be suppressed due to event rejection by the anti-coincidence system. For many gamma-rays of interest, there are more events contained in the first and second escape peaks than the photopeak itself. Rejecting these events would result in an unacceptable loss in sensitivity. To recover this information, the GRS keeps watch for any coincident events with shield energies falling near the 0.511 MeV or 1.022 MeV lines and produces two additional spectra containing only these types of events. These coincidence modes are somewhat vulnerable to false detections arising from galactic cosmic-rays and Compton scattering, but extensive testing and calibration have demonstrated the overall improvement in detection using this technique (Trombka et al., 1996).

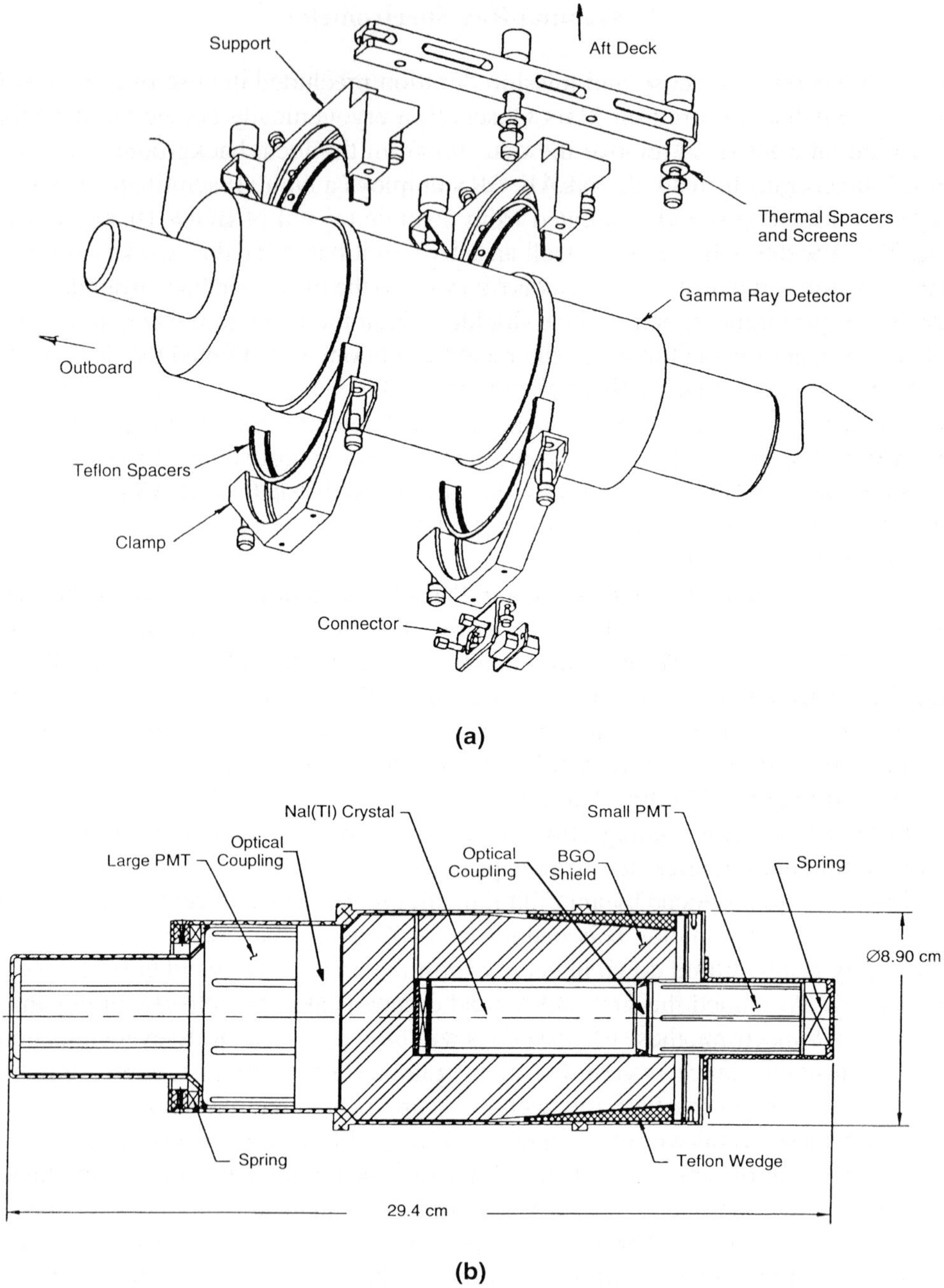

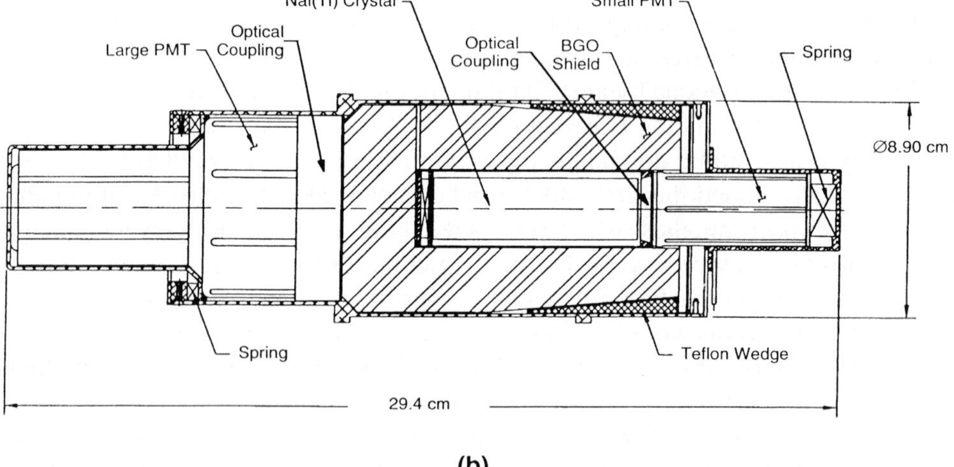

Figure 6. NEAR gamma-ray detector: (a) sensor assembly; (b) detector cross section.

7.1. FLIGHT GAMMA-RAY DETECTOR

The gamma-ray detector was designed and built by EMR Photoelectric, a division of Schlumberger. Figure 6(a) shows an outline drawing of the detector assembly and Figure 6(b) provides a cross sectional view. The central detector assembly is based on a ruggedized NaI(Tl) unit used in their oil logging operations. The BGO cup was designed especially for NEAR and is fabricated from a single crystal. The NaI(Tl) scintillator is a 2.54 cm × 7.62 cm right circular cylinder and couples to a 3.17 cm diameter metal ceramic photomultiplier tube. The BGO scintillator cup has outside dimensions of 8.9 cm × 14 cm. The base of the cup has a thickness of 3.1 cm and the sides of the cup have a maximum thickness of 2.6 cm with a slight taper towards the front to direct light back towards the base. The base couples to a 7.6 cm diameter metal ceramic photomultiplier. The mass of the gamma-ray detector assembly is 6 kg.

Scintillation detectors produce a light intensity proportional to the energy absorbed with a wavelength that is characteristic of the detector material. The NaI(Tl) and BGO crystals each couple to a photomultiplier tube (PMT) with optimized photocathodes. Crystal and PMT combinations were carefully chosen by EMR to achieve the best energy resolution. The measured energy resolutions for the NaI(Tl) and BGO detectors mounted in the flight configuration are 8.7% and 14% fwhm, respectively @ 662 keV. The energy range for both detectors is 0.1 to 10 MeV. The presence of the small PMT between the NaI crystal and the gamma-ray source introduces an overall loss in efficiency for the NaI detector which is a function of both energy and angle (Trombka et al., 1996). The flight spare exhibits nearly identical performance and will be used extensively to further characterize GRS performance and develop a detector model.

The light output of the NaI(Tl) and BGO scintillators strongly depends on temperature. The temperature coefficient for the NaI(Tl) scintillator is roughly 0.2%/°C and the BGO can approach 2%/°C. Temperature fluctuations during asteroid operations would adversely affect detector calibration. For this reason, the detector assembly is thermally isolated with active control to stabilize the gamma-ray detector temperature to within 0.25 °C. The temperature setpoint is commandable to allow its optimal value to be chosen during operations at Eros.

The electron gains of the PMTs are not particularly sensitive to temperature, but are sensitive to voltage variations. The PMT high voltage power supplies must be stable to a fraction of a volt over temperature so as not to adversely affect GRS calibration. To meet this stability requirement, the XGRS uses an external feedback control system to produce ultra-stable high voltage outputs from ordinary high voltage power supplies.

     J. O. GOLDSTEN ET AL.

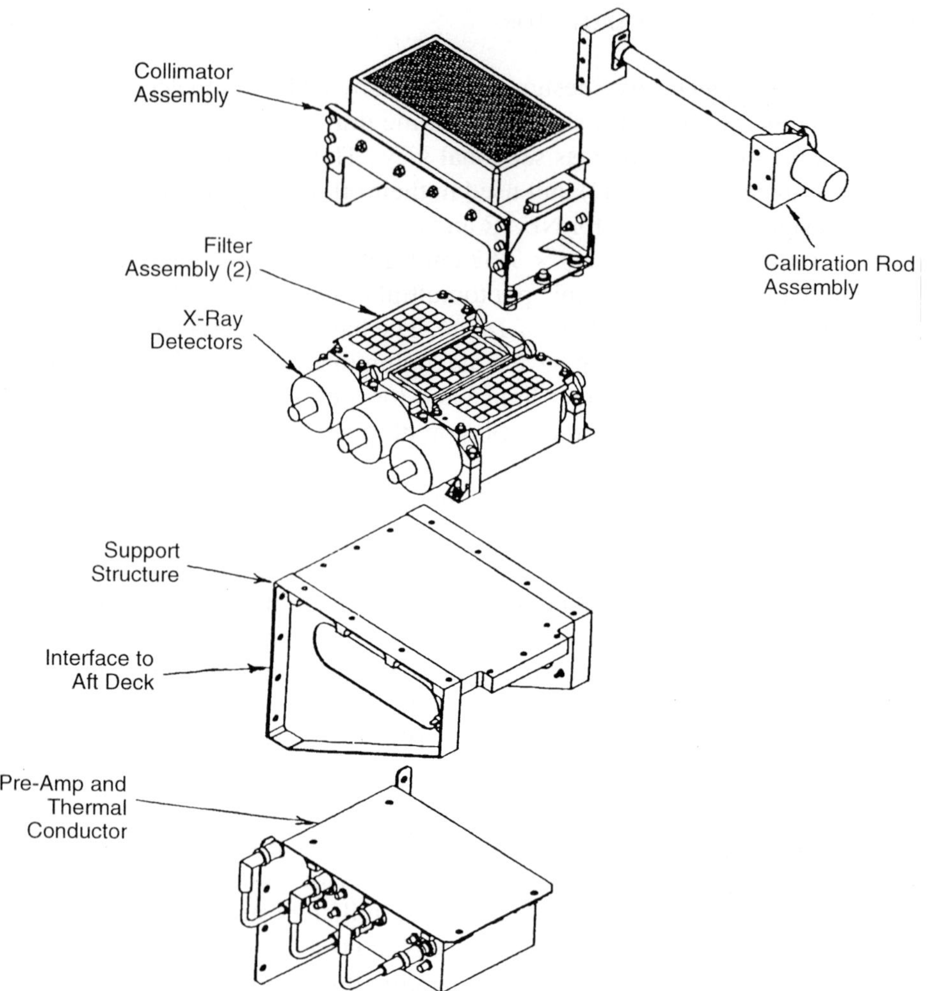

Figure 7. NEAR X-ray sensor assembly.

8. X-ray Spectrometer

The X-ray Spectrometer (XRS) consists of three identical gas-filled proportional counters designed and built by Metorex International Oy. Gas tubes of this type provide the large active area and therefore sensitivity required for remote sensing. Similar detectors have been flown on lunar orbital missions (Mandel'shtam et al., 1968) and most recently on Apollo missions 15 and 16 (Adler et al., 1972a, b). The XRS uses an updated design and achieves an energy resolution of 850 eV fwhm @ 5.9 keV. The energy range of the detectors is 0.5 to 10 keV. Figure 7 shows an exploded view of the major mechanical assemblies of the XRS sensor.

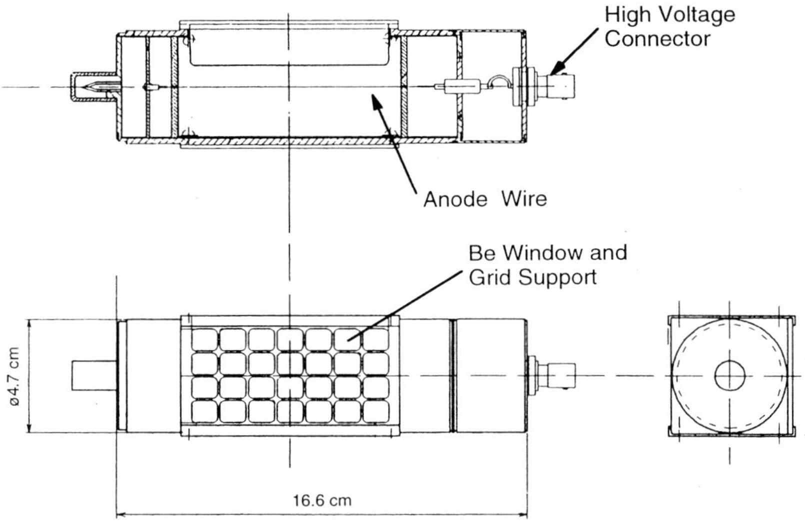

Figure 8. NEAR X-ray detector.

8.1. FLIGHT DETECTOR

Figure 8 shows outline and cross-sectional views of the flight detector. The detectors have a rectangular aperture with an active area of 25 cm^2. The window material is 25 μm Be foil supported by an external grid made of 2 mm solid Be. The grid panes are slightly rectangular. The detector housing is stainless steel with a 0.5 mm thick Be liner to absorb fluorescence X-rays arising from the housing. The chamber diameter is 42 mm and is filled with P-10 gas, a mixture of 90% argon and 10% methane, to an absolute pressure of 1200 mbar. The anode wire is gold-plated tungsten and has a diameter of 13 μm.

The gas proportional counters work in the following manner. An incoming X-ray photon penetrates the thin window of a sealed gas-filled chamber and interacts with the gas producing an energetic photoelectron (the photoelectric effect). The kinetic energy of the electron is then progressively absorbed in the gas leaving an ionization trail. The free electrons from the ionization are attracted to a thin wire anode at high potential stretched down the center of the chamber. In the region of high electric field strength, very close to the wire, the electrons reach sufficient energy to liberate even more electrons in a stable multiplication effect providing signal gain. This signal gain helps to overcome the preamplifier noise such that the energy resolution of the gas tubes is primarily limited by the Poisson statistics of the ionization process.

Unlike the gamma-ray detectors, the X-ray gas tubes are not particularly sensitive to temperature and may be operated over a wide temperature range, since the multiplication effect has more to do with the number of gas molecules rather than the gas pressure. Under cold conditions, spacecraft survival heaters keep the tubes above $-30\,°C$ to prevent gas condensation within the tube.

The gain in the gas tubes is even more sensitive to voltage variations than the photomultiplier tubes in the GRS and is roughly 1%/V. To maintain calibration, the XRS shares with the GRS, the feedback control system which maintains ultra-stable high voltage outputs.

The relatively large size of the proportional counters (as compared to a solid-state detector) increases the number of interactions with the cosmic-ray background. These interactions usually occur at the walls of the detectors and leave long ionization trails that exhibit characteristically longer charge-collection times. The XRS employs rise-time discrimination circuitry to reject up to 70% of these background events.

8.2. X-RAY FILTERS

The energy resolution of the gas tubes is not sufficient to resolve the closely spaced Mg, Al, and Si lines. To separate these lines, the XRS uses balanced filters (Trombka et al., 1996). The two outer detectors have thin absorption filters mounted externally. A Mg filter on one detector attenuates the Al line, and an Al filter on the other detector attenuates the Mg and Si lines. The very steep absorption edges of the filters makes the separation of the lower energy lines possible. At higher energies, the filters are essentially transparent and the Ca and Fe lines are resolved directly by the detectors. The center detector has no filter.

The filters are 8.5 μm thick. They are bonded to a copper wire mesh and then sandwiched between two flat pieces of Be-Cu which have a grid structure that precisely co-aligns with the support grid for the detector window. This mesh and grid approach ruggedized the filters while preserving active area. No problems were encountered in acoustic tests.

8.3. X-RAY COLLIMATOR

The XRS requires a collimator with a 5-deg acceptance angle to provide spatial resolution and eliminate unwanted X-ray sky background. A pure Be collimator was considered too massive and costly. Instead, the XRS uses a compact, honeycomb structure fabricated from Be-Cu foil. Copper is considered an acceptable material for the collimator because any X-ray lines excited in the collimator itself will not interfere with line emissions from the asteroid surface. The inclusion of Be (2%) added the necessary stiffness to ruggedize the collimator.

To make the collimator, over 50 individual Be-Cu foils 0.05 mm thick were first corrugated and then hand assembled using an epoxy containing only light elements. The honeycomb section was then mounted in a sheet metal frame. The hexagonal cells measure 3.2 mm across the flats and the collimator is 76 mm thick. The collimator achieves better than 99% transmission, and its small-cell design eliminated the need for precise alignment with the detectors. Angular and energy response measurements confirm predicted performance (Trombka et al., 1996).

8.4. X-RAY CALIBRATION ROD

A Vespel rod containing three Fe-55 radioactive sources spans the X-ray detector assembly near the base of the collimator. The bottom edge of the collimator is beveled to allow the rod to rotate the sources into view of the detectors for periodic in-flight calibration without sacrificing active area. An outer copper tube shields the sources when not in use. A stepper motor mechanism at one end directly drives the rod and a pair of optical fiducial holes report its position.

The radioactive source holders are embedded axially in the rod, one at each detector location, and at 45 deg from each other so that only one detector can be in calibration at a time. Pinholes in the outer copper tube restrict the count rate and also act as crude collimators to minimize source crosstalk between detectors. This ensures maximum science return in the event of a failed mechanism.

8.5. SOLAR MONITORS

Two sunward-pointing X-ray detectors positioned on the forward deck of the space-craft monitor the incident solar flux. The spacecraft uses a fixed position high-gain antenna generally pointed toward Earth. As a result, the sun-angle changes signific-antly during the mission. The solar monitors must possess a wide 60° field-of-view to observe the Sun at all times. The two solar monitors are mounted close together about 10 cm above one corner of the deck to ensure an unobstructed view.

The asteroid-pointing detectors require large active areas to collect the weak emissions from the surface. The solar monitors experience the opposite conditions of very strong X-ray emissions directly from the Sun, especially during solar flares. For a maximum count rate of 10,000 counts s^{-1} under M-level solar conditions, the active area for a solar monitor needs to be only 1 mm^2 as compared to the 2500 mm^2 active area of a single gas tube.

The requirement for a small active area allowed the use of a new miniature X-ray detector developed and built by AMPTEK, Inc. (Figure 9). The compact sensor consists of a Si-PIN photodiode mounted on a miniature thermo-electric cooler in a hermetic package 15 mm in diameter. Also mounted on the cooler are the input FET and RC feedback components to the charge-sensitive preamplifier, and an integrated circuit temperature monitor. A 75 μm thick Be window rejects the intense solar flux below 1 keV. The Si-PIN solar monitor achieves an energy resolution of 600 eV fwhm @ 5.9 keV. AMPTEK has recently improved the performance of their detector to better than 250 eV using transistor feedback, but the new design was not available in time for the NEAR mission.

The NEAR mission is the first to fly this new X-ray detector technology and therefore reliability was a concern. Early tests indicated that exposure to ionizing

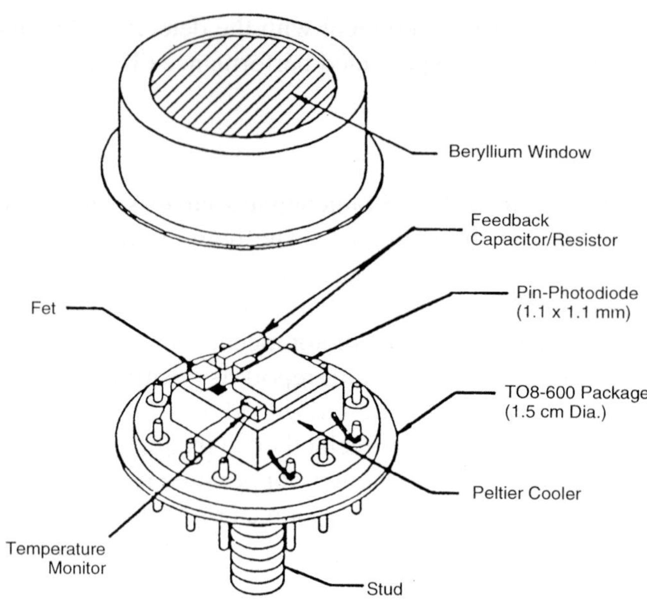

Figure 9. Miniature X-ray detector used to monitor solar X-ray flux.

radiation could produce significant damage in the detector. However, subsequent testing demonstrated that the detector will anneal almost completely when heated to 100 °C for 24 h. To address the possibility of radiation damage during the mission, the XGRS contains circuitry to monitor the leakage current of the Si-PIN detector and, if necessary, to reverse the operation of the thermo-electric cooler to provide in-flight annealing.

Reliance on this new Si-PIN X-ray detector for such a critical function as the solar monitor was judged to incur too much risk. Therefore, a gas proportional counter, identical to the asteroid viewing ones, is also mounted on the sunward deck. In this case, the large active area of the gas tube needed to be severely restricted. Rather than use a simple pinhole, the gas tube is equipped with a specially designed graded shield to enhance the detector sensitivity at higher energies. At low energies the shield has an aperture size of roughly 1 mm^2. With increasing energy, the graded shield becomes increasingly transparent, opening up the effective area of the detector and increasing its sensitivity. Such a modified energy response is better matched to the solar spectrum that follows a power-law and may drop four orders of magnitude in intensity from 1 to 10 keV region.

Construction of the graded shield is shown in Figure 10. The innermost shield is a 1.58 mm thick layer of Delrin with a 1.36 mm diameter pinhole as shown. The next layer is Be foil, 0.1 mm, with a smaller pinhole (0.95 mm dia.) directly above the first. The next layer is an Al plate 0.38 mm thick with a rectangular aperture 7.5 mm × 30 mm. The outermost layer is a thin (0.025 mm) sheet of Kapton which

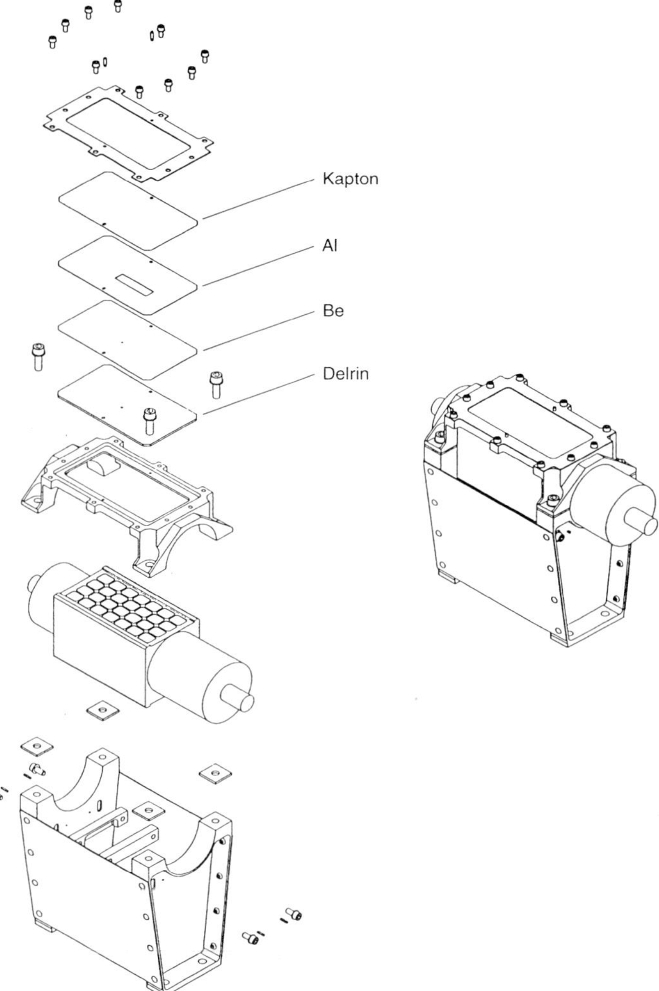

Figure 10. Assembly drawing of NEAR proportional counter solar monitor showing graded shield design. This detector configuration is designed to operate over four orders of magnitude in solar intensity, from quiescent to M-type flares.

helps reduce solar heating as well as absorbing solar flux below 1 keV. A discussion of the graded shield design may be found in (Trombka et al., 1996).

Telemetry data may be selected from either solar monitor. An additional 'toggle mode' alternates the data return from the two solar monitors. In this mode, the instrument provides nearly continuous observations from both solar monitors and allows close cross-calibration between the two detectors without increasing the overall instrument data rate. The solar monitors will rely on prominent line emissions from the Sun for in-flight calibration.

9. Electronics Design

To minimize the development time of the flight electronics, the XGRS makes use, where possible, of functional elements common to both the XRS and GRS portions of the instrument. Where needed, the design is augmented or adapted to implement those elements which are unique to each experiment.

9.1. DETECTOR ELECTRONICS DESIGN

Refer to Figure 5 for the block diagram of a generalized XGRS detector processing channel. The detector produces an output signal proportional to the energy absorbed in the detector. The output signal is generally in the form of a charge pulse which may last from a few nanoseconds to several hundred nanoseconds. A charge-sensitive preamplifier collects the charge from the detector and produces a voltage step proportional to the total amount of charge. A linear filter network shapes the preamplifier signal to reduce noise and provide a signal more suitable for pulse height analysis.

The shaping network produces both an energy signal and an accompanying timing signal. The shaped energy signal is best described as a pseudo-gaussian pulse with a peaking time of 5 μs. The timing signal has a bipolar-shaped characteristic whose zero-crossing precisely coincides with the peak of the energy signal. The logic portion of the pulse height analyzer (PHA) circuitry uses this zero-crossing to command an analog-to-digital converter (ADC) to hold and convert the peak amplitude of the energy signal.

In general, when detectors of this type are exposed to a monoenergetic source (discrete line), the measured spectrum exhibits a characteristic 'line spread' related to the system resolution. The spectrum may also contain other various features due to device physics such as Compton scattering or incomplete charge-collection, but ultimately, all of these responses must be 'convolved' with the system resolution. The system resolution is generally due to a combination of the inherent detector resolution and the electronics noise set by the preamplifier and shaping network. The number of spectral bins is chosen to be enough to sufficiently describe the characteristic line spreads, but not so many as to waste telemetry with excessive 'oversampling.'

The ADC is of the successive approximation type, 11-bits plus sign, and has a fixed conversion time of about 8 μs. The ADC produces more channels than is required for either the X-ray or gamma-ray experiments and so only the requisite bits are used in the binning process. However, the extra precision in the ADC improves the differential linearity of the PHA. Although the linearity is not quite as a good as that which can be achieved using a Wilkinson type ADC, the XGRS ADC is a monolithic device which greatly simplified the PHA design. With its on-chip reference circuit and built-in sample-and-hold function, the ADC exhibits excellent stability and comparable linearity. The device also draws relatively little power and has a moderately fast conversion time.

A logic section controls the operation of the ADC. Two conditions must be met before a new conversion can occur. The energy signal must rise above the threshold of a lower-level discriminator (LLD), and the pulse pile-up rejection logic must indicate the absence of pulses above threshold for at least 25 μs prior to the event. The function of the LLD is to discriminate against electronics noise and prevent 'swamping' of the PHA channel by non-photon events. The function of the pile-up rejection (PUR) logic is to assure that the measurement of a given energy pulse is not corrupted by the tail of a previous pulse. A 25 μs minimum 'dead time' guarantees that a gaussian-shaped pulse with a fwhm of 5 μs will return to less than 0.1% of its peak amplitude. In the event of an overload signal, the PUR logic first waits for the channel to recover from its saturated condition before applying the fixed dead time. This adaptive approach minimizes data corruption, due to pile-up, of X-ray and gamma-ray events in the desired energy range by events with energies beyond the desired range (e.g., cosmic-ray background).

In general, incoming photons are random, independent events. The time between events follows an exponential distribution with a characteristic mean related to the intensity. The pile-up rejection logic restarts a dead time counter every time the energy signal crosses a lower-level discriminator. This produces a waiting time that is extended until the time between two events exceeds the minimum dead time. Consequently, the percentage of missed events increases as the input rate rises (the mean time between pulses shortens). For a given shaping time, this leads to a selection of dead time that is a compromise between maximizing data throughput and minimizing data corruption.

A proper quantitative analysis requires that the measured spectra be corrected for losses due to pile-up rejection. For a single detector system, the exact relationship of the output rate to the input rate can be predicted by analysis (Goulding and Landis, 1982) or easily characterized by measurement. However, the XGRS system is complicated by the use of a single DPU to service multiple detectors. The throughput for a particular detector will be dependent on the activity of all the others. A bank of hardware counters, located in the DPU, count LLD triggers providing a best estimate of the input rate to each detector. By measuring the input and output rates directly, each detector may be accurately normalized for dead-time losses.

The overall performance of the pile-up rejection system is limited by the pulse pair resolution of the shaping network. Multiple pulses occurring within the shaping time can generally not be discriminated. The XGRS shaping networks employ a 5-pole design (Mosher, 1976) which produces a pulse with a fwhm about equal to its peaking time (5 μs). To first order, the fwhm is a good approximation for the mean pulse pair resolution. The probability for one or more pulses occurring within this 'window of vulnerability' relates to the input rate. If the mean time between pulses is large relative to the pulse pair resolution, then the probability may be approximated by the simple product of the count rate and the pulse pair resolution. The maximum counting rate requirement for a given XGRS detector is 10 kHz which bounds the data corruption to 5%. The nominal count rates from the asteroid will generally be below 1 kHz which corresponds to less than 0.5% data corruption. The general appearance of data corruption in the spectra due to pile-up is a low-level continuum 'filling in' the regions between prominent lines.

A special operating feature of the XGRS PHA allows the DPU to 'force' a given ADC to sample and convert its input regardless of signal conditions. The DPU samples each detector channel in this way once per second. At the end of each integration period, the DPU computes the mean and standard deviation of the samples. Because the ADC is configured for bipolar operation, it can measure the zero offset even when negative. Wild points are assumed to be samples of energy signals and are discarded so as not to bias or distort the computations.

The mean provides a direct measure of the system offset or 'energy zero'. Such a measurement simplifies energy scale calibration by separating scale and offset terms which usually requires at least a two-point energy calibration. Furthermore, the energy zero can be tracked over count rate, temperature, and age. This function is equivalent to the operation of a gated baseline restorer, but has the added benefit of including the offset and drift within the ADC itself.

The standard deviation provides a measure of the electronic noise in the channel and should directly relate to the noise threshold or the energy resolution if the system is preamp noise limited (e.g., the Si-PIN detector). The measurement is also an excellent diagnostic. A rise in the standard deviation may indicate a noisy preamp or possibly an interference problem. As an example, this parameter was critical in tracking down a microphonics problem in the gas tubes resulting from coupled vibrations from the spacecraft reaction wheels.

9.2. HIGH VOLTAGE POWER SUPPLIES

Historically, high voltage supplies have been one of the least reliable components in space-flight science instrumentation. To reduce this risk, the XGRS includes a separate high voltage power supply (HVPS) for each of its seven detectors. An additional benefit to this approach is that the individual supplies may be adjusted as necessary to equalize the gains of the detectors and to counter any aging effects. The supplies were designed and built by K and M Electronics, Inc., with design heritage from units developed for the SSUSI instrument, a UV spectrometer to fly on a series of DMSP satellites for the Air Force.

The mechanical housings of the supplies have an outer lip to allow vertical stacking. On NEAR, up to three supplies are stacked together. To reduce mass, only the critical high voltage sections within the supplies are potted rather than the complete unit. A single supply has approximate dimensions $14 \times 8 \times 3$ cm and a mass of 300 g.

The electrical design of the supplies is identical except for the two units designated for the gamma-ray detector. These generate negative rather than positive voltages. Each supply provides a regulated voltage output 0 to 2000 V for loads up to 50 μA. The high voltage supplies operate from a regulated ± 15 V power supply and can be adjusted via an analog control voltage. Each HVPS also provides a high voltage monitor and an internal temperature sensor. Power dissipation is about 300 mW.

9.3. HIGH VOLTAGE CONTROL ELECTRONICS

Both the GRS and the XRS require ultra-stable high voltages for their detectors. The electron gain of the gamma-ray detector PMTs varies roughly 0.5%/V at a nominal operating voltage of 1300 V. The X-ray gas tubes exhibit an electron gain variation approximately 1%/V at a nominal operating voltage of 1150 V. The GRS stability requirement is actually the more stringent of the two because gamma-ray spectra are binned to 1024 channels as compared to only 256 channels for X-ray spectra. To maintain energy scale calibrations to within one spectral bin would require a high voltage power supply to be stable to better than 80 ppm over temperature – not easily achieved.

The stability of a high voltage supply ultimately relates to its ability to precisely measure the high voltage output. This is usually accomplished using a simple resistive voltage-divider. The problem is that the temperature coefficient of the high voltage divider inside the HVPS may approach 80 ppm/°C and the spacecraft deck temperature may vary 60 °C. Rather than attempting to thermally isolate and regulate the temperature of the three stacks of high voltage power supplies, the XGRS contains separate electronics to make precise, independent measurements of each high voltage output and feed back corrections via the existing HVPS control input.

The High Voltage Control Electronics (HVCE) achieves the necessary precision by locating seven independent voltage dividers on a single printed circuit board which has its own built-in thermal control. The response of the secondary control loop is purposely slow so as not to interfere with the primary control loop within each HVPS. The HVCE board also contains precision digital-to-analog converters for commanding the high voltages and the HVCE temperature setpoint. The housing for the HVCE is completely blanketed except for a passive radiator to space. The radiator is sized such that the heat loss through the radiator dominates any leakages through the many connecting cables. Measured performance of the HVCE during spacecraft-level testing demonstrates internal temperature stability to better than 2 °C over a 25 °C range. The high voltage divider resistors are specified to have a maximum temperature coefficient of 25 ppm/°C resulting in an overall HVPS stability better than 100 ppm. The HVCE adds 1.1 kg of mass to the XGRS.

10. X-ray Electronics

10.1. X-RAY PREAMPS

Three X-ray preamp channels are located on a single printed circuit board in a preamp assembly mounted on the back of the XRS. The preamps are positioned to minimize connecting cable capacitance to the detectors. A preamp channel consists of a high voltage bias network, a fast, low-noise charge-sensitive preamplifier, and a buffer/driver stage. The charge-sensitive preamplifier is a modified industry standard AMPTEK A250 with the internal feedback capacitance reduced to 0.4 pF and the feedback resistor raised to 2 GΩ. The external JFET to the charge-sensitive preamp was selected from a special wafer run by INTERFET for use in AMPTEK miniature X-ray detectors. The low leakage, low capacitance devices exhibit high transconductance and are well suited for XGRS. Devices procured for the XGRS underwent additional high reliability processing by the manufacturer.

10.2. SOLAR MONITOR PREAMPS

The Si-PIN detector is mounted on a top corner of the solar monitor electronics box. The discrete FET and RC feedback components to the charge-sensitive preamplifier are both located inside the miniature detector hybrid. This approach minimizes the stray capacitance and improves the characteristics of the FET. The feedback capacitor is about 0.3 pF and the feedback resistor is an extremely high 80 GΩ.

Connections to the rest of the preamp are made through the cover plate which also serves as a thermal sink for the thermo-electric cooler (TEC). One side of the electronics assembly is not blanketed and serves as a passive radiator. The preamp consists of a standard AMPTEK A250 followed by an inverting buffer/driver stage. A high voltage zener diode protects the PIN detector from an overvoltage.

The preamp for the gas tube solar monitor is the same design as that used for the asteroid-pointing portion of the XRS, but repackaged to fit in the solar monitor electronics box. The input connector is positioned to minimize cable capacitance to the gas tube detector assembly which mounts separately to the deck.

10.3. X-RAY SHAPER

The X-ray Shaper board accepts inputs from all four gas tube detector preamps (3 asteroid + 1 solar). Each shaping channel produces three related signals: an energy signal, a timing signal, and a fast signal for rise-time discrimination (RTD). Figure 11 shows a simplified schematic. The design uses passive filter networks to preserve linearity and provide long-term stability. The energy signal shaping network implements a 5-pole pseudo-gaussian shaper with a peaking time of 5 μs (Mosher, 1976). The fwhm of the pulse is approximately equal to the peaking time. The accompanying timing signal has a bipolar-shaped characteristic whose zero-crossing time coincides with the peak of the energy pulse. The RTD shaping

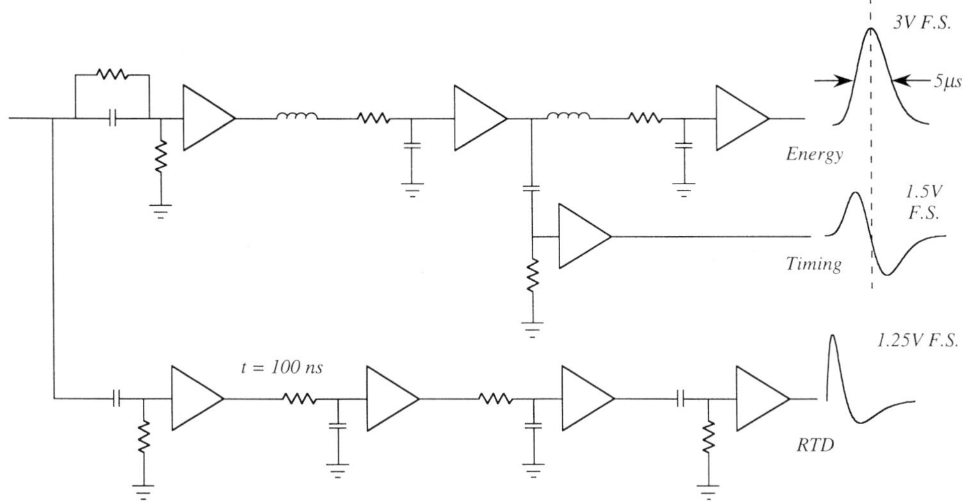

Figure 11. Simplified schematic of X-ray spectrometer shaping networks.

network is a double-differentiated, double-integrated type with a time constant of 100 ns. For the RTD shaper, fast pulse amplifiers are used to preserve bandwidth.

The preamplifier noise contribution to the energy channel is less than 200 eV fwhm. The preamplifier noise term determines the minimum energy threshold of the system. The RTD channel, with its wide bandwidth and subsequently higher noise, has a minimum noise threshold of about 1 keV and exhibits reliable RTD operation down to about 2 keV. The system energy resolution is determined by the preamp noise term added in quadrature with the inherent detector resolution. For the XRS, the preamplifier noise is always small compared to the signals produced by the gas tubes, which possess electron gain. Therefore, the energy resolution is primarily limited by the Poisson statistics within the gas tube.

10.4. X-RAY PHA

An X-ray PHA board contains two complete PHA channels. Two boards are required to accept all of the signals from the four-channel X-ray shaper board. The X-ray PHA design is based on the generalized scheme described above with the addition of rise-time discrimination capability (Figure 12).

The RTD portion of the PHA produces a flag if the preamp signal exhibits a characteristically slow rise time. Two voltage comparators with fixed-level thresholds fire on the positive and negative lobes of the fast-shaped RTD signal. This compensating arrangement minimizes amplitude walk thus producing a logic pulse whose width is only related to the preamp rise-time. The logic pulse is fed into a 2 μs integrator which acts as a simple time-to-amplitude converter (TAC). The TAC signal is then compared to a rise-time threshold. If the TAC signal rises above this

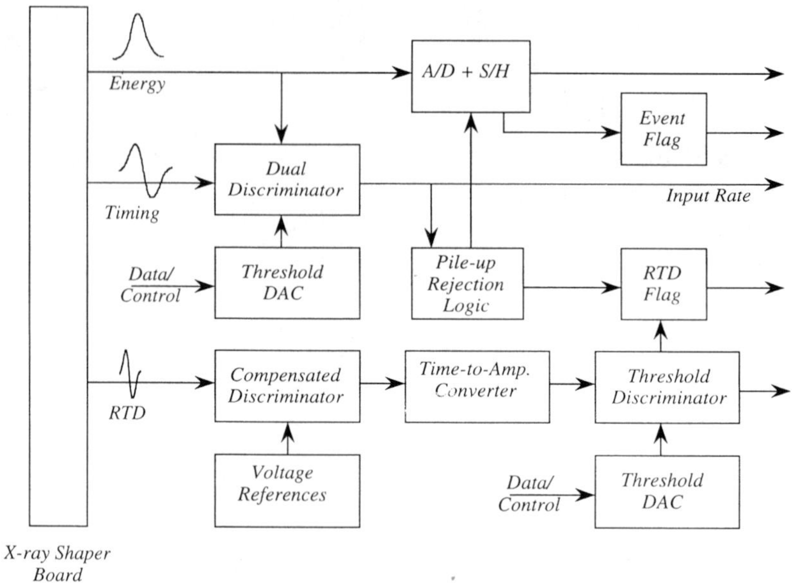

Figure 12. X-ray processing electronics block diagram.

threshold, an RTD flag signals the DPU to optionally reject the event. To minimize false rejections, the DPU will only reject slow rise-time events if the corresponding measured energy is higher than a threshold set in software, nominally about 2 keV.

The LLD and the RTD each have commandable thresholds. The signals from the LLD and RTD discriminators both pass to the DPU for count-rate monitoring.

10.5. Sɪ-PIN PHA

The Si-PIN PHA board is dedicated solely to processing signals from the Si-PIN solar monitor. This partitioning was in part the result of an early effort to 'decouple' the electronics associated with this new technology detector from the rest of the XGRS. In this way, any uncertainty in its inclusion would not impact the overall instrument development. The board is comprised of a shaping network and PHA similar in design to the other X-ray channels but without the need for rise-time discrimination. Some additional amplifier gain is provided because a solid state detector lacks the internal multiplication gain of a gas proportional counter. For the same reason, the energy resolution of the Si-PIN detector is preamplifier noise limited. The miniature size and cooling of the Si-PIN detector greatly reduces the preamplifier noise and as a result the Si-PIN solar monitor achieves better energy resolution than the gas tube solar monitor at energies above 3 keV. However, the Si-PIN solar monitor does exhibit a higher noise threshold, about 1.2 keV.

11. Gamma-Ray Electronics

11.1. GAMMA-RAY PREAMP

The GRS shares the preamp assembly with the XRS. The extra cable length to the detector is not critical because the PMT's provide such large gain. In fact, the system is practically 'noiseless', i.e., the energy resolution of the detectors is not degraded by electronics noise. This obviates the need for a low noise preamp. Of more importance is large signal linearity and overload recovery. The gamma-ray preamp consists of a FET-input operational amplifier configured as a charge-amp with a 470 pF feedback capacitor and a fall time of 47 μs. Diode clamping circuitry turns on for large overload signals and reduces the fall time by a factor of ten to hasten recovery. Back-to-back high conductance diodes protect the front-ends from damage by electrical discharge. The preamps for the NaI(Tl) and BGO detectors are identical. A Be-Cu ground shield reduces crosstalk from the gamma-ray preamps to the sensitive X-ray preamps.

11.2. GAMMA-RAY PHA

The pulse shaping networks for the NaI(Tl) and BGO channels reside on the Gamma-ray PHA board. The shaping networks are similar in design to the X-ray shapers, but do not need to be very low noise and instead focus more on baseline recovery and linearity. The amplifier gain of the gamma-ray shapers is much lower than that for the X-ray shapers because of the large input and high PMT gain of the gamma-ray detectors. The lower amplification requirement allows the shaping networks to be d.c. coupled throughout which improves baseline stability at high count rates. The gain for the BGO channel is slightly higher than that for the NaI(Tl) to account for its lower light yield.

The basic gamma-ray PHA design is similar to the X-ray PHA with some important differences (Figure 13). The coincidence arrangement of the NaI(Tl) and BGO detectors requires their PHA channels be closely linked. The two PHA's are linked together through a common pile-up rejection circuit, i.e., neither PHA can analyze an event unless the baselines are clear for both channels. This produces the highest quality data by minimizing data corruption of coincident events and minimizing false detections of anti-coincident events. A consequence of this approach is that the throughput of anti-coincidence events in the NaI(Tl) is subject to suppression by the higher expected rates in the BGO shield. For nominal count rates in each of the detectors the suppression is less than 10%.

The zero-crossing detectors which signal the NaI(Tl) and BGO channel ADCs to convert also drive the coincident logic. Bipolar signals when coupled to zero-crossing detectors do not exhibit the characteristic 'walk' with pulse amplitude of leading-edge discriminators and thus are ideal for coincidence timing. This feature allows a coincident window of 500 ns to work reliably with 5 μs pulses. Stability is

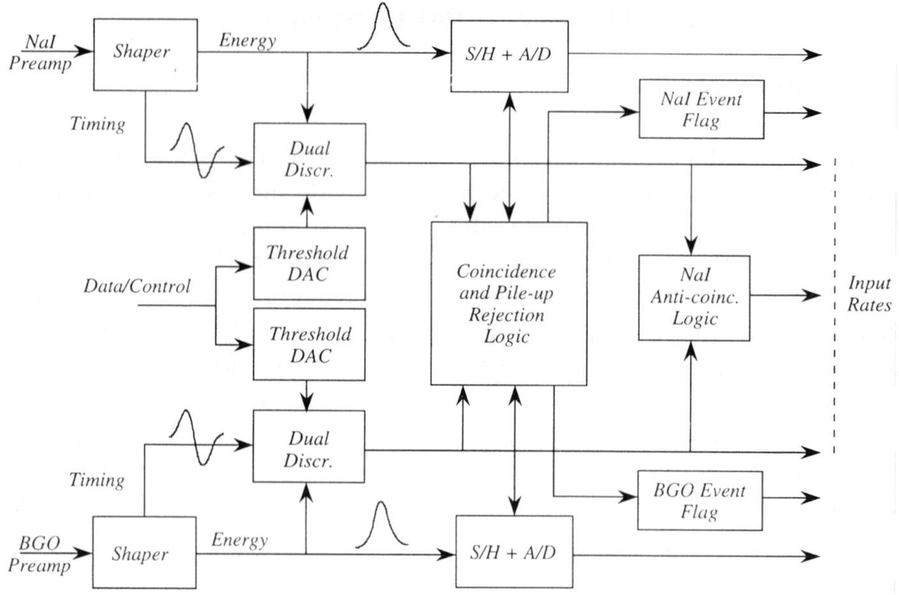

Figure 13. Gamma-ray processing electronics block diagram.

also improved because any perturbations on the size of a small coincident window should have little effect on the overall operation.

The state of the NaI(Tl) and BGO event flags indicates to the DPU the type of gamma-ray event: NaI(Tl) only, BGO only, or coincident event. The DPU bins the events to form the corresponding spectra. A DPU read operation of the BGO re-arms both PHA channels so the NaI(Tl) must be polled first.

The LLD outputs for NaI(Tl) and BGO pass directly to the DPU counter bank. An additional logic circuit passes pulses for anti-coincident events in the NaI(Tl). Two separate software counters in the DPU keep track of the number of overload events in both detectors.

12. Data Processing Unit

A single DPU services all seven XGRS detectors and performs all X-ray and gamma-ray event processing. Each PHA retains an event until polled and read by the DPU. The DPU read operation re-arms the given PHA to accept new events. The DPU performs several tests to determine the nature and validity of the event and then bins the value into the appropriate spectrum. At the end of each integration period, the DPU formats the different spectra into science telemetry packets. The DPU also samples the event counters and controls operation of the high voltage power supplies, operational heaters, TEC, and the calibration rod. Other functions include instrument housekeeping, command handling, and autonomy.

12.1. DPU DESIGN

To reduce cost and development time, a common DPU design was developed to support NEAR instruments. Functions were modularized so that a DPU could be assembled to meet a variety of system requirements. Core functions include the central processing unit (CPU), interface to the spacecraft, and instrument housekeeping. Additional interface board(s) are customized for each instrument. A separate section of the DPU houses power conditioning and a motor controller. The selection of power converters is tailored to each instrument.

The CPU board is based on a Harris RTX2010RH, a radiation-hardened, reduced instruction set (RISC) microprocessor designed for the direct execution of Forth. The processor board contains 4K words of programmable read-only memory (PROM), 96K words of random access memory (RAM), and 32K words of electrically erasable PROM (EEPROM). The processor first boots from PROM, then copies the operational software from EEPROM to RAM, and then executes out of RAM to maximize speed. A shared memory space of 32K words of RAM facilitates data transfer between the DPU and the CTP. The processor clock rate is 6 MHz. Other features of the processor board include a watchdog reset, a low-voltage reset, an EEPROM write-protect circuit, and an expansion bus to communicate with peripheral boards. An external serial test port allows for easy downloading and troubleshooting of software.

The spacecraft interface board operates as a remote terminal on the spacecraft MIL-STD-1553 bus, a high-speed serial communications bus used in military avionics. The board is built around the United Technologies Microelectronics Center (UTMC) SuMMit chip, a radiation-hardened monolithic device which handles all of the 1553 protocols. Data are transferred to and from the processor board through 32K words of shared memory.

The housekeeping board contains multiplexers, amplifiers, and a 12-bit ADC to gather housekeeping telemetry data from a variety of sources. The board contains channels to measure instrument currents, single-ended and differential voltages, and temperatures (Analog Devices AD590s). The XGRS uses 24 of the 36 housekeeping channels.

The XGRS interface board supports all of the digital interfaces to the instrument. PHA data is read via a 16-bit read bus. Thresholds are controlled via an 8-bit write bus. The two interfaces share a 5-bit address bus. A separate digital interface goes to the High Voltage Control Electronics. The event counters reside on a single chip comprised of 16 24-bit counters. The device was developed at Goddard Space Flight Center using a radiation-hardened UTMC gate array. The event flags for the seven detectors are grouped together into a single byte.

The XGRS power board provides conditioned power to the instrument at several standard voltages (±15 and ±5 V). Post-regulators in the Detector Electronics supply ±6 V volts to the critical analog sections. Separate DC/DC converters supply power to the high voltage power supplies and the thermo-electric cooler to allow on/off control. The heater within the high voltage control electronics taps power from the detector electronics and can be disabled via its temperature setpoint. The GRS operational heater receives power directly from the spacecraft bus via a power-switching relay under processor control.

The power motor board supplies the motor drive circuits for the XRS calibration rod, supplies the interfaces to the calibration rod fiducial sensors, and provides conditioned power to the DPU itself. The motor is a two-phase stepper motor with bipolar drive. To minimize power dissipation on the power motor board, the current to each motor phase is regulated using pulse width modulation and is commandable to 8 levels. Interface circuits to the fiducial sensors provide both commandable brightness and receiver thresholds.

13. Software

13.1. OVERVIEW

The XGRS DPU software has three components: a Forth language and operating system, DPU common software, and XGRS application specific software. The Forth system and DPU common software comprise a kernel which is common to all the NEAR instrument DPUs.

13.1.1. *NEAR Forth System*
The Forth system used for the NEAR instrument DPUs is an ANS Forth compliant system which is implemented for the RTX-2010 processor. The NEAR Forth system contains a Forth development environment including an interactive Forth interpreter and compiler which are accessed via a test port UART and concurrent programming extensions. The interactive interpreter and compiler support application debug and monitoring. The concurrent programming extensions support the concurrent, real-time requirements of the NEAR instrument software.

13.1.2. *Concurrent Programming Extensions*
The concurrent programming extensions which are included in the NEAR Forth system support a prioritized, concurrent process software architecture and address mutual exclusion, event notification, and resource allocation between the processes. Higher priority processes can preempt lower priority processes based on hardware or software events. The support for mutual exclusion, event notification, and resource allocation is based on wait and signal synchronization operations applied to software semaphores.

The NEAR Forth concurrent programming extensions allow applications to build and manage a dynamic, prioritized process list. Associated with each process is the code which it executes (typically in the form of an infinite loop in the NEAR instrument applications), a software process state (waiting or runnable), an RTX processor state, and dedicated data and return stacks which Forth uses when the process runs. All processes (runnable or waiting) are linked together in a single process list with process priority from top (highest priority) to bottom (lowest priority). When a multi-process application runs, generally, the highest priority runnable process will have control of the processor at any given time. A running process continues running until (1) it 'waits' on an event which has not yet occurred or for a resource which is not yet available or (2) a higher priority process becomes runnable. A higher priority process may become runnable when (1) the current process 'signals' that an event has occurred or that a resource it was using is available or (2) an external hardware interrupt occurs and the RTX interrupt handler signals a semaphore on which the higher priority process is waiting.

The RTX supports single level interrupt handlers. While an interrupt handler is executing, it cannot be preempted by a Forth process. An interrupt handler can signal a software semaphore to notify a Forth process waiting for that interrupt to proceed with its response to the interrupt. If the running Forth process is interrupted due to a hardware interrupt, when the interrupt handler completes, the Forth system will resolve which process to run based on the new semaphore state. Forth software processes may also make judicious use of disabling interrupts.

13.2. DPU COMMON SOFTWARE

The DPU common software is the software in addition to the NEAR Forth system which all NEAR instrument DPUs require. The functionality provided by the DPU common software includes CCSDS standard telemetry and command handling which occur over the 1553 communications bus between the DPU and the Command and Telemetry Processor (CTP), real-time housekeeping collection, housekeeping A/D board interface, time management, watchdog timer control, and initialization. The DPU common software also implements a set of core commands (including memory uploads to EEProm or RAM, program control and reset commands, macro commands, memory dump commands, configure telemetry commands, and time-tagged commands) and forms the basic real-time housekeeping telemetry which are required for all NEAR instrument DPUs.

Much of the DPU common software functionality is provided as basic services to the DPU application specific software. The application specific software generally uses the common software services by registering for those services. In registering for a common software service, the application specific software specifies various parameters of the service which customize the service to the needs of the application. In some cases the application may also specify code which augments a service's behavior as a further means of customization. For instance, the telemetry service is registered in the application software by specifying its various CCSDS parameters including record length and id. Application code registers command services by specifying with each command opcode a reference to command handling code which the common software then executes each time it receives a command with that opcode. The timing services of the common software allow the application specific software to readily access current Mission Elapsed Time (MET) and to flexibly schedule tasks.

13.3. XGRS SOFTWARE

Utilizing the combined functionality of its three components: the NEAR Forth system, DPU common software, and XGRS application specific software, the XGRS software performs a multitude of tasks relating to instrument configuration and housekeeping, data collection and formation of spectra, autonomy, and commands and telemetry. The XGRS software process architecture is based on the concurrent processing extensions built into the NEAR Forth system. Figure 14 illustrates that process architecture. The process priorities are from top (highest priority) to bottom (lowest priority). The Forth process is implemented in the Forth system software. The telemetry, command, and dump processes are implemented in the DPU common software. The motor, 1 Hz, and background sensor processes are implemented in the XGRS specific application software. The XGRS specific application software has the final say on process priorities to meet its requirements.

13.4. XGRS SHARED RESOURCES

The XGRS software uses a number of mechanisms to control access to its shared hardware and software resources. These mechanisms include the wait and signal synchronization on software semaphores provided in the Forth system as well as interrupt disabling. Examples of XGRS shared resources are the sensor hardware interface, spectral buffers, and telemetry packet buffers. The sensor hardware interface and spectral buffers are each shared by the background sensor process which uses them for reading sensor data/integrating spectra and the 1 Hz process which uses them for sensor configuration/science record formation. Telemetry packet buffers are shared between the telemetry process and the 1 Hz process or the dump process.

13.5. SOFTWARE MEMORY USAGE AND METRICS

Because of the desire to design redundancy into the DPU program memories and the implementation of the boot procedures in the NEAR Forth system, only 28K words are readily available for total application code (not including boot code). Fortunately, additional memory for various buffers needed by the application code is provided in separate remote memory 'pages' which have a slight additional access overhead compared to the program memory page. Thus, in order to preserve program memory space, the XGRS software has been designed so that most large buffers are located on these remote memory pages and primarily the actual code definitions reside on the program memory page. Specifically, the large spectral accumulation buffers, macro, and various command and telemetry buffers are located on remote memory pages. In addition to optimizing program memory usage

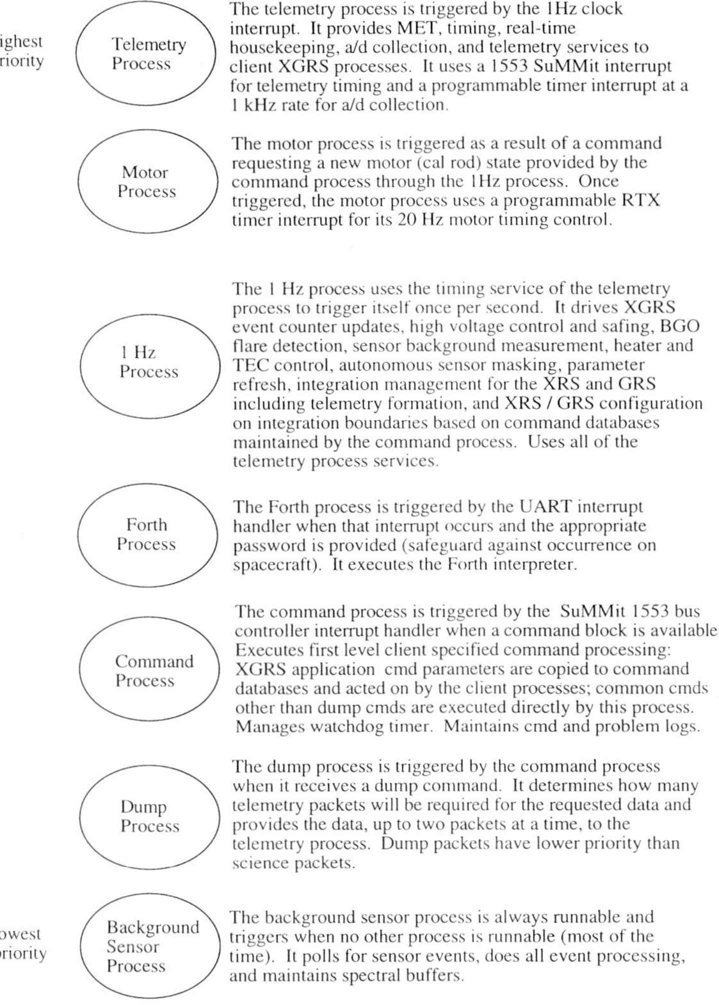

Figure 14. X-ray/Gamma-ray Spectrometer Software Process Architecture.

by using remotely allocated data memory, the XGRS code has been optimized for memory usage by removing a portion of the Forth code 'headers' which allow Forth code to be executed by symbolic reference (as opposed to by address or opcode) in the Forth interpreter.

XGRS software version 5 requires approximately 9000 lines of source code; the corresponding object code requires 27.8K words out of the total available 28K words of program space. The breakdown of these numbers by software component is provided in Table III. An additional 36.6K words of remote data buffers are used by the XGRS software as also indicated in Table III.

Table III
X-ray/Gamma-ray Spectrometer software components

XGRS source code lines	
Forth system	2000 (~20%)
DPU common software	2000 (~20%)
XGRS application software	5000 (~60%)
Total	9000
XGRS object code size	
Forth system	6.2K words (~20%)
DPU common software	5.3K words (~20%)
XGRS application software	16.3K words (~60%)
Total	27.8K words
XGRS remote buffer sizes	
Spectral buffers	6.4K/16K words (p. 2)
Macro buffers	8K/8K words (p. 2)
Time tagged command buffer	8K/8K words (p. 2)
Telemetry packet buffers	9.4K/16K words (p. 4)
1553 bus buffers, etc.	4.8/16K words (p. 4)
Total	36.6K words

The average number of Forth symbols (a Forth symbol is a symbolic reference such as a number, variable name, Forth word name) in a line of source code in the XGRS source code is 3.1 symbols per line. Note the general rule that the Forth code symbol count equates to the number of words required in the object code (provided large memory buffers are separately accounted for).

14. Software Functionality

14.1. Spectral integration

Spectral integration takes place in the background sensor process. The software polls hardware event flags set by the sensor channels and circulates event processing among those that are unmasked. The X-ray and gamma-ray integration periods operate independently. Both are controlled by the 1-Hz software process which is tied to a 1-Hz interrupt synchronized with the Mission Elapsed Time (MET). The overall instrument telemetry rate may be raised or lowered by commanding shorter or longer integration periods. An integration interval for the XRS or GRS normally terminates in one of two ways. Either the natural end of the commanded integration period is reached or the XGRS DPU receives one of several commands which cause the integration period to end immediately. End integration commands are used to synchronize the start of an integration period with a particular event or time and also to ensure that the instrument configuration is stable for each telemetered spectrum. An integration interval may also be extended if, for example, the CTP delays in picking up the science telemetry packets from the previous integration period.

14.2. SPECTRAL BIN OVERFLOW

For lengthy integration periods or in the event of bright spectral features, the binning process may cause one or more of the 16-bit spectral bins to overflow. XGRS software provides two instrument modes to handle spectral bin overflows. In 'rollover' mode, the bins are simply allowed to roll over and continue counting. In 'normal' mode, if a spectral bin reaches its maximum value, then event processing for that sensor is suspended for the remainder of the integration period. The highest energy bin is exempted as it contains the sum of all events beyond full scale and is likely to overflow. For the XRS, non-overflowing sensors will continue to integrate with the actual integration time for each sensor reported separately. An overflow in either of the gamma-ray detectors suspends integration for all gamma-ray spectra. In either rollover or normal mode, any bin overflows that occur are reported by simple flags in the telemetry.

14.3. SENSOR MASKING

Sensors may be excluded from event processing through a process called sensor masking. Masking occurs in the software either as a result of direct command or by autonomous action by the DPU. Commanded masking provides for ground selection of sensor processing, such as the selection of the active solar monitor, while autonomous sensor masking ensures that spectra are not corrupted with data collected under transitional or aberrant conditions. The DPU may autonomously mask a sensor for the following reasons: sensor safing, a condition in which a high input rate is detected; high voltage ramping, due to a commanded change in the high voltage set point or as an artifact of safing; the corresponding science record telemetry is commanded off; proton flare event, in which an elevated count rate is detected in the BGO shield; or a bin overflow occurs while in normal mode as described above. All spectra appear in the telemetry regardless of mask state with the exception of solar spectra. To save telemetry and to reduce the processing load, the DPU provides spectra from only one of the two solar monitors at a time. The unmasked solar monitor is referred to as the 'active solar monitor'. In 'toggle' mode, the DPU alternates between the two solar monitors. The actual integration time for each spectrum is reported in the science telemetry and generally corresponds to the period of time during the integration period for which that sensor was unmasked for data collection.

14.4. SCIENCE EVENT COUNTERS

The DPU maintains a total of 24 32-bit software event counters which relate to collected spectra. Of these, 12 are based on values read from the 24-bit hardware event counters and include the number of input events for each detector, the number of rise-time rejection events for the gas proportional counters, and the number of NaI anti-coincidence events. The DPU reads and resets each 24-bit hardware event counter once per second and adds its value to the corresponding 32-bit software counter. The remaining 12 counters are strictly software. They measure the number of valid events processed, the number of gamma-ray events above 10 MeV, and the number of low-energy X-ray events with rise-time rejection flags. All counter values are reported in the science telemetry regardless of the sensor mask state. Note that purely software counters will have zero values when the corresponding sensors are masked for the entire integration by command or autonomously. The relationship of the various event counters are depicted in Figure 15.

14.5. SENSOR SAFING

An important function of the raw input rate counters is to enable autonomous sensor safing. The DPU monitors the raw input rates once per second and if a rate exceeds a safing threshold, the DPU takes action to quickly reduce the high voltage to that sensor. Such conditions might occur in the event of a strong solar flare and could persist for several hours. After a commandable waiting period, the DPU attempts to restore a sensor to nominal operation by slowly ramping up its high voltage while checking its input rate. If this test fails, the DPU waits and tries again. If a sensor cannot be successfully restored after a commandable number of attempts, the sensor is placed in 'level two safing'. At this level, the affected sensors remain masked with their high voltages set to zero volts and must wait for ground command to be reset.

14.6. SOLAR FLARES

Much of the prime science for the X-ray experiment will occur during solar flares which excite strong emissions from the asteroid surface. Some of these flares may be followed by a period of energetic solar protons causing elevated count rates in the BGO shield. While this count rate should remain well below the gamma-ray safing level, the increased burden on the DPU might suppress processing of prime events in the X-ray sensors. For this reason, a special 'proton flare' mode was created. If the input rate to the BGO shield exceeds a commandable threshold, then the BGO sensor is immediately masked. When the BGO rate falls below this threshold, BGO processing resumes at the start of the next gamma-ray integration period. In this way, the instrument can maximize the science return from the X-ray Spectrometer without having to alter the stabilized voltage to the BGO detector.

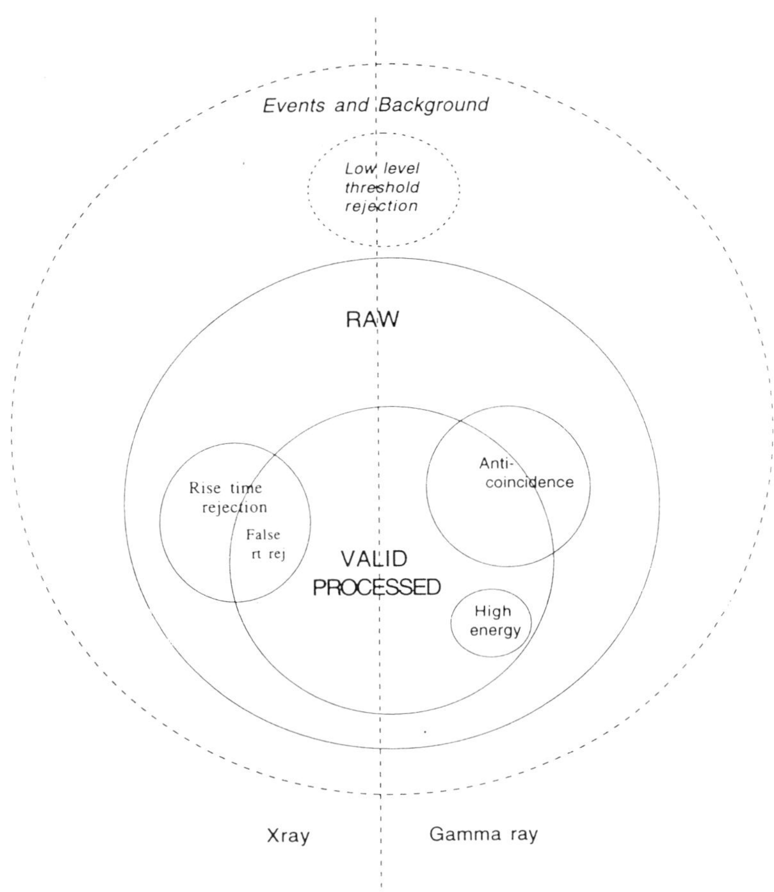

Figure 15. Venn diagram showing relationship of XGRS event counters.

In addition to reading the hardware counters, the DPU forces each PHA channel to sample its input once per second regardless of signal conditions. The DPU computes the mean and standard deviation of these samples each integration period. The majority of these samples will contain only the effects of electronics noise and offset. Individual samples which lie beyond roughly 3 sigma are assumed to contain signal and are discarded so as not to bias or distort the computations. The DPU maintains a separate threshold for each channel. Each threshold value is dynamically linked to the value of the corresponding lower-level discriminator whose setting generally provides a rough indication of the electronic noise level for that channel. In this way, the value of the sample reject threshold is somewhat adaptive and may be indirectly controlled via the commandable discriminator setting.

The DPU also provides one second control of the gamma-ray operational heater, the TEC for the PIN solar monitor, and all detector high voltage ramping. The

DPU also refreshes critical hardware parameters at this rate to guard against any possible catastrophic damage due to a Single Event Upset (SEU). This functionality is provided in the 1-Hz process as described below.

14.7. GAMMA-RAY HEATER CONTROL

Once per second, the DPU measures the temperature of the gamma-ray detector assembly and decides whether to turn the gamma-ray heater on or off. Both the temperature set point and hysteresis level are commandable. Nominal settings provide a low limit of 20 °C with a delta of 0.25 °C. The DPU provides an estimate of the heater duty-cycle by simply counting the number of seconds the gamma-ray heater is on during an arbitrary 100 s period. The most recent measurement is reported in the telemetry. Individual duty-cycle measurements tend to exhibit a large variation, but this variation indicates proper regulation and a long term average of these samples indicates the level of margin.

14.8. TEC CONTROL

The DPU does not need to regulate the cooling temperature of the TEC because the gain of the Si-PIN detector is stable for temperatures below -20 °C. However, the DPU does monitor the TEC temperature to ensure the detector temperature does not get too cold and suffer undue thermal stress. The DPU will turn the TEC off as necessary to keep the detector temperature above a commandable lower limit. The DPU performs a similar function to keep the detector from getting too hot if the detector is placed in its annealing mode. If the detector temperature gets too hot while in cool mode, a fault is assumed and the DPU turns off the TEC. The DPU also checks if the state of the cool/heat relay does not match the expected TEC state. For either of these error conditions, the DPU reports a problem code in the DPU portion of the real-time housekeeping telemetry and turns the TEC off until re-enabled by ground command.

14.9. VOLTAGE RAMPING

To reduce the risk of damage to the detectors by rapid changes in high voltage, the DPU provides controlled ramping between high voltage settings. The XGRS software ramps the voltages up or down at a rate of 100 V every 20 s. The 20 s stabilization period allows the X-ray preamps sufficient time to come out of saturation and report a valid input rate, which is especially important for the sensor safing algorithm to work properly.

14.10. CALIBRATION ROD CONTROL

The XGRS motor process controls the stepper motor for the X-ray calibration rod. The DPU can step the motor forward or backward at a rate of 20 steps per second.

A total of 200 steps are required to rotate the calibration rod a full 360°. Two Light Emitting Diode (LED) and photodiode receiver pairs are located at the normal or 'home' position, and 180° from the normal position. The software has the ability to search for the home fiducial to establish a reference. The software then uses dead reckoning from that point to move to the three different X-ray calibration positions at step 75, 100, and 125. Once the motor has reached its goal position, the motor power is turned off. The stepper motor has sufficient off torque to hold the calibration rod in place.

14.11. THROUGHPUT

Due to efforts in code optimization, the DPU achieves an event processing throughput of greater than 30,000 events s^{-1}, despite its myriad of tasks. The DPU's high throughput capability allows full use of the available sensitivity of the XGRS sensors. Weak line emissions from the asteroid surface will not be suppressed by the presence of high activity on the complement of XGRS detectors.

15. Data Products

Full details to recover and assimilate science data from the X-ray/Gamma-ray Spectrometer may be found in the applicable Johns Hopkins University Applied Physics Lab documents, specifically the XGRS software requirements (Schneider, 1995) and the DPU common software requirements (Hayes, 1995).

15.1. GAMMA-RAY DATA PRODUCTS

The GRS produces a collection of data products every integration period: five 1024-channel spectra, two 21-channel spectrum segments, and seven scalar counts. Spectral bins contain 16 bits. Scalar counts are 32 bits. The five primary spectra are: the raw NaI(Tl) spectrum (all processed events); the raw BGO spectrum; the NaI(Tl) anti-coincidence spectrum (events in NaI(Tl) detector only); the NaI(Tl) single escape spectrum (events coincident with the BGO 0.511 MeV line); and the NaI(Tl) double escape spectrum (events coincident with the BGO 1.022 MeV line). In addition, the BGO coincidence spectrum around 0.511 MeV and the BGO coincidence spectrum around 1.022 MeV are collected with a variable size window of up to 200 keV width which is reported in 21-channel spectrum segments. Along with the spectra, the following scalar counts are reported: NaI(Tl) raw counts (number detected); BGO raw counts; NaI(Tl) valid counts (number processed); BGO valid counts; NaI(Tl) anti-coincidence counts (number detected); NaI(Tl) counts above 10 MeV (number processed); and BGO counts above 10 MeV.

A full GRS science record, including housekeeping parameters which reflect the actual integration time, background statistics, and instrument configuration state over the integration, requires 31 telemetry packets. The data rate is roughly

70 bits s^{-1} for a nominal integration period of 1200 s. The data rate may be raised or lowered by commanding a shorter or longer integration period.

15.2. X-RAY DATA PRODUCTS

The XRS produces four X-ray spectra and 17 scalar counts every integration period. All spectra contain 256 channels. Spectral bins contain 16 bits. Scalar counts are 32 bits. The spectra correspond to the unfiltered gas tube detector, the Mg-filtered detector, the Al-filtered detector, and the active solar monitor. The XRS scalar counts report the following: the number of events detected by each of the five detectors; the number of events processed for each of the four active detectors; the number of rise-time rejection events detected by each of the four gas tube detectors; and the number of rise-time rejection events processed with energies below 2 keV.

A full XRS science record, including housekeeping parameters which reflect the actual integration time, background statistics, and instrument configuration state over the integration, requires 7 telemetry packets. The data rate is roughly 180 bits s^{-1} for a nominal integration period of 100 s. The data rate may be raised or lowered by commanding a shorter or longer integration period.

15.3. SUMMARY DATA PRODUCTS

An additional summary science mode provides a reduced set of spectra from the X-ray and gamma-ray experiments and combines them into a single data stream. The summary packet contains all of the instrument housekeeping telemetry and all of the scalar counts. For the GRS portion, compressed versions of the raw NaI(Tl) and BGO are included. The spectral resolution is reduced from 1024 to just 16 channels and the channel accumulations are quasi-log compressed from 16-bits to 8-bits. X-ray spectra are reduced from 256 channels to 16 channels and the channel accumulations are quasi-log compressed. Transmission of summary data is tied to the X-ray integration period. For an X-ray integration period of 300 sec, the summary science data rate is approximately 10 bits s^{-1}.

16. Instrument Calibration

An extensive calibration program has been undertaken for both the X-ray and gamma-ray spectrometers using flight systems and flight spares. Detector response and efficiency as a function of energy and angle have been measured for both detectors. For the X-ray spectrometer, additional work was done to determine filter transmissions as a function of energy and the efficacy of the rise time discrimination circuitry. For the gamma-ray spectrometer, the anti-coincidence rejection efficiency as a function of energy was measured and tests were conducted to determine the coincidence efficiency of the annihilation line recovery logic. Details of the many

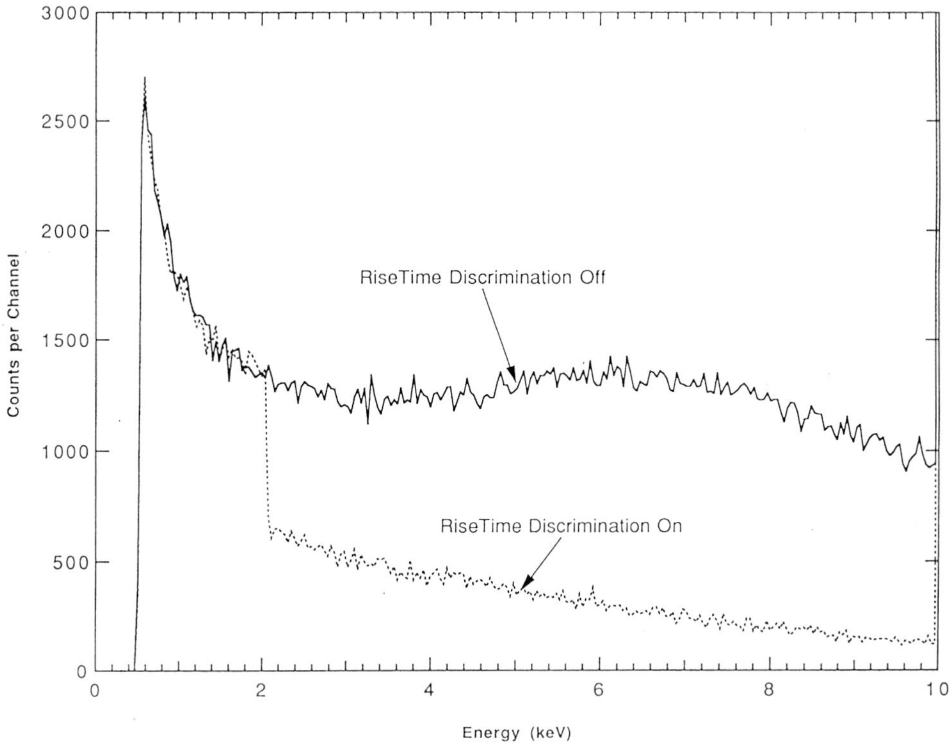

Figure 16. Background spectra obtained in flight with the Mg-filtered proportional counter. Spectra with and without rise time discrimination are shown. The break in the spectrum at 2 keV clearly shows where rise time discrimination is in effect.

calibration measurements taken on the ground are described in Trombka et al. (1996). Here we present calibration measurements using in-flight data.

16.1. X-RAY SPECTROMETER

As mentioned in a previous section, the X-ray proportional counters use rise time discrimination to reduce the background due to cosmic rays and gamma rays. At energies less than 2 keV the rise time discriminator cannot distinguish between background and true events. Therefore, rise time rejection is used only above 2 keV. Since background reduction is most useful at higher energies, where the signal-to-background ratio is poor, the lack of rise time rejection below 2 keV will have a minimal impact on asteroid measurements. Figure 16 shows background spectra obtained in flight with the Mg-filtered proportional counter. Spectra with and without rise time discrimination are shown over the energy region from about 0.7 to 10 keV. The break in the spectrum at 2 keV clearly shows where rise time discrimination is in effect and the reduction in background above this energy can be seen. A pulse height spectrum of the Mg filtered proportional counter with the

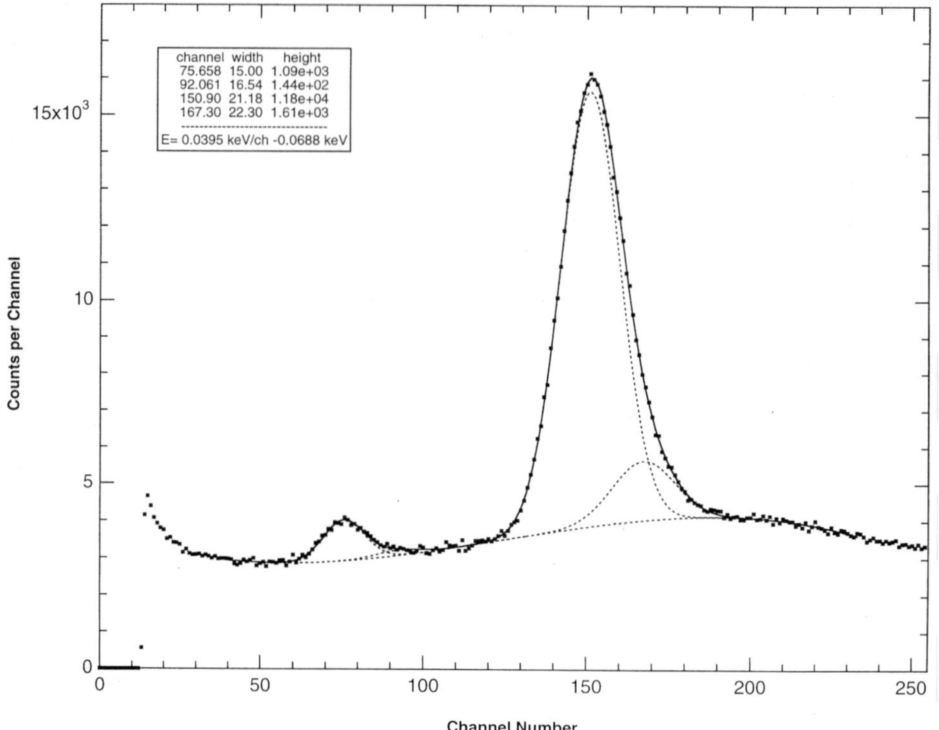

Figure 17. Pulse height spectra of the Mg filtered proportional counter with the Fe-55 source in the field of view. The fit to the spectrum (solid line) is accomplished by including the Kα (5.89 keV) and Kβ (6.54 keV) lines in both the full energy and escape peaks (dashed lines) plus a polynomial for the background (dashed line). The fit for the full energy Kα line gives a fwhm of 840 eV.

Fe-55 source in the field of view is shown in Figure 17. The fit to the spectrum (solid line) is accomplished by including the Kα (5.89 keV) and Kβ (6.54 keV) lines in both the full energy and escape peaks (dashed lines) plus a polynomial for the background (dashed line). The fit for the full energy Kα line gives a fwhm of 840 eV.

Figure 18 shows the pulse height spectrum obtained with the solar monitor proportional counter and the PIN detector during a period of low solar flare activity. The proportional counter, which has a much larger volume than the PIN, is dominated by the cosmic ray background when the Sun is quiet, thus displaying a fairly flat energy spectrum. The PIN spectrum is also flat above about 3 keV, but shows its response to the quiet Sun X-ray flux below this energy. The spectrum turns over at about 1.5 keV because of the 3 mil Be window which is used to decrease the flux in the lower energy region thus facilitating measurements over several orders of magnitude. The high peak at low energy is due to detector noise.

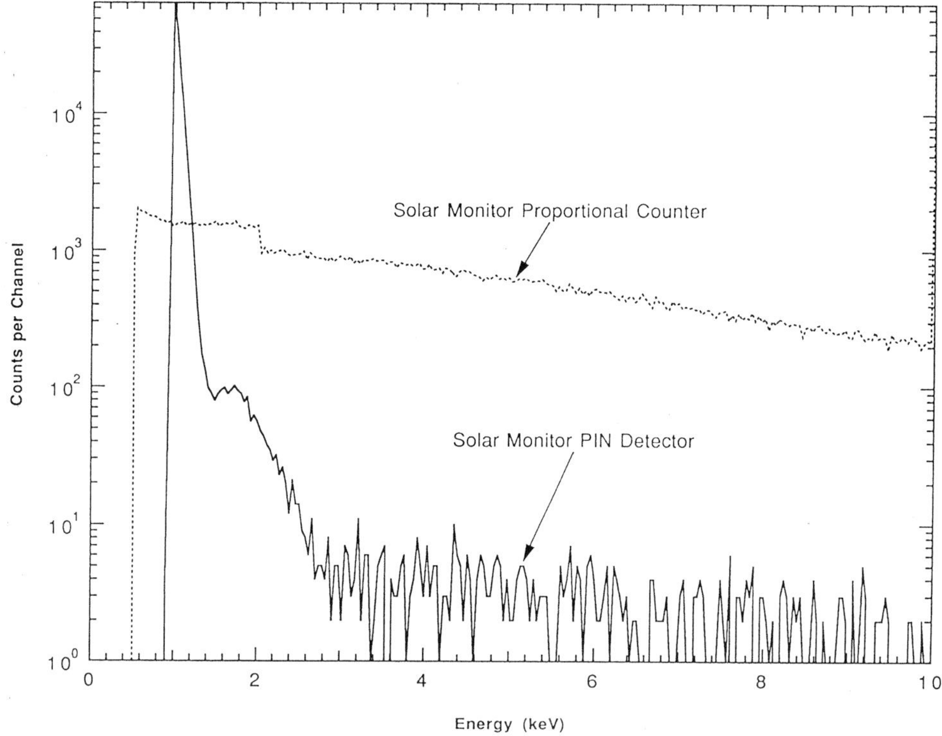

Figure 18. Measurement of the quiescent solar X-ray spectrum with both the solar monitor proportional counter and the solar monitor PIN detector.

16.2. GAMMA-RAY SPECTROMETER

Many days of BGO and NaI detector flight spectra have been obtained. A BGO background spectrum is shown in Figure 19. Some of the lines that have been identified are at 0.511 MeV (e^+-e^- annihilation), 0.844 MeV (Al-27), 1.809 MeV (Mg-26), 2.223 MeV (H) 4.438 MeV (O), and 6.129 MeV (O). The four NaI background spectra shown in Figure 20 are the raw spectrum, in anti-coincidence with the BGO, in coincidence with the 511 keV line in the BGO, and in coincidence with the 1022 keV line in the BGO. The raw spectrum is dominated by cosmic-ray interactions above 3 MeV and is seen to flatten and begin to rise at higher energies. As with the ground tests, the anti-coincidence spectrum clearly shows the effectiveness of the shield in reducing background. Above 3 MeV the BGO shield reduces the background by two orders of magnitude. Some of the lines that have been identified are at 0.175 MeV (Ge-71 – anti-coincidence spectrum), 0.511 MeV ($e^+ - e^-$ annihilation – raw and coincidence spectra), 1.369 MeV (Na-24 – anti-coincidence spectrum), 1.779 MeV (Al-28 – anti-coincidence spectrum), 2.614 MeV (Bi – anti-coincidence spectrum) 4.438 MeV (O – anti-coincidence

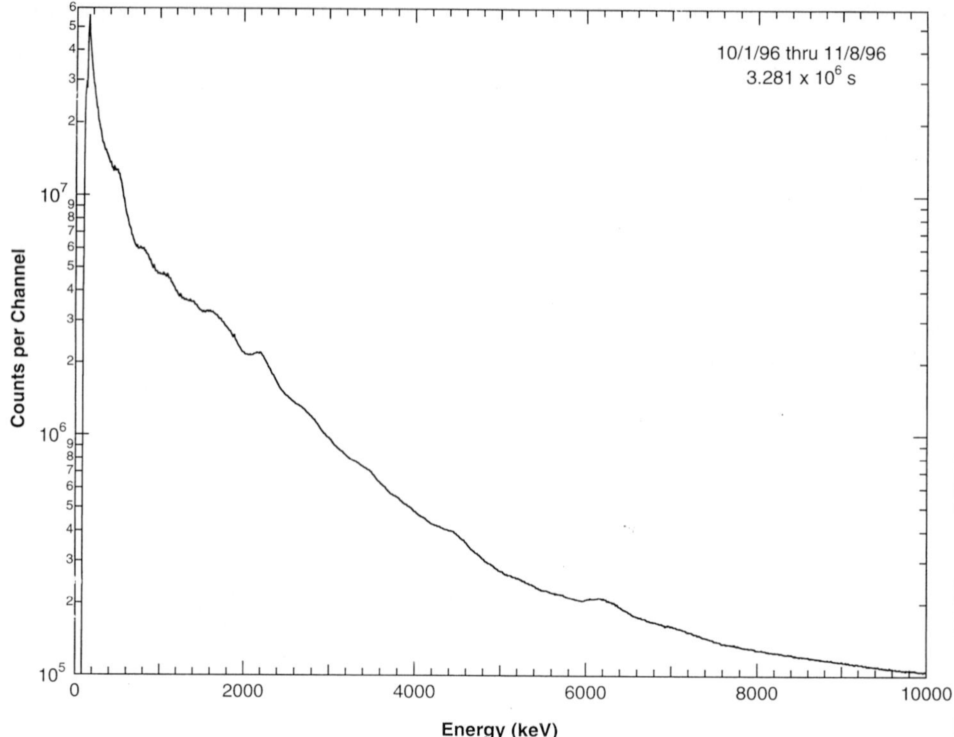

Figure 19. Background spectra obtained in flight with the NEAR BGO detector. The accumulation time is approximately 38 days. Some of the lines that have been identified are at 0.511 MeV ($e^+ - e^-$ annihilation), 0.844 MeV (Al-27), 1.809 MeV (Mg-26), 2.223 MeV (H) 4.438 MeV (O), and 6.129 MeV (O).

and coincidence spectra), and 6.129 MeV (O – anti-coincidence and coincidence spectrum).

Acknowledgements

The authors would like to thank all of the XGRS team members for their dedicated efforts in developing this instrument and making the X-ray/Gamma-ray Spectrometer a success. We also wish to express our gratitude to the Technical Services Department of the Johns Hopkins Applied Physics Laboratory for meeting a very demanding schedule, the NEAR Mission Operations team for assisting in the collection of many hours of in-flight data, and to the NEAR program manager, T. B. Coughlin. The authors would also like to acknowledge those organizations who worked closely with us to develop the flight detectors: S. Vajda at EMR for the gamma-ray detector; T. Andersson at Metorex International for the X-ray detectors; and J. Pantazis at AMPTEK, Inc. for the miniature X-ray detector. K and

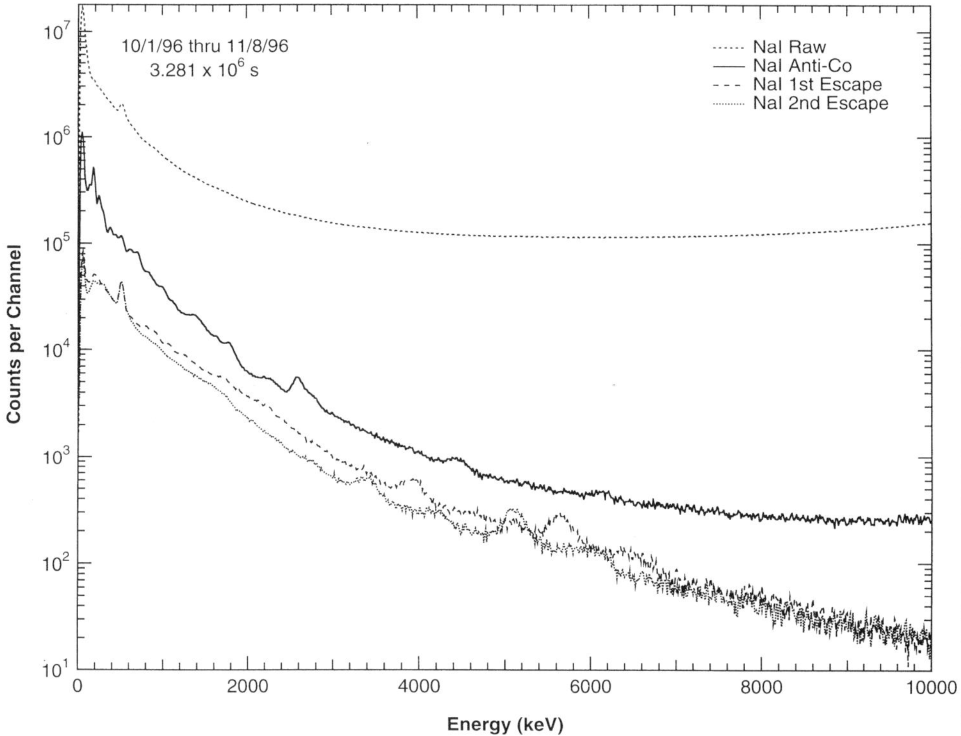

Figure 20. Background spectra obtained in flight with the NEAR NaI detector. The accumulation time is approximately 38 days. Four spectra are shown. Pulse height spectrum without anti-coincidence; pulse height spectrum with anti-coincidence; pulse height spectrum with 511 keV coincidence; and pulse height spectrum with 1022 keV coincidence. The anti-coincidence spectrum clearly shows the effectiveness of the shield in reducing background.

M Electronics provided the high voltage power supplies. We also thank the staff and management of Paradise Animal Hospital of Catonsville, MD and, especially, Dr N. Auclair for arranging the purchase and acquisition of an X-ray detection screen on short notice. Many thanks also go to: J. Buttermore and R. E. Thompson for their fine work in assembling the instrument; P. Bierman for his dedicated effort on the X-ray collimator; D. Fort for his work on rise-time discrimination; J. D. Boldt and P. Eisenreich for providing the DPU; A. A. Chacos for providing superb instrument GSE; and finally E. H. Darlington for his general expertise in scientific instruments and helpful nature.

References

Adler, I., Trombka, J., Gerard, J., Lowman, P., Schmadebeck, R., Blodget, H., Eller, E., Yin, L., Lamothe, R., Gorenstein, P., and Bjorkholm, P.: 1972a, *Science* **175**, 436.

Adler, I., Trombka, J., Gerard, J., Lowman, P., Schmadebeck, R., Blodget, H., Eller, E., Yin, L., Lamothe, R., Osswald, G., Gorenstein, P., Bjorkholm, P., Gursky, H., and Harris, B.: 1972b, *Science* **177**, 256.

Boynton, W. V., Trombka, J. I., Feldman, W. C., Arnold, J. R., Englert, P. A. J., Metzger, A. E., Reedy, R. C., Squyres, S. W., Wänke, H., Bailey, S. H., Brückner, J., Callas, J. L., Drake, D. M., Duke, P., Evans, L. G., Haines, E. L., McCloskey, GF. C., Mills, H., Shinohra, C., and Starr, R.: 1992, *J. Geophys. Res.* **97**, 7681.

Evans, L. G., Starr, R., Trombka, J. I., Brückner, I., Bailey, S. H., Goldsten, J. O., and McNutt, R. L., Jr.: 1996, 'Performance of the Gamma-Ray Detector System for the NEAR Mission and Application to Future Missions',*Preprints of the Second IAA International Conference on Low Cost Planetary Missions, IAA-L-0905P*, Laurel, Maryland, U.S.A., April 16–19.

Evans, L. G. and Squyres, S. W.: 1987, *J. Geophys. Res.* **92**, 9153.

Farquhar, R. W., Dunham, D. W., and McAdams, J. V.: 1995, *J. Astron. Sci.* **43**, 353.

Gaffey, M. J., Bell, J. F., and Cruikshank, D. P.: 1989, 'Reflectance Spectroscopy and Asteroid Surface Mineralogy', in *Asteroids II*, University of Arizona Press, Tucson, pp. 98–127.

Gaffey, M. J., Burbine, T. H., and Binzel, R. P.: 1993, *Meteoritics* **28**, 161.

Goulding, F. S. and Landis, D. A.: 1982, *IEEE Trans. Nucl. Sci.* **29**, 1125.

Hawkins III, S. E., Darlington, E. H., Murchie, S. L., Peacock, K., Harris, T. J., Hersman, C. B., Elko, M. J., Prendergast, D. T., Ballard, R. W., Gold, R. E., Veverka, J., and Robinson, M. S.: 'Multi-Spectral Imager on the Near Earth Asteroid Rendezvous Mission', *Space Sci. Rev.*, this volume.

Hayes, J. R.: 1995, 'NEAR Instrument DPU Common Software Requirements Specification', JHU/APL, pp. 7358–9001.

Mandel'shtam, S., Tindo, I. P., Cheremukhin, G. S., Sorokin, L. S., and Dmitriev, A. B.: 1968, *Cosmic Res.* **6**, 100.

McFadden, L. A., Gaffey, M. J., and McCord, T. B.: 1984, *Icarus* **59**, 25.

Mosher, C. H.: 1976, *IEEE Trans. Nucl. Sci.* **23**, 226.

Pieters, C.: 1976, *Icarus* **28**, 105.

Santo, A. G., Lee, S. C., and Gold, R. E.: 1995, *J. Astron. Sci.* **43**, 373.

Schneider, S. E.: 1995, *X-ray/Gamma-ray Spectrometer DPU Software Requirements Specification*, JHU/APL , pp. 7358–9002.

Trombka, J. I., Boynton, W. V., Brückner, J., Squyres, S. W., Clark, P. E., Starr, R., Evans, L. G., Floyd, S. R., Fiore, E., Gold, R. E., Goldsten, J. O., McNutt, R. L., Jr., and Bailey, S. H.: 1996, 'Compositional Mapping with the NEAR X-ray/Gamma-ray Spectrometer', *J. Geophys. Res.*, submitted.

Warren, J. W., Peacock, K., Darlington, E. H., Murchie, S. L., Oden, S. F., Hayes, J. R., Bell III, J. F., and Krein, S. J.: 1997, 'Near Infrared Spectrometer for the Near Earth Asteroid Rendezvous Mission', *Space Sci. Rev.*, this volume.

Wetherill, G. W.: 1985, *Meteoritics* **20**, 1.

Wetherill, G. W. and Chapman, C. R.: 1988, in J. F. Kerridge and M. S. Matthews (eds.), 'Asteroids and Meteorites', *Meteorites and the Early Solar System*, University of Arizona Press, Tucson, pp. 35–67.

THE NEAR-EARTH ASTEROID RENDEZVOUS LASER ALTIMETER

T. D. COLE, M. T. BOIES, A. S. EL-DINARY and A. CHENG
Space Department, The Johns Hopkins University Applied Physics Laboratory Laurel, MD
20723–6099, U.S.A.

M. T. ZUBER
Department of Earth, Atmospheric and Planetary Sciences, 54-518 Massachusetts Institute of
Technology, Cambridge, MA 02139, U.S.A.

D. E. SMITH
Laboratory for Terrestrial Physics, NASA/Goddard Space Flight Center, Greenbelt, MD, U.S.A.

(Received 9 September, 1996)

Abstract. In 1999 after a 3-year transit, the Near-Earth Asteroid Rendezvous (NEAR) spacecraft will enter a low-altitude orbit around the asteroid, 433 Eros. Onboard the spacecraft, five facility instruments will operate continuously during the planned one-year orbit at Eros. One of these instruments, the NEAR Laser Rangefinder (NLR), will provide sufficiently high resolution and accurate topographical profiles that when combined with gravity estimates will result with quantitative insight into the internal structure, rotational dynamics, and evolution of Eros. Developed at the Applied Physics Laboratory (APL), the NLR instrument is a direct-detection laser radar using a bistatic arrangement. The transmitter is a gallium arsenide (GaAs) diode-pumped Cr:Nd:YAG (1.064-μm) laser and the separate receiver uses an extended infrared performance avalanche-photodiode (APD) detector with 7.62-cm clear aperture Dall–Kirkham telescope. The lithium-niobate (LiNbO3) Q-switched transmitter emits 15-ns pulses at 15.3 mJ pulse^{-1}, permitting reliable NLR operation beyond the required 50-km altitude. With orbital velocity of 5 m s^{-1} and a sampling rate of 1 Hz, the NLR spot size provides high spatial sampling of Eros along the orbital direction. Cross-track sampling, determined by the specific orbital geometry with Eros, defines the resolution of the global topographic model; this spacing is expected to be <500 m on the asteroid's surface. Combining the various sources of range errors results with an overall range accuracy of 6 m with respect to Eros' center-of-mass. The NLR instrument design, perfomance, and validation testing is decribed. In addition, data derived from the NLR are discussed. Using altimetry data from the NLR, we expect to estimate the volume of 433 Eros to 0.01% and its mass to 0.0001% accuracies; significantly greater accuracies than ever possible before NEAR.

1. Introduction

On 17 February 1996, the Delta-II rocket launch of the Near-Earth Asteroid Rendezvous (NEAR) spacecraft began NASA's first Discovery mission. The objective for the NEAR mission is to answer questions relating to the nature and origin of near-Earth asteroids (NEAs). NEAs are of interest from two perspectives. The study of asteroids and meteorites is critical to understanding the formation and early evolution of the solar system. Asteroids, in particular, may contain clues to the nature of the building blocks from which the inner planets, including the Earth, were formed. Certainly interest in NEAs has been further heightened by the realization of their potential for catastrophic terrestrial impacts. NEAs are a significant source of large bodies that have repeatedly collided with the Earth and are

© 1997 *Kluwer Academic Publishers. Printed in Belgium.*

considered highly responsible for the cataclysmic collision ending the Cretaceous age 65 million years ago.

Knowledge about asteroids has been derived from Earth-based observations, two *Galileo* spacecraft flybys, selected data from the asteroid-like Martian moons, analyses of relationships to meteorites, and theoretical modeling of asteroidal dynamics, structure, and thermal evolution. For asteroids with diameters less than 100 km, shape is believed to be controlled by collisions and impacts, with only a minor contribution due to self-gravitation. The modest size of even the largest NEA dictates that details of an asteroid's surface structure cannot be unambiguously resolved from Earth. Consequently, a variety of novel techniques have been developed to infer their sizes and gross shapes. Sizes and shapes of asteroids contain important information concerning thermal, collision and dynamic histories of these bodies and their internal structures. Detailed numerical representations of the shapes of the best-studied asteroids employ limb, terminator and photogrammetric measurements derived from astronomy and radar.

The asteroid to be investigated by the NEAR mission is 433 Eros, one of the largest and most observed NEA since its discovery in 1898. In the particular case of Eros, size and shape have been estimated from measurement of angular dimensions, stellar occultations, color and polarization observations, radar measurement, and thermal radiometry. All of these studies yielded average radii at photometric maxima in the 10 to 20-km range. The rotational period and albedo for Eros are 5.27 h and 0.15, respectively (Binzel et al., 1989). None of these observations, however, rule out the possibility that this asteroid may actually be a 'rubble pile' consisting of two or more small, gravitationally bound bodies.

Figure 1 illustrates the NEAR spacecraft orbiting 433 Eros. The spacecraft's non-articulating solar arrays must point towards the Sun while the parabolic dish is kept within $1°$ of Earth. For the NEAR spacecraft, this limits the Earth–spacecraft–Sun angle, θ, to $\leq 30°$. Throughout an orbit, the spacecraft is rolled using momentum wheel action to maintain nadir-pointing orientation, the direction all NEAR instruments view. This restriction on spacecraft orientation precludes large excursions in scanning the instruments across Eros' surface; therefore, global coverage of the asteroid by the NEAR instruments depends upon spacecraft orbit and Eros' self-rotation beneath the spacecraft.

2. Mission Objectives for the Near-Laser Rangefinder

The Near-Earth Asteroid Rendezvous (NEAR) Mission provides an opportunity to dramatically improve our understanding of a near-Earth asteroid that has been well studied by Earth-based observations. In particular, the objective of the NEAR-laser ranging investigation is to obtain sufficiently high resolution and accurate profiles, that when combined with gravity estimates, provide quantitative insight into the internal structure, rotational dynamics, and the evolution of NEAs. Using altimetry

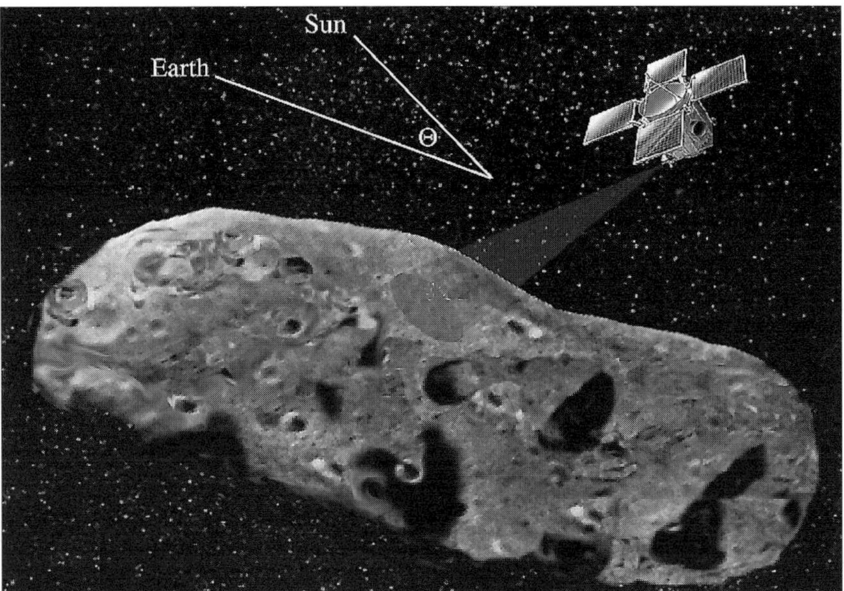

Figure 1. Geometry of the NEAR spacecraft as it orbits 433 Eros. Optical instruments, including the laser radar, are continuously pointed towards the asteroid's surface.

data and orbital tracking data, we expect to resolve the volume of 433 Eros to 0.01% and its mass to 0.0001% (Zuber et al., 1997). Comparison between the NLR dataset with the predetermined gravity field (estimated through Eros' shape) permits correlation of surface topography to the local gravity field. Although the measurement of 433 Eros' mean density will be limited by the accuracy of the topographic field, volume and mass will be estimated with significantly greater accuracy than that for any other asteroid.

To achieve this desired resolution in volume and mass, single-shot range resolution of the NLR must not exceed 6 m and range data accuracy must be $\sim$ 6 m with respect to the asteroid's center-of-mass (COM). The along-track spatial resolution of the instrument will vary from 4 to 11 m depending upon spacecraft altitude above Eros. Cross-track spacing, determined by the final orbit parameters and asteroid rotation, is expected to be $\sim$ 500 m.

Figure 2 summarizes the geometry associated with the altimeter's footprint. All NLR measurements will be in an absolute COM reference frame allowing precise registration with data from other NEAR sensors. Determination of the gravity and topography fields will reveal any appreciable offset between the center-of-mass (COM) and center-of-figure (COF) for Eros. This COM/COF relationship will provide information as to whether internal density differences are distributed uniformly and will help verify if Eros consists of two or more gravitationally bound bodies.

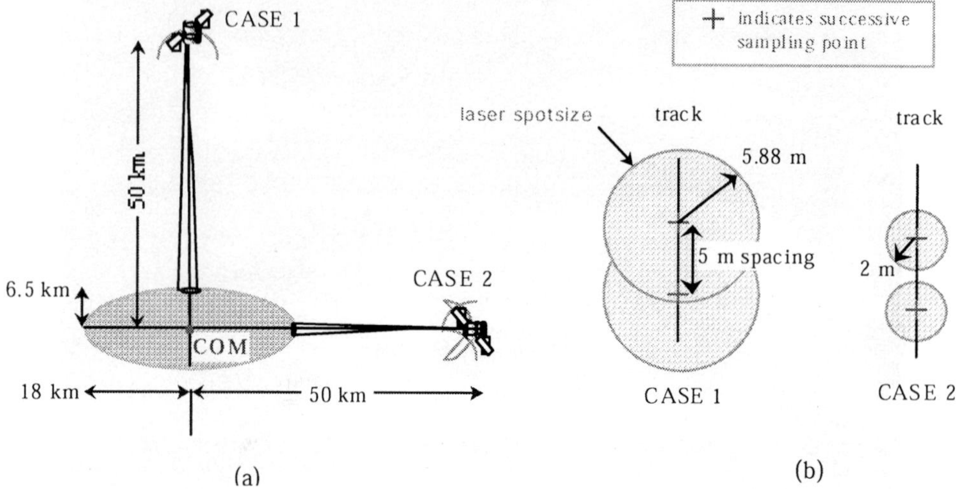

Figure 2. Mission geometry of the NEAR spacecraft during encounter with the near-Earth asteroid, 433 Eros assuming a ground-track velocity of 5 m s^{-1} (actual velocity of NEAR relative to Eros may range from 3 to 10 m s^{-1}). Two cases of orbiting Eros are indicated demonstrating different perspectives: (a) orbit position relative to Eros' major axis, and (b) ground-track sampling. Case 1, orbiting about the minor axis, shows significantly overlapped sampling, whereas Case 2, orbiting about the major axis, has near-contiguous sampling. Estimates of 433 Eros dimensions are from Earth-based observations (Binzel et al., 1989).

2.1. THE NLR MISSION SPECIFICATIONS

Table I summarizes mission specifications for the NLR instrument. These specifications were derived from altimetry data objective, spacecraft constraints, and estimates concerning Eros. The tabular values, determined through direct measurement or inferred from related measurements, are presented for the integrated instrument. Where values are not available, 'N/A' is indicated.

2.2. NLR DESIGN IMPLEMENTATION

The NLR is a bistatic configuration of a diode-pumped Cr:Nd:YAG transmitter and a 7.62-cm aperture Dall–Kirkham Cassegrain receiver. Direct-detection was selected using time-of-flight (TOF) measurements between transmission of 15-ns laser pulses and detection of Eros' surface backscatter (Cole et al., 1996). No waveform processing is performed, instead, the receiver uses leading-edge detection based on Neyman–Pearson thresholding. Figure 3 presents the integrated NLR instrument with additional details on each of the major subsystems.

Our goal in designing the NLR altimeter was to maintain simplicity, motivating our choice for direct detection of the return pulse s leading edge. Leading-edge altimeters are relatively simple to implement and their design reduces range scintillation resulting from target range extent exceeding transmitter pulse width, a

Table I

Altimetry mission requirements imposed on the NLR instrument

Parameter	Specification	Measurement[a]/estimate[b]
Max. range (altitude)	50 km	>160 km[b]
Range accuracy	$\leq$ 6 m	<6 m[b]
Range resolution	$\leq$ 1 m	31.22 cm[a]
Instrument weight (max.)	5 kg	4.9 kg[a]
Instrument power (avg.)	<22 W	15.1 W[a]
Sample (grid) spacing	contiguous	contiguous (overlapped)[a]
Data rates	<56 bps	variable; 6.4 to 51 bps[a]
Eros' albedo, Pb	0.10 to 0.22	0.15[b]
Operational period	1 yr (continuous),	N/A,
Lifetime	4 yr	N/A

possible condition when ranging to irregular surfaces as expected with an asteroid. The selection of a bistatic configuration permitted parallel development of the transmitter (TX) and receiver (RX). The transmitter (TX) was designed and fabricated by McDonnell-Douglas (MDC) of St. Louis, and the RX was designed and fabricated by APL. The RX design was based on previous APL efforts that used APD-based detection circuits and APL developments in TOF implementation using GaAs applications-specific integrated circuitry (ASIC).

2.3. SYSTEM DESIGN

A block diagram of the NLR instrument is presented in Figure 4. Eight subsystems comprise the NLR instrument: laser resonator assembly (LRA), laser power supply (LPS), fiber-optic delay assembly (FODA), optical receiver, analog signal section (detector and processor boards), digital processing unit (DPU), low voltage power supply (LVPS), and medium voltage power supply (MVPS). The LRA, LPS, and FODA together form the NLR TX subsystem. The optical receiver, analog section, DPU, LVPS, and MVPS are collectively referred to as the NLR RX subsystem. The TX and RX subsystems were integrated at APL to form the NLR instrument and was subsequently tested and flight-qualified at the instrument and spacecraft levels. In the figure, 'Htr' represents heater power and AD590 indicates feedback control used to maintain survival and operable temperatures. The TX contains optical elements sensitive to thermal gradients and therefore must be controlled within ± 2 °C of its operational temperature, which is 20 °C. The analog signal section has two boards, a detector board containing an avalanche photodiode (APD) detector and an analog processing board that contains a video amplifier (gain $= K_0$), a matched filter (Bessel-type), and a threshold comparator (C). The DPU uses a 480 MHz oscillator, an APL-designed TOF application-specific integrated circuit (ASIC), an Actel 1280 field-programmable gate array (FPGA), an RTX2010 microcontroller,

Figure 3. The integrated NLR instrument showing key subsystems. The detector assembly (*upper-left*) is an enhanced avalanche photodiode (APD) hybrid mounted at the receiver focal plane (APD detector is not visible; it is located on the side opposite that shown). The laser resonator and associated laser power supply (*lower-left*) are based on previous designs. Receiver optics (*upper-right*) is an all-reflective, lightweight Dall–Kirkham made of aluminum (athermal design). Receiver electronics (*lower right*) is shown mounted in the NLR chassis. Both the digital processing unit (lower board) and the low-voltage power supply are shown. The calibration spool, which is a 109.5-m fiber optic spool, is shown on the transmitter in the center photo. Not shown is the APD power supply that was remotely located from the NLR receiver to reduce electrically induced interference.

an analog-to-digital converter (ADC), and a 1553-bus chip set. Memory units installed on this single digital board include: 2 K × 16 PROM, 32 K × 16 EEPROM, and 32 K × 16 SRAM. The DPU proceses time-of-flight measurements through digitalization of altimetry data and formatting of this data in real-time.

NLR operation can be traced through use of Figure 4. A FIRE command is issued by the DPU at the selected pulse repetition frequency, PRF. The FIRE command enables the TX, fires the diode-pump, and produces a laser optical pulse $\sim$192 μs subsequent to TX enable. As this optical pulse is emitted from the TX, a portion is simultaneously directed into the FODA for calibration. There are eight selectable threshold levels for RX operation, ranging from 0 (corresponding to a threshold voltage of 16 mV) through 7 (2.028 V). With the START signal from the TX, two TOF counters (range and calibration) are started. This START signal is produced by the optical detection of the optical pulse as it leaves the TX. The START signal is immediately converted into an electrical signal by a photodiode (PIN) within the TX prior to its transfer to RX electronics. During ranging operation, the receiver sees two optical pulses per transmitted pulse. The first, arriving approximately 558

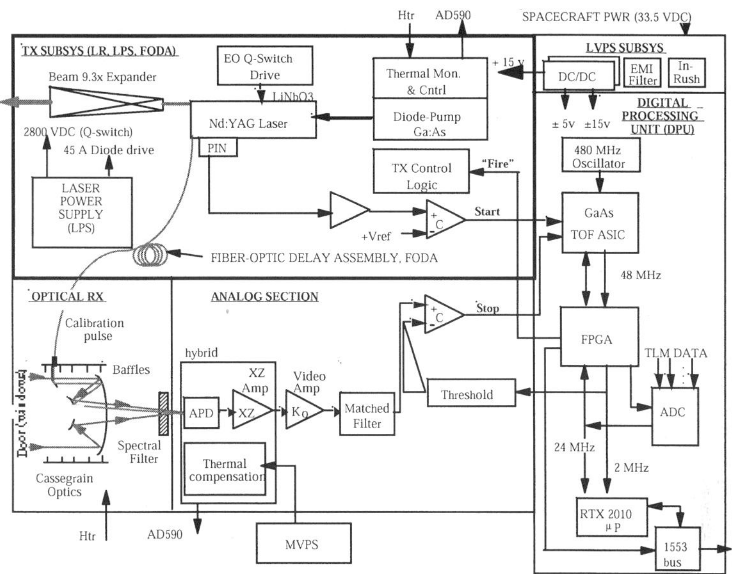

Figure 4. Block diagram of the NLR instrument.

ns after laser firing, is the calibration pulse routed through the FODA. Detection of this pulse by the receiver halts the calibration counter. The second return results from backscatter off the asteroid, which halts the range counter. All optically produced signals are compared to a threshold level to eliminate noise triggering and to improve rejection of false range values. Threshold levels are set by command from the ground or through the NLR auto-acquisition sequence. The level for the calibration signal is set through the use of neutral density filters, using prelaunch tests, to trigger the receiver detection process at the highest receiver threshold level. After both TOF counters stop, the DPU reads the TOF counter values and formats them into science data packets for transmission over the 1553 bus as requested by the spacecraft data collection process.

2.4. TRANSMITTER DESIGN

The LRA is based on a proven polarization-coupled U-cavity design (Culpepper et al., 1995). Figure 5 illustrates the laser resonator design for the NLR. An internal aperture reduces higher-order modes to provide well-behaved, far-field characteristics, and an external beam expander (9.3× Galilean telescope) reduces angular divergence. The resonator uses a 20-bar GaAs diode array to optically side-pump a Cr:Nd:YAG zigzag slab with the opposite side used for heat removal. An antireflection coating is applied to the long dimension of the slab to efficiently couple the optical pump. A high reflectance coating is applied to the side attached to the heat sink to reflect diode-pump energy into the YAG slab. Through this zigzag pumping, optical energy is uniformly distributed throughout the YAG medium, thereby

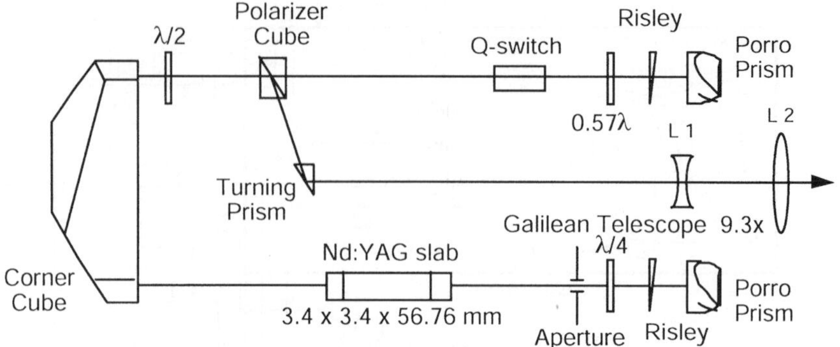

Figure 5. NLR laser resonator cavity design.

increasing absorption path length. At both slab faces, an antireflection coating is applied to reduce optical losses for rotated polarization states. In the slab leg, a quarter waveplate is used to maintain polarization. The slab is located in a cross-porro cavity that provides boresight stability. Each porro prism assembly includes a Risley wedge to allow for beam alignment of the standing-wave cavity during assembly and to remove beam offsets produced by the prisms.

The laser is operated in Q-switch mode, which alters cavity loss by changing the polarity state of return light incident on the polarizing cube. Polarization compensation is accomplished in the Q-switch leg by a half waveplate. The Q-switch (Pockel cell) material is lithium niobate, $LiNbO3$. Unfortunately, $LiNbO3$ is pyro-electric which requires avoidance of thermal gradients while operating in vacuum. Placement of the Q-switch in the output arm places the $LiNbO3$ element in the low circulating power portion of the resonator thus minimizing potential laser damage; but the concern for thermal sensitivity of this element is resolved only through stringent operation and survival control of the element as discussed.

The laser resonator was built on an aluminum structure which is thermally controlled using redundant heaters and temperature monitors (AD590). Figure 6 is a photograph of the LRA optical components; note the lightweight structure used to mount the elements for the Galilean beam expander. The laser YAG slab is mounted onto a separate aluminum optical bench to preserve mechanical and thermal stability. Thermal gradients are further minimized in the slab through use of the zigzag pump arrangement. Finally, to thermally isolate the LRA, it is mounted to the NLR housing using Vespel shoulder washers. This permits the LRA to operate separately from the thermal environments associated with spacecraft interface for the NLR receiver subassembly.

We measured typical pulses having pulse widths of 15 ns (FWHM) as presented by the data captured in Figure 7. Although the far-field beam pattern approached single-mode characteristics, it still contained multiple longitudinal modes as exhib-

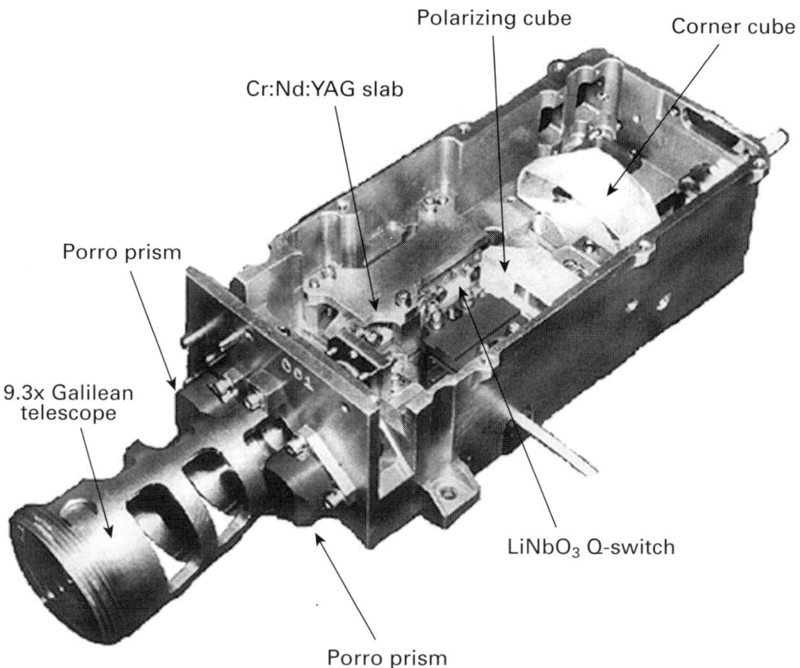

Figure 6. LRA optical component configuration.

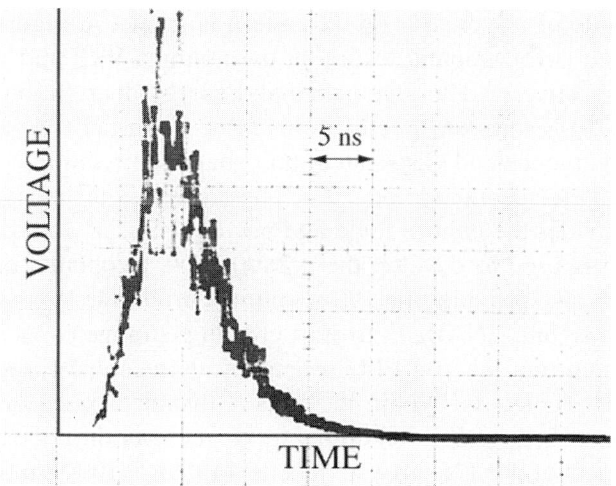

Figure 7. Temporal response of LRA optical pulse.

ited by this waveform. Figure 7 is a single NLR laser pulse waveform measured using a fast (picosecond response) detector; note the longitudinal mode structure.

The temporal, spatial (near- and far-field), and pulse energy characteristics of the laser were measured at various stages of system development. Pulse energy

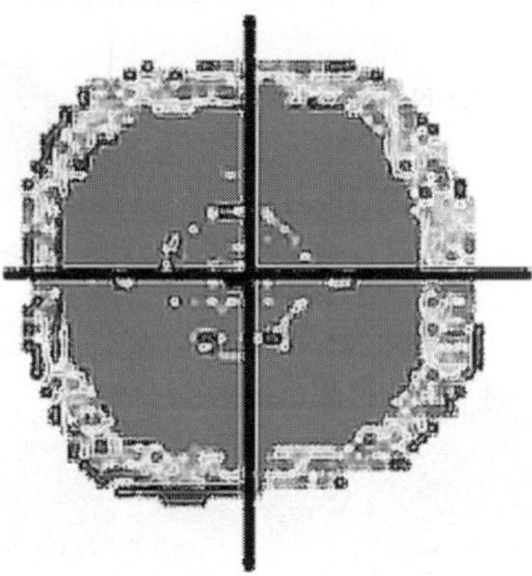

Figure 8. Ambient (atmospheric pressure) far-field pattern of a NLR output laser pulse (135 ± 2 μrad at $1/e^2$ points) with the NLR operating at 1 Hz.

measurements of the transmitter beam and FODA port output were measured using NIST-traceable equipment accurate to $\pm 5\%$. Transmitter optical energy of 15.3 mJ was measured consistently throughout testing whereas FODA pulse energy of 7.2 pJ was measured at the LRA output coupler. Characterization of the near-field included measurement of beam diameter ($1/e^2$), modal structure, and energy distribution (top-hat). Far-field testing consisted of beam divergence, jitter, and wander. A sample of a typical far-field pattern is shown in Figure 8. (The beam pattern presented is for ambient testing at the nominal PRF and voltage of 1 Hz and 33.5 V, respectively.) The divergence and energy distribution of the TX met instrument and NLR mission specifications of $\leq$ 300 μrad (235 μrad measured) output beam divergence and Gaussian output shape (correlation coefficients, both X- and Y-axes, were specified to be >0.9. These coefficients measured 0.91).

The LPS provides the control logic and power necessary to enable and fire the LRA. A hybrid is used to develop the +2800 VDC to operate the resonator Q-switch. The LPS also provides the 190-μs pump current (45 A) to the diode array. The LPS uses a signal, ENABLE, to start charging storage capacitors to produce the diode drive current when a FIRE command is received from the DPU. The switching DC/DC converter within the LPS is disabled based on the maximum ranging duration to reduce noise coupling to the receiver during expected returns.

The final aspect of the TX subsystem is the FODA, a 109.5-m fiber-optic delay line that serves as an in-line system timing and, to a lesser extent, output TX power calibration capability. Calibration is accomplished by optically sampling each transmitted pulse at the laser output and injecting it into the single-mode fiber-optic. This fiber-optic is connected to the receiver telescope via a 45° elliptical mirror. The fiber spool, made of fused silica, provides a constant optical delay of 529.2 ns (measured). (Once the fiber-optic was integrated into the NLR system, overall delay was measured resulting with 558.33 ns, indicative of a 29.13-ns systematic delay.) Having such a constant 'range target', we perform end-to-end calibration for each emitted pulse by detecting range-walk due to threshold-level changes or from oscillator drift.

The NLR uses a leading-edge detection scheme wherein the receiver threshold determines if a valid return detection is present (i.e., an altimetry measurement). As this level is altered, the timing position where the leading edge crosses the threshold also changes which corresponds to earlier, or later, range values, an error condition known as 'range-walk'. Operating at a fixed threshold, our nominal operating mode, the long-term stability of the NLR oscillator can be estimated by assuming oscillator drift during any one calibration measurement (~ 558.33 ns) is insignificant. Comparing successive calibration intervals, oscillator drift can be extracted, through judicious filtering of calibration data, and subsequently removed from NLR altimetry data.

To implement this calibration capability, the APL-designed GaAs TOF chip implemented two range counters: one for calibration (11-bits) and one for the range measurements (21-bits, with 1-bit overflow). Explicit range gates were not implemented in the NLR, instead, counter overflow conditions provided the NLR with an effective range gate of 2.181 ms for the return pulse (equivalent to one-way range of 327 km). (The calibration counter does not implement an overflow indication, instead, it continues to 'wrap-around' as long as the main counter continues to increment – once the range counter stops, the contents of the calibration counter is a modulo of 11-bit 'range' values, or 4.267 μs.) Observing both counters during testing allowed us to determine quality of the system based on threshold level, power levels, operational mode, or environmental conditions. This feature proved invaluable for debugging throughout integration and test and has also allowed us to evaluate NLR functionality during the cruise phase.

2.5. RECEIVER OPTICAL SUBSYSTEM

A significant aspect of the NLR configuration is the fact that the receiver optics act as a direct-detection 'photon bucket' – this drastically simplified the design and development of the optical receiver (Boies et al., 1996). Since imaging is not required, the optics could tolerate optical aberrations provided the resultant spot size fit within the physical and alignment bounds of the APD detector (Burns, 1994). This attribute provided us the freedom to select the receiver optical design best suited for low weight (our telescope weighs 167.4 g) and manufacturing ease. An all-reflective approach from the Cassegrain family was indicated; specifically, a Dall–Kirkham design because of its simple manufacturing process, acceptable paraxial aberrations, and compact geometry. Figure 9 is an illustration of the RX telescope design. The NLR aberration allowance was sufficient to allow for diamond-turned mirror surfaces, a very economical manufacturing approach. Considering our needs, we did not pursue additional polishing of the surfaces as acceptance testing of this telescope indicated that 98% of the focused energy is

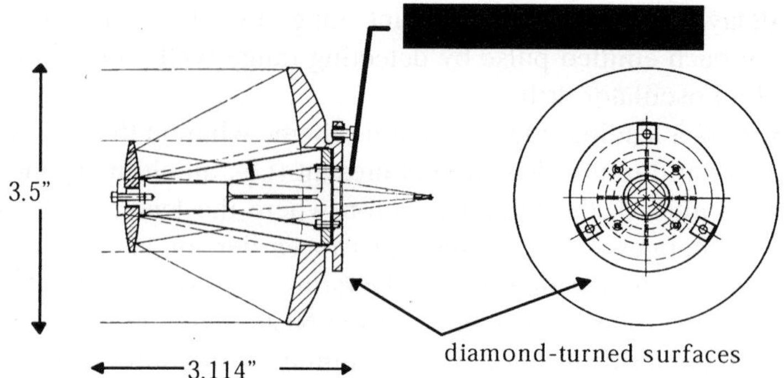

Figure 9. Dall–Kirkham design (scaled) for NLR $f/3.4$ receiver telescope.

located within a 100-μm Airy disk. This 100-μm spot size is easily accommodated on the 700-μm active diameter of the APD detector.

To reduce optical background noise, we inserted a 7-nm FWHM spectral filter in the convergence cone of the RX telescope, a location selected to reduce weight and cost. This matched our laser line stability, ± 1 nm, and allowed operation over our operating angles, 4 to 9°. Passband for spectral filters are quite sensitive to their thermal environment, however, thermal influence on our filter's passband attenuation were minimized by selecting a deposition process (Swenson, 1994) that reduces passband shift by a factor of 10.

To minimize contamination, the NLR receiver uses a one-time deployable door over its entrance aperture. Our major concern was with contamination from propellant burns during the NEAR 3-year cruise phase. In addition, we were interested in reducing contamination and providing mechanical protection throughout the prelaunch period. In addition to the door, both the TX and RX subsystems were under a positive purge using research-grade nitrogen throughout development, test, and prelaunch; only hours before launch was the purge line removed from the instrument.

The door-release mechanism uses redundant pyrotechnic wire cutters with a tempered beryllium-oxide (BeO) wire. In the rare event of a door-release failure, the door contains six 1-inch diameter windows (refer to Figure 3) to provide 50% of the collection area achieved if the door failed to open; this reduced collection area still allows successful NLR operation at our operational range of 50 km. To maintain sufficient purge flow, the RX door contains a one-way, filtered purge valve. The door was successfully opened during flight towards Eros on September 24, 1997.

Telescope baffling of the NLR receiver is required to reject stray sunlight from focusing onto the detector. During operation, the Sun angle may vary and come within 30° of the optical axis of the receiver. Placement of that solar energy on the spectral filter or detector surface would severely degrade system performance (e.g., thermal heating, APD saturation and/or surface ablation). In addition, solar input

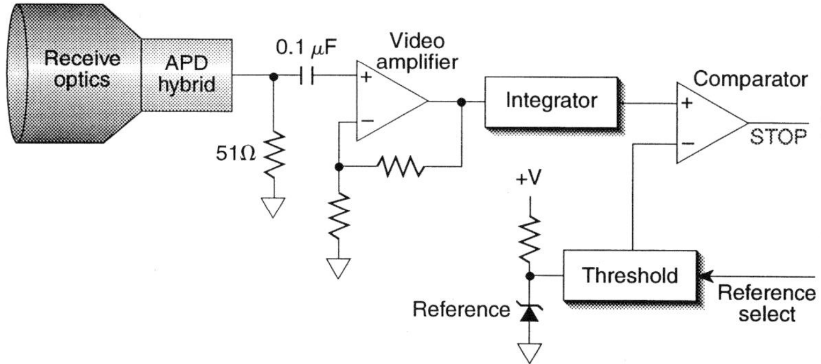

Figure 10. Analog signal section block diagram.

greatly affects thermal balance and may cause the instrument to thermally expand and exceed design (focus) conditions. To minimize the solar background, a baffle assembly was designed using the method of shadowing. This method ensures that no ray entering the telescope at an off-axis angle could reach the primary mirror in less than two reflections. The advantage of the method is that it reduces the number of baffles and hence the weight. To further reduce scattering, all baffles were knife-edged to minimize direct reflections into the telescope.

2.6. RECEIVER ANALOG SIGNAL SECTION

The purpose of the analog signal section is to convert laser backscatter from the asteroid into a digital 'stop' pulse to compute round-trip time measurement (range). The NLR analog signal section, depicted in Figure 10, is implemented in four fundamental stages: (1) APD detector, (2) video amplifier, (3) integrator stage (Bessel-type filter), and (4) comparator. The APD used in the analog receiver is a hybrid that combines an enhanced silicon APD (with active thermal compensation circuitry) with a transimpedance amplifier. Our particular hybrid is a wide-FOV detector having a measured bandwidth response of 50 MHz and responsivity of 770 kV W^{-1}. Although this particular hybrid provides <5% gain variation over a -8 to $+40\,°$C temperature range, we thermally controlled the RX housing to $+20\pm$ 10 °C. Optical energy inputs (peak power) expected during measurement operations will range from our minimally detectable 9 nW, to a maximum (saturation) of 0.5 W. This dynamic range of power levels assumes a 10-ns pulsewidth at 1.064 μm.

The video amplifier stage consists of a wideband op-amp to provide a voltage gain of $+50$ at 75-MHz bandwidth. The integrator stage is a seven-pole lowpass Bessel filter with 3-dB cutoff of 30 MHz; it is used to integrate return pulses that are significantly time dilated in response to target-surface topology. The integrator optimizes the probability of detection for anticipated surface slopes through such temporal integration while also eliminating high-frequency noise through the analog signal section.

The comparator stage determines whether a backscattered signal generates a STOP signal; which is generated if the return signal exceeds the threshold level set by command or auto-thresholding (these eight threshold levels were described in the System Design section). Threshold values were determined by setting the lowest threshold (TH $= 0$) below the measured receiver noise floor and the highest setting, TH $= 7$, just below the signal strength associated with the calibration input signal so that calibration signals are never suppressed. Threshold levels increment as $2^n V_{th}$, where V_{th} is the minimal threshold voltage (16 mV), resulting in a 24-dB dynamic range of signal input levels ($n = 0, \ldots, 7$). The comparator device exhibits very low propagation delay (<2 ns) and has very little overdrive dispersions making it ideal for precise timing applications (Reiter, 1993).

2.7. DPU

To save weight and minimize the number of high-speed lines to the TOF chip, the DPU uses a single rigid-flex multilayered board (Rodriguez, 1994). Functionality of the DPU board is presented in Figure 11. The board contains the GaAs TOF ASIC, radiation-hardened FORTH-language (RTX-2010) microcontroller, memory (RAM, PROM, EEPROM), A/D converter and multiplexer, redundant channel 1553-bus chip-set and transformers, ECL/HCMOS dual-output 480 MHz oscillator, CMOS FPGA chip, and associated control logic and compensation components. The RTX-2010 is a parallel 16-bit microcontroller (μC) operating at 1 MHz to conserve power. The microcontroller chip contains three on-chip timers, a dual-stack architecture, and an interrupt controller with the ability to handle five external interrupts.

The 1553-bus interface consists of two redundant channels including two bus transformers, a 1553 dual bus transceiver, and a dual-bus protocol controller. The controller responds to bus commands sent by the spacecraft command telemetry processor (CTP) and handles data transfers, commands, and telemetry, that is sent to the NLR and the CTP. Data bus arbitration is handled by the Actel 1280 FPGA (refer to Figure 4), which simplified digital hardware design and fabrication by incorporating several functions: address decoding, RAM arbitration, clock generation, receiver range-gating, laser transmitter firing (T-0) masking, and transmitter and receiver configuration control. T-0 noise is that interference that occurs simultaneously with the formation of the laser optical pulse responsible for the starting of the range counters.

The oscillator produces two output frequencies: 480 MHz $\pm0.01\%$ (ECL) and 48 MHz $\pm0.01\%$ (HCMOS), with short-term (100-μs) frequency stability of 10^6 ppm. The FPGA uses the 48 MHz clock and produces 24-MHz, 2-MHz, and 500-kHz output clocks for the 1553-bus controller, microcontroller, and FPGA-internal mask-counters, respectively.

One RX control line (STOP) is used to indicate a detected optical pulse above the set threshold. This stop pulse terminates one of the TOF counters (calibration

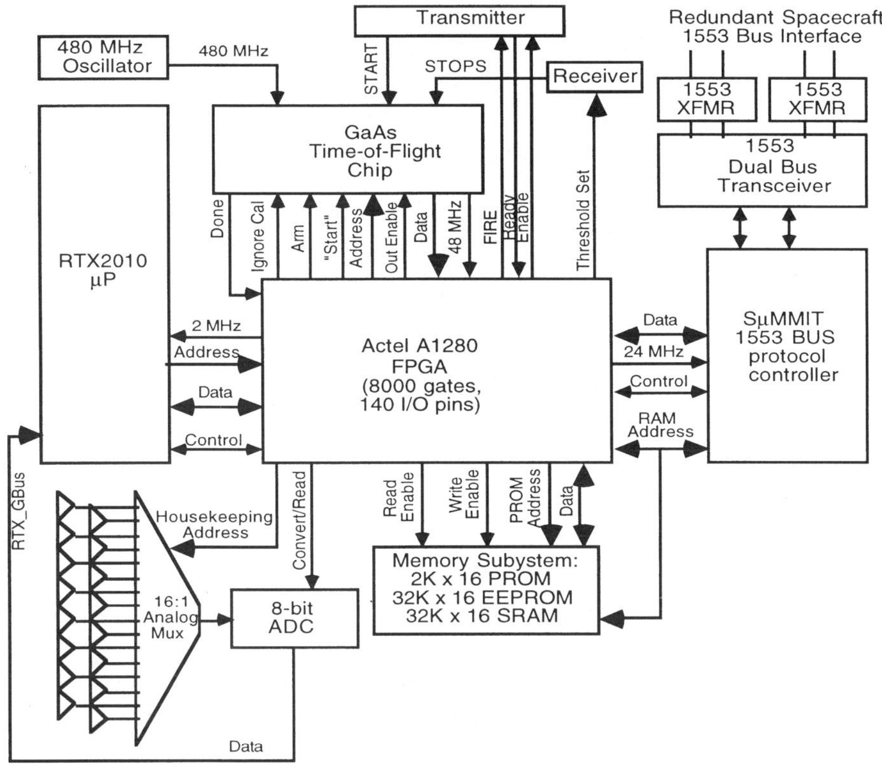

Figure 11. DPU single-board implementation (Penn, 1994).

or range), depending on system configuration. The second control line (GATE) is used to disable the RX comparator during TX firing. During TX firings, the analog section is momentarily disabled via a programmable range gate to prevent firing noise from the switching power supply (LPS) or from the laser pulse generation itself (T-0) from prematurely triggering the receiver and halting the TOF counters. This range gate is a 10-bit programmable timer implemented in the FPGA with a timing resolution of 41.67 ns and a maximum delay time of 42.628 μs.

Figure 12 is a photograph of the actual flight DPU board mounted in the NLR chassis alongside the LVPS. The DPU is electrically isolated from the LVPS by design of a tongue-and-groove structure as part of its Mg chassis and through use of EMI filters at the intervening bulkhead.

The NLR software was programmed in JH-FORTH for the RTX-2010 micro-controller (μC). This software was required to interface with the spacecraft CTP via the 1553-bus protocol, execute commands from the ground control, format science telemetry and instrument housekeeping, and control operation of the instrument. The software was implemented in a multitasking environment running four differ-ent software processes (Moore, 1994; El-Dinary et al., 1996): NLR_PROCESS to

Figure 12. Flight DPU configuration showing LVPS and flight chassis and connectors.

handle TX/RX control and data formatting, TELEMETRY_PROCESS to handle 1553 protocol for data transfers, COMMAND_PROCESS to handle incoming instrument commands, and DUMP_PROCESS to allow dumping blocks of memory through the telemetry stream.

The NLR_PROCESS is the main controlling task for the instrument. This task has the highest priority and requires 5 ms to execute. Once every minor frame interrupt (spaced at intervals of 125 ms), the algorithm determines when to fire the transmitter, initialize parameters (T-0 timing gate mask, threshold, and range gate), and read and format calibration and range. Range data, spacecraft time, and instrument configuration parameters are formatted in a science packet for each transmitter shot, and a science packet accumulates data from 56 shots (112 shots for PRF = 2 Hz) for transmission via the downlink. Transmitter firing rates are determined through uplink commands, and timing is controlled by NLR software to insure transmitted pulses are coincident with 1553-bus minor frames. An internal microcontroller timer and interrupt maintain minor frame counting with periodic updates from the 1553 'Sync-With-Data' pattern, which occurs at 1 puse sec^{-1}. For selected PRF, NLR transmitter firings occur in the following minor frames:

PRF (Hz)	Spacecraft minor frame(s)
$\frac{1}{8}$	4
1	4
2	0, 4
8	0–7

The NLR can be commanded into the 8-Hz mode for a 2-s burst followed by 14-s quiescent period to reduce thermal stresses within the TX. Following the 14-s pause, the NLR automatically resets itself to the nominal 1-Hz rate. At all other PRF rates, the NLR can operate indefinitely.

The NRL_PROCESS is programmed with several calibration algorithms. In the FAILSAFE mode, the TOF counter is started by a delayed FIRE command as opposed to an electronic trigger from the TX photodiode. The FAIL-SAFE mode configures the DPU gate array to detect the FIRE signal and send a delayed version to start the TOF chip. This mode will be commanded if the START pulse from the TX fails. The FIRE command is delayed because of a variable delay in the TX optical output, nominally 192 μs. This delay is determined by initiating an AUTO_CALIBRATION function, which varies the T-0 mask through a 10-bit programmable counter until a reasonable calibration range is detected. Since the counter has a 500-ns resolution, the maximum range error due to a failed TX_START pulse would be 150 m (500 ns), well above NLR 6-m specification, therefore, the bias must be estimated and removed from the data – this is readily perfomed by having the calibration signal available. The T-0 mask, along with the range gate, is configurable using uplink commands. A One-Stop mode exists to configure the TOF chip to stop both counters at the first detected return pulse. The NLR will be commanded to this mode if the calibration pulse is nonexistent (e.g., broken fiber-optic). In this case, the returned range will be the single stop pulse.

An AUTO_ACQUIRE sequence exists to allow autonomous noise-floor measurements for the analog section. This algorithm configures the NLR to step through each threshold level and collect 16 samples of range and calibration data at an 8-Hz rate. Collecting range data at 8 Hz minimizes the influence of relative movement between the NLR and asteroid terrain lending to high correlation between data points. This is important as range rate is calculated for each new sample at each threshold and is compared to a predetermined difference (sent as a command argument) to determine the system noise floor, and hence, the operating threshold for reliable range returns.

2.8. SUMMARY OF NLR SPECIFICATIONS

Table II summarizes the parameters describing the NLR design. Analytic evaluations used these values to examine NLR performance.

3. Performance Analysis of the Near-Laser Radar

Determination of the adequacy of the NLR design required evaluation of performance given parameter values describing Eros and the NLR mission (Table I), and the design implementation (Table II). The process we used to evaluate the performance of the NLR is given in this section.

Table II
NLR performance specifications

Parameter	Specification	Measurement[a]/estimate[b]
Transmitter pulse energy	>5 mJ @ 1.064 μm	15.3 mJ[a]
Transmitter energy jitter	<10%	<1%[a]
Transmitted pulse width, tpw	10 ns < t_{pw} < 20 ns	15 ns[a]
Transmitted pulse-width jitter	<2 ns	0.82 ns[a]
Transmitter wavelength broadening	$\pm$3 nm	$\pm$1 nm[a]
PRF rates	$\frac{1}{8}$, 1 (nominal), 2, 8 Hz	$\frac{1}{8}$, 1, 2, 8 Hz[a]
T-0 event mask (resolution)	T-0 mask required	0.00–511.5 μs (500 ns)[a]
TEM00 mode (% Gaussian)	>90%	91%[a]
Divergence ($1/e^2$)	<300 μrad	235 μrad[a]
Beam waist (near-field)	N/A	22.93 $\pm$ 0.12 mm[a]
Beam centroid jitter (shot-to-shot)	<50 μrad	16.31 $\pm$ 24.39 μrad[a]
Beam centroid wander	<300 μrad	4.81 $\pm$ 31.25 μrad[a]
Calibration power jitter	$\pm$5%	<5%[a]
Calibration timing jitter	<1 m	<31.22 cm[a]
Thermal control	$\pm$2 °C	$\pm$2 °C[a]
Shots (lifetime)	>31.5 M	>1 B[b]
Effective aperture, f/#	N/A	7.62 cm, f/3.4[a]
Spectral receiver bandwidth	<10 nm	7 nm[a]
Temporal receiver bandwidth	$\leq$ 100 MHz	30 MHz[a]
APD dark voltage	195 μVrms (24 MHz)	150 μVrms (24 MHz)[a]
APD responsivity	770 kV W^{-1}	770 kV W^{-1} [a]
Optical receiver FOV	>900 μrad	2.9 mrad[b]
Threshold levels	N/A	8 ($2^n \times$ 16 mV)[a]
Data rates	51 bps, 6.4 bps	variable, incl. 51 and 6.4 bps[a]
TX-to-RX alignment shift	<1100 μrad	345.0 μrad[a]
		(pre-to-post tests of vibration)

N/A: not applicable.

3.1. THE APD PHOTOELECTRON OUTPUT MODEL

An APD provides internal gain for primary signal photons that may otherwise have been partially obscured by thermal noise associated with the load resistor and amplifier. The internal gain is a random variable as the impact ionizations occur at random, generating output signal randomness characterized by a factor, F,

$$F \equiv E\{m^2\}/E^2\{m\} , \tag{1}$$

where $E\{\ \}$ is the expectation operator and m is the number of photoelectrons produced at the output of the APD. The probability density of the output photo-electrons based on the number of absorbed primary photons, n, has been described by the Conradi distribution (Cole, Davidson et al., 1996), which uses average APD gain, G, and the ratio of ionization coefficients, k_{eff}

$$F = k_{\text{eff}}G + (2 - 1/G)(1 - k_{\text{eff}}) . \tag{2}$$

The probability density function for the integrated output is $p(x|\lambda\tau_p)$, where λ is the photon absorption rate and τ_p is the effective integration period. We are assuming a well-matched filter receiver where integration period, τ_i, is approximately equal to the effective pulse width of return pulse, τ_p. Secondary electrons, m, are Gaussian distributed and describe the variance in the preamplifier noise and APD surface leakage current noise,

$$\sigma = \left(2eI_s + \frac{4kT}{R_L}\right) B_N \tau_p^2 , \tag{3}$$

where e is the electronic charge; k, the Boltzman's constant; T, equivalent noise temperature; I_s, APD leakage current; and B_N, noise bandwidth.

With the mean number of absorbed primary photons, $n = \lambda\tau_p$, and $m > \lambda\tau_p$, the Conradi distribution can be closely approximated by the Webb density. Webb's approximation approaches a Gaussian density with a mean value described by the product, $G\lambda\tau_p$, and having variance of $FG^2\lambda\tau_p$ for values of m close to its mean. However, this approximation departs from Gaussian behavior at large values of secondary electrons (m) which defines the contribution to false alarms. The photon arrival rate, λ, is given by total photon rate from backscattered return signal (λ_s), background photons (λ_b), and contributions from bulk leakage current (I_b).

The APD used in the analog receiver is a hybrid part that combines an enhanced silicon APD with bias voltage thermal compensation and transimpedance amplification. Table III presents characteristics associated with this hybrid. This unit is designated as a wide-FOV detector with a measured bandwidth response of 50 MHz and responsivity of 770 kV W^{-1}. Given these values, and an estimate of the noise floor, energy inputs (peak power) may range from a minimally detectable 9 nW to a maximum (saturation) of 0.5 W assuming a 10-ns pulse width at 1.064 μm.

3.2. Radiometric Performance

Using standard radiometric relationships (Jelalian, 1992), the background photon rate, impinging the detector, is given by

$$\lambda_b = \left(\frac{\eta_{\text{APD}}}{1.19 \times 1.602 \times 10^{-19}}\right) (I_{\text{Sun}}\Delta\lambda\eta_{RX}\pi) \left(\frac{\theta_{\text{FOV}}}{2}\right)^2 \left(\frac{\rho_b}{\pi}\right) \left(\pi\frac{D_0^2}{4}\right) , \tag{4}$$

where $I_{\text{Sun}} = 230$ W m^{-2} μm^{-1} (solar irradiance at Eros), $\Delta\lambda = 7$ nm (spectral bandpass), $\eta_{RX} = 0.8$ (receiver efficiency, or, transmissivity), $\theta_{\text{FOV}} = 3$ mrad (field-of-view of the receiver), ρ_b is Eros' albedo, and $D_0 = 7.62$ cm (receiver clear diameter).

Geometric albedo, defined as the ratio of asteroid brightness at zero phase angle to the brightness of a perfectly diffusing disk having the same apparent size and

Table III

APD hybrid characteristics used in the NLR instrument

Parameter	Value
Quantum efficiency, h	0.35
Surface leakage current, I_s	20×10^{-9} A
Bulk current, I_B	50×10^{-12} A
Average gain, G	100
Ionization coefficient ratio, k_{eff}	0.0065
Preamplifier feedback resistance, R_L	22 kΩ
Preamplifier equivalent noise temperature, T	750 K
Signal bandwidth, B_s	50 MHz

position as the asteroid, provides the best estimate for reflectivity (ρ_b) at our laser wavelength, 1.064 μm. We compared this value to reflectivity measurements of similar materials at the 1.064 μm-laser which varied from 0.15 (loam, rusty sandy type) to 0.62 (Guthrie fine silt loam). In addition, photometric measurements of several asteroids indicate spectral albedo values vary from 0.12 to 0.55 displaying a range similar to different soils. By selecting $\rho_b = 0.15$, our performance evaluation should reflect a conservative scenario.

Evaluating Equation (4) yields 4.833331×10^9 background phots s^{-1}. This calculation assumes solar noise dominates; zodiacal light as well as emission from stellar field or celestial bodies (e.g., Moon) are not included as their contributions are insignificant. Detected signal photons, N_s, is determined using

$$N_s = \eta_{\text{APD}} \left(\frac{E_p}{1.17 \times 1.602 \times 10^{-19}} \right) \left(\frac{\rho_b}{\pi} \right) \left(\pi \frac{D_0^2}{4R^2} \right) \eta_{RX} , \qquad (5)$$

where E_p is the laser transmitter's energy/pulse. Equation (5) is evaluated for various values of slant range, R; at $R = 50$ km, $N_s = 2246.2$ phots. For our instrument, performance is defined as the probability of correct detection as a function of range, $P_d(R)$ given signal and noise current levels at the APD output and a false alarm probability, P_{FA}. The NLR mission requires $P_d \geq 0.95$ with <1 false alarm per sampling period, or false alarm $<10^{-4}$. The mean output signal (electrons) of the APD can be expressed as,

$$\mu_0 = (\lambda_B \tau_p X) + \frac{I_B}{1.602 \times 10^{-19}} \tau_p , \qquad (6)$$

where X is the mismatch of the filter to the return pulse and is calculated by the ratio of integration time to the return pulse width ($0 \geq X \geq 1$). I_B is the bulk current within the device (see Table III). For the NLR, μ_0 evaluates to 154.375 electrons. With a signal present, the APD output is modified to become $\mu_1 = (\mu_0 + N_s X)$.

To compute P_d and P_{FA}, the variance (σ^2) of preamplifier noise and APD leakage current noise must be included in the Neyman–Pearson interpretation. The variance depends upon equivalent preamplifier temperature, T, and the surface leakage current, I_s. The overall variance at the output of the APD requires combining σ^2 with the variance associated with the APD-induced variance, σ_0^2. F and μ_0 are required for the calculation of σ_0^2. F is determined using Equation (2) and μ_0 depends Equation (6). Now we can employ Webb's approximation for noise-only (μ_0) and signal-plus-noise (μ_1) conditions. To account for the influence of speckle, the noise factor for signal and noise is augmented as

$$F^*(N_s) = k_{\text{eff}}G + (1 - k_{\text{eff}})\left(2 - \frac{1}{G}\right) + \frac{N_s X}{M} , \tag{7}$$

where M is the number of speckle patterns seen by the APD. The noise variance for the signal plus noise case is then computed similarly as for the noise-only case except F^* and μ_1 are used rather than F and μ_0.

Integration of P_{FA} and P_d based on the threshold setting, N_{TH}, given in terms of noise standard deviations, provides the estimate for performance of the NLR as a function of slant range. The integrals are expressed in terms of the variable, $\xi \equiv (\mu - G\mu)/\sigma$,

$$P_{FA}(N_{TH}) = 1 - \exp\left(-\frac{R_g}{\tau_p}\int_{\xi_0}^{\xi_1} p(m|\lambda_n \tau_p)\, \mathrm{d}\xi\right) , \tag{8}$$

where R_g is the functional range gate, and $p(m|\lambda \tau_p)$ is Webb's approximation; $\xi_0 = (\mu_0 - G\mu_0)/\sigma_0$ and $\xi_1 = 100$. The functional range gate for the NLR is determined by the mode of operation. Nominally, the range gate is completely open awaiting the expected return signal to halt the counters. For the calibration signal, the gate is effectively 558 ns whereas for the actual range data from Eros, it would be $\Delta t = 2R/c$, with $c =$ velocity of light. Without a target, the TOF chip runs until overflow occurs, with 21 bits, at $R \sim 327$ km. For our calculations, this overflow range value is used for R_g. Similarly, for probability of detection given threshold (N_{TH}) and signal (N_s) photoelectrons,

$$P_d(N_{TH}, N_s) = \int_{\xi_2}^{\xi_3} p(m|\lambda_{s+n}\tau_p)\, \mathrm{d}\xi . \tag{9}$$

The integrands associated with Equations (8) and (9) incorporate normalized parameters based on the variances of the noise sources involved; Figure 13 plots the results of these equations for the NLR. Given the requirements for $P_d \geq 0.95$ with $FA \leq 10^{-4}$, we observe that the NLR should be capable of satisfactory operation at $R = 50$ km.

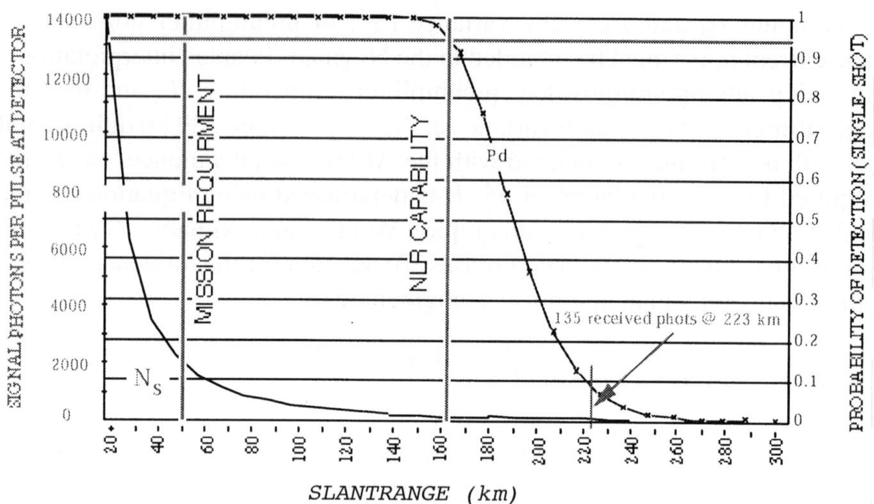

Figure 13. Plot of performance for the NLR instrument based on parameter values describing Eros, NEAR mission geometry, and the NLR design. Use of Webb's approximation for APD output current was made. P_d indicates adequate operation beyond 50 km. At 223 km, this analysis indicates ~135 equivalent photons arriving at the detector surface ($P_d = 0.1$). Statistics assumed single-shot operation. False alarm was 0.00057785 using threshold (TH) set at $7\sigma_0$.

3.3. RANGE MEASUREMENT ADJUSTMENTS

Profiles and topographic grids using range measurement from an orbiting spacecraft require corrections. The largest correction expected is associated with uncertainty in the knowledge of the spacecraft orbit. The orbital accuracy of the NEAR spacecraft around Eros should be recoverable to the 5-m level with respect to the COM.

Because the NLR uses leading-edge detection, pulse dilation due to interaction with the surface will introduce range error and can lead to performance loss due to receiver filter mismatch. Pulse dilation is defined as the temporal stretching that occurs due to timing differences between the initial and final arrival of the backscattered photons. Range errors occur because spreading of the return pulse delays the time at which accumulated photons exceed a prescribed threshold, thus causing an error in the time-of-flight counter. Within a footprint, a sloped surface or significant surface roughness will lead to pulse dilation.

The interaction of the altimeter pulse with the surface is described using Figure 14 where h_0 = altitude (nadir), r_0 = minimum detected range, r_1 = maximum detected range, $\theta_d = 1/e^2$-points full beamwidth (divergence), θ_s = average slope deviation from horizontal, θ_p = NLR (spacecraft) pointing angle, x = intersection of beam with surface, and Z is the dilation-induced range error. The full-width, half-maximum (FWHM) pulse width is t_p and average range (R) is shown in Figure 14(b).

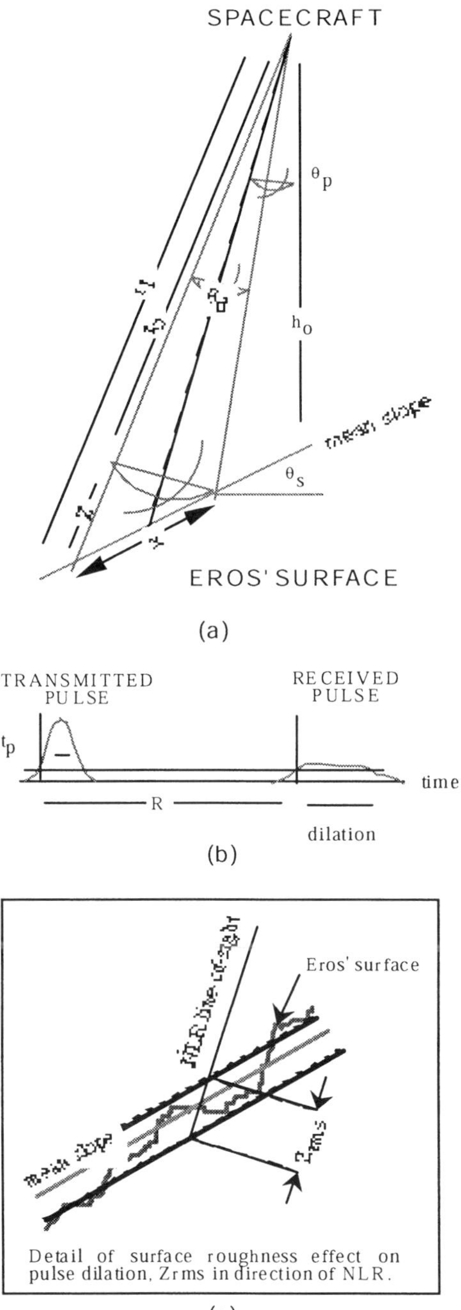

Figure 14. Geometry describing NLR pulse interaction with the surface. (a) terms describing the interaction of the pulse as it strikes Eros' surface, (b) temporal spreading of return pulse based on threshold crossings as compared to transmitted pulse, (c) description of surface roughness indicating an r.m.s. deviation, $Z_{r.m.s.}$, from the mean surface in the direction of the NLR.

$Z_{\text{r.m.s.}}$ represents the r.m.s. deviation from the mean slope in the direction of the rangefinder and is expected to contribute <1 m error. The total amount of pulse spreading is also dependent on receiver resolution and biases due to curvature of the beam wavefront (<1 m), pointing jitter ($\ll 1$ m), and variations in surface reflectivity (~ 1 m). Contribution of pointing jitter to range error is negligible based on performance of the NEAR spacecraft. Variations in surface reflectivity can produce pulse dilation due to contributions of the return signal from varying parts of an illuminated footprint. If the illuminated footprint is uniformly bright, then the return pulse will be dilated only by range increments from the closest point (center if pointing nadir). However, if the footprint has irregular returns, due to reflectivity variations, more complicated (in regards of time description) return signals would appear. Additional range errors can arise from electronic delay, timing errors, and incorrect correlation of attitude with range sample. Root-sum-squaring all of these various sources of range error are expected to result in a topographic field that is accurate to approximately 6 m with respect to Eros' COM.

The 3-axis angular stability of the NEAR spacecraft is anticipated to be 50 μrad over 1 s. Spacecraft 3-axis attitude control and knowledge are expected to be 1.7 μrad and 50 μrad, respectively. The presence of glint (strong scatterer) is not considered; however, if present, operational range could be drastically increased albeit just for those surface areas containing such glint. At our clock frequency of 480 MHz, the NLR provides a temporal resolution of 2.08 ns, or 31.22-cm range resolution. An estimate to dilation (ignoring the $Z_{\text{r.m.s.}}$ term) from geometric considerations, is provided by

$$Z \equiv r_1 - r_0 = h_0 \sin(90 + \theta_s) \times$$

$$\times \left\{ \frac{1}{\sin\left(90 - \theta_s - \theta_p - \dfrac{\theta_d}{2}\right) - \sin\left(90 - \theta_s - \theta_p + \dfrac{\theta_d}{2}\right)} \right\}. \tag{10}$$

Using the maximum radius of Eros as 18 km, θ_p can range from $0°$ to $\pm 19.8°$ prior to having the beam walk off the asteroid. In Equation (10), pulse dilation, Z, identifies the extent of range error expected assuming sufficient backscatter from all points along the intercepted surface, x. Figure 15 is a surface plot of dilation, $Z(\theta_s, \theta_p)$ for the NLR mission parameters, where $\theta_d = 235$ μrad and $h_0 = 50$ km. The surface curve, $Z = 6$ m, is shown as it represents maximum allowable range error.

Given the size (curvature) of Eros, as pointing (θ_p) approaches $19.8°$, slope angles become significantly influenced by the curvature of Eros' surface restricting availability to the area where $Z \leq 6$ m. Using the best case for maximum θ_p, the spacecraft would orbit at a point perpendicular to the largest Eros dimension (Case 1 in Figure 2). Here, $Z = 6$ m at $\theta_p \sim 10°$. For Case 2, orbiting perpendicular to the shortest axes would restrict pointing angles to $< 3.5°$. This does not indicate that

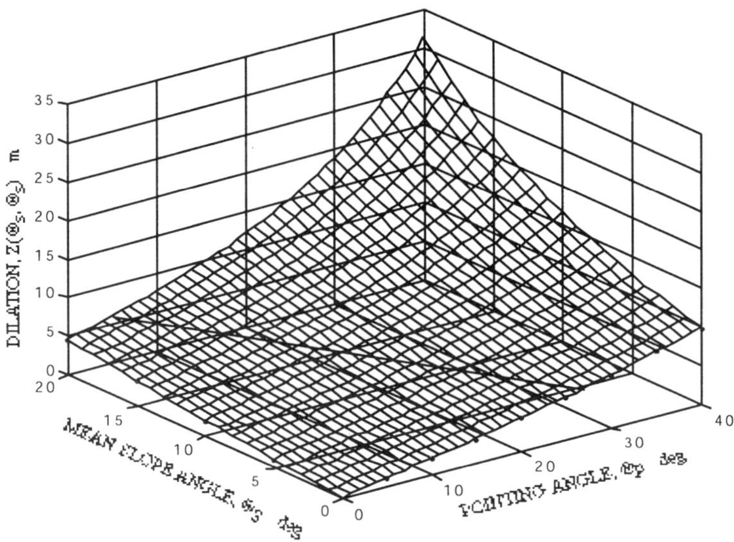

Figure 15. Surface plot of dilation, $Z(\theta_s, \theta_p)$ with locus of points describing $Z = 6$ m shown to indicate maximum permissible Z for NLR mission.

the NLR would not receive usable returns from angles beyond these limits, rather, it indicates that the range error would become excessive at such pointing angles. Also of consideration is the cosine loss of the illuminated area, $(\pi R^2 \theta_d^2/4)\cos(\theta_p)$, due to Eros' curvature that would reduce return signal strength (where R is the slant range from the NLR to the illuminated area).

4. Performance Testing of the Near Laser Radar

The NLR was subjected to numerous tests to satisfy both instrument and spacecraft requirements. At the instrument level, the NLR went through design verification testing. These tests ensured that the instrument components met interface and performance requirements. The NLR was environmentally tested to evaluate operation and performance at various operating environments. During these tests, characterization of the NLR instrument took place as data were required to determine if any shifts developed in the NLR performance.

The door to the NLR receiver had an optic (GRIN-lens) fitted to a window to support a fiber-optic cable connection. The TX also had a cap that supported a fiber-optic connection. Through these optics and fiber-optic connections, the NLR could be operated safely and reliably throughout system functional and qualification testing without restricting personnel. We could directly simulate range effects to perform end-to-end system performance testing. This was possible by converting the optical pulses from the capped TX, transmitted through a fiber-optic cable, to an electrical pulse within the high-speed photodiode trigger unit. This electrical

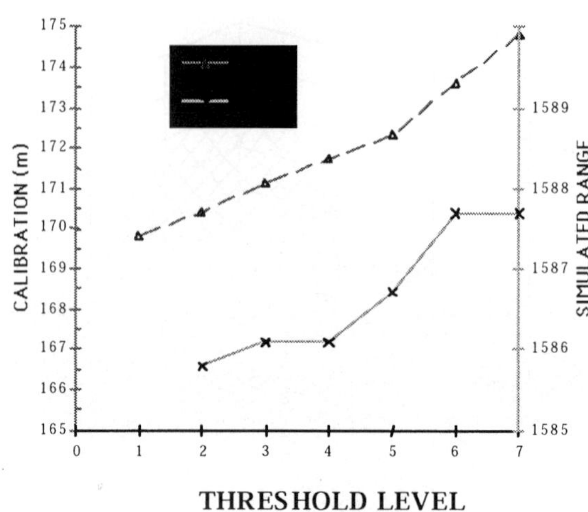

Figure 16. Functional test of NLR using simulated backscatter (level, delay, and dilation). The ordinate reads CALIBRATION and simulated range values in equivalent meters (optical path); actual numbers are counts that must be converted using 2.1 ns count^{-1} velocity of light (c). For the fiber-optic calibration values, use of light velocity (c/n) in fiber optic is required (where n is the fiber-optic index of refraction).

pulse was then used to trigger the laser-pulse generator to provide an optical pulse to the RX, via fiber optic, with a variable delay to simulate round-trip time for satellite-to-asteroid ranging. Prior to going to the RX, the optical output pulse from the pulse generator could be attenuated through a programmable attenuator. In addition to simulating range, this system could also be used to simultaneously measure power of the TX using an optical splitter with a power-meter head.

Figure 16 presents the data for a functional test performed on the NLR prior to environmental tests (16 July, 1995). The receiver was sequenced through the range of threshold levels, 0-to-7, and range counts were averaged over multiple laser firings to produce this graph. Threshold values correspond to comparison voltages on the analog comparator. For each increase in threshold level, comparison voltage is doubled. In Figure 16, calibration ranges are denoted by triangles whereas simulated ranges are represented using squares. Both curves illustrate range-walk, typical behavior of a leading-edge detection scheme.

As threshold is increased, correspondingly more energy is required to signal the arrival of a return pulse. The calibration plot (Figure 16) indicates a range-walk of ∼4 counts while operating the NLR between thresholds of level 2 to 7. The corresponding walk seen for simulated range is ∼2 m. The difference in walk is a result of calibration range being a one-way value whereas simulated range takes the time of flight and divides by two to obtain range to target. In this particular test, calibration data was unreliable for level 0 and was not plotted. For the particular values of simulated range, measurements overflowed our NLR counters for levels 0 and 1, thereby indicating insufficient power level (actually, signal-to-noise) for signal detection. Once we selected level 2, we obtained reliable range measurements (lower plot in Figure 16).

4.1. NLR ENVIRONMENTAL TESTS

Each NLR component was subjected to environmental qualification (El-Dinary, 1995). Performance data was collected following each axis of vibration and instrument alignment was verified before and after vibration using two optical methods. To verify that the NLR would survive the launch on a Delta-II, vibration tests were conducted to the limits given in Table IV. These vibration levels were set to protoflight levels, +3 dB above actual flight, based on modeling of the Delta-II launch environment.

The first method measured the relative change between optical reference cubes mounted on the TX, RX, and instrument base. The alignment of these cubes was measured using a Theodolite with 2 arc-sec precision. TX-to-RX alignment was also determined by observing the TX beam in the RX far-field field of view (FOV). This was performed by mapping the TX central lobe to the electronically determined FOV of the RX.

The operating temperature of the spacecraft is projected to be between $-29\,°C$ and $+55\,°C$ and survival range is projected to be between $-34\,°C$ and $+60\,°C$. However, NLR subassemblies had to maintain at different temperature limits based on specifications from the manufacturers, which was achieved through blanketing and heater control. To test the adequacy of the NLR thermal design, the thermal ranges given in Table V were used.

To evaluate the NLR over these ranges, the entire integrated instrument was placed into a thermal-vacuum (T-V) chamber and operated over several thermal cycles which lasted five days (three days of actual T-V testing). Special T-V chamber fiber optics were employed to route the TX signal out of the chamber to the test equipment rack and to route the resulting test signal back into the chamber to the RX. This permitted continuous performance evaluation using actual altimetry measurements over temperature.

Figure 17 presents the thermal-vacuum test profile followed for the qualification testing of the NLR. Six cycles were implemented with temperature ranging from $-29\,°C$ to $+55\,°C$ in approximately 1.5 h.

The opposite ramp ($+55$ to $-29°$) required approximately 2 to 3 times longer due to thermal latency in the test configuration. To ensure reliable operation of the high-voltage power supplies, the NLR was soaked in vacuum ($<10^{-5}$ T) for 36 hours during the startup of the test to prevent corona. No anomaly was observed with the hardware during thermal-vacuum testing. At each thermal plateau, and abbreviated performance (limited performance, LP) test was conducted in situ using the NLR test equipment. Figure 18 present calibration data obtained during the thermal-vacuum test for NLR. For this graph, we again plotted counts rather than range to highlight slight thermal dependencies. Although range would have

Table IV

Vibration levels for testing the NLR instrument for both sinusoidal and random vibration sensitivities. (A) and (B) indicate the sinusoidal component in the thrust and lateral axes, respectively. (C) indicates the test levels for random vibration levels having an overall amplitude of 13.6 g.

(A)

Z axis (thrust)

Frequency (Hz)	Acceleration
10–30	0.33 in. (double amplitude)
30–40	15.0 g
10–100	2.0g

(Rate: 4 octaves min^{-1})

(B)

Lateral axes

Frequency (Hz)	Acceleration
10–15	0.65 in. (double amplitude)
15–25	7.5 g
25–100	1.4 g

(Rate: 4 octaves min^{-1})

(C)

Frequency (Hz)	PSD level
20–40	$0.01 \text{ g}^2 \text{ Hz}^{-1}$
40–140	$+6.0 \text{ dB octave}^{-1}$
140–300	$0.12 \text{ g}^2 \text{ Hz}^{-1}$
300–400	$+3.0 \text{ dB octave}^{-1}$
400–830	$0.16 \text{ g}^2 \text{ Hz}^{-1}$
830–2000	$-6.0 \text{ dB octave}^{-1}$
2000	$0.028 \text{ g}^2 \text{ Hz}^{-1}$

Duration: 60 s each axis.

Overall r.m.s. amplitude: 13.6 g.

been adequate, the use of counts avoided systematic biases associated with software updates that were occurring during initial T-V testing.

For Figure 18, several hundred (>200) points were collected at each threshold for each of three plateau temperatures: $-29\,^\circ$C (diamond), $+25\,^\circ$C (triangle), and $+55\,^\circ$C (square). Results from altering ambient temperature from hot ($+55\,^\circ$C) to

Table V

Thermal environment for the NLR subsystems, including both operational and survival test limits

Operational limits	0 °C to +20 °C	(TX) with blankets and heaters
	+10 °C to +30 °C	(RX) with blankets and heaters
	−29 °C to +55 °C	(DPU)
Survival limits	−5 °C to +25 °C	(TX) with blankets and heaters
	−34 °C to +60 °C	(RX and DPU)

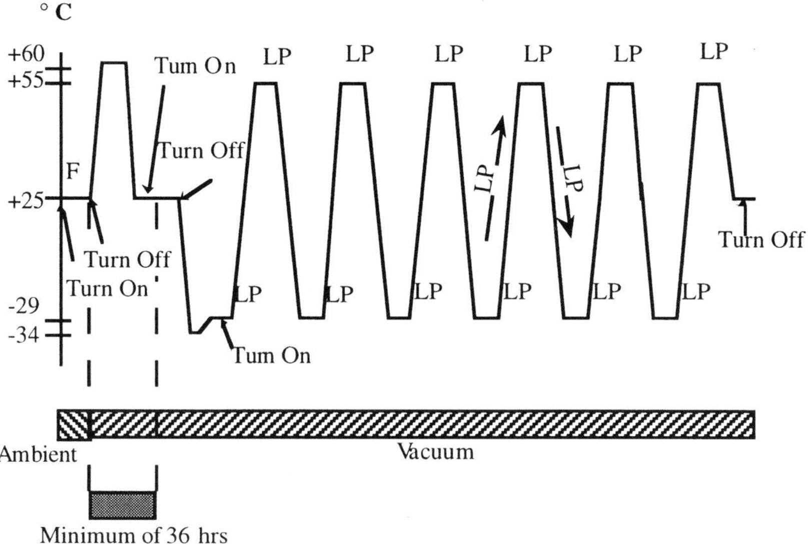

Figure 17. Thermal-vacuum profile used for qualification of the NLR instrument. Note the prolonged vacuum ($<10^{-5}$ T) soak period of 36 hr. LP indicates limited performance test.

cold ($-29°C$) indicate thermal influence on NLR performance as being relatively benign. Threshold-induced range walk is observed, but performance appears to be relatively insensitive to thermal effects as represented by this thermal-vacuum test scenario (Figure 18).

Finally, we observed variance of these data to be quite small, approximately 1 to 2 counts (refer to error bars in graph). No anomalies were observed with NLR operation or data during the thermal-vacuum test period. However, we did observe a decrease in signal levels of approximately 1 dB during $-29°$ to $+55°C$ gradients. The cause of this decrease was suspected attenuation by the test fiber-optic cable inside the T-V chamber due to the large thermal difference between the chamber walls and the NLR during these transient periods. At all thermal plateaus or during negative transients, this 1-dB loss was not observed. Verification of performance prediction was only partially completed due to schedule, however, numerous functional tests were conducted that provide information regarding design adequacy.

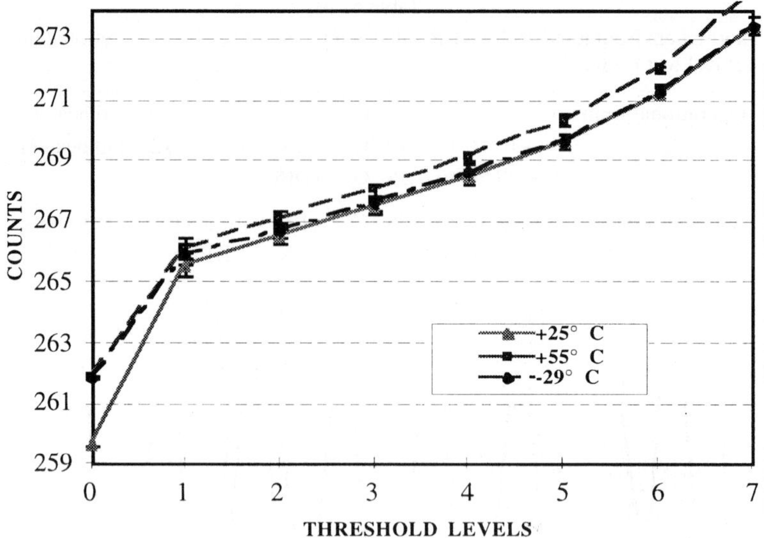

Figure 18. Functional test of NLR using simulated backscatter (level, delay, and dilation).

Relating test results to performance predicts is possible by observing threshold levels used. As stated, the lowest threshold level, TH = 0, is below the receiver noise floor. Throughout testing, neither calibration nor simulated range measurements were obtained at this level.

4.2. END-TO-END SYSTEM TEST

Following environmental qualification, a comprehensive, end-to-end system test was conducted to verify that the NLR operated as a ladar system. The test characterized range walk of the system and provided data useful to estimate the maximum operational range (>100 km). This test was conducted in a moderately controlled test area, a 216.4-m (710-foot) hallway, using a sand-blasted aluminum plate (Lambertian scatterer) and a silicate rock as ranging targets. Range simulation, based on return energy levels, was accomplished using neutral density filters, an aperture stop, and the receiver door in the closed position (windows uncovered) for a total attenuation of 71 dB. (The closed door provided mechanical and contamination protection to the receiver optics.) The NLR was configured to fire at 1 Hz, and data were collected at multiple ranges to both targets. These ranges varied from 600 to 690 ft and actual ranges were determined using a NIST-traceable 300-ft surveyor's rule. A composite of the targets used during this test along with test results is presented in Figure 19.

The graphs (Figure 19) have range data as expected given the attenuation used for threshold (TH) levels 2 through 5; noise was seen at TH = 0, 1, and no returns were detected for TH = 6 or 7. The abscissa on both graphs represents elapsed

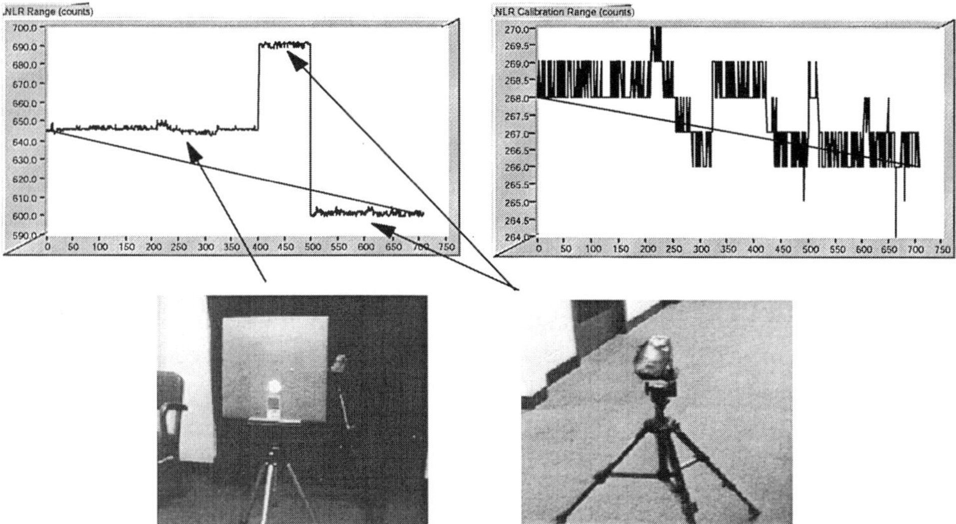

Figure 19. Free-air test (700-ft hallway) within APL campus; evaluation and calibration of NLR along with photos of targets used (sand-blasted aluminum plate, left, and granite rock, right). Note the presence of the laser beam on the targets due to the infrared sensitivity of these 'photos'.

time into the test and ordinate for the graphs plots NLR range in meters (leftmost plot) and calibration in counts (rightmost graph; refer to Figure 16 for definition of counts). Occasionally, calibration signals would be correctly detected by TH = 1; this resulted from having a very short calibration signal range gate. The arrows in Figure 19 indicate which target was illuminated during the data collection. Data were collected at one range for the aluminum target and two separate ranges for the rock target. Of note is the illuminated spot on the targets indicating where the YAG beam was impinging the target during data collection. (An 8-mm camcorder recorded these images using a typical intrinsic silicon CCD array that is sensitive to the near infrared.)

Considering the successful detections, an estimate of the performance of the NLR was determined from this test. Infrequently we would obtain good data at TH = 1 indicative of this threshold level being close to the system noise floor. At TH = 5, we received all data expected but at TH = 6, no returns were observed indicating signal strength below threshold. Given the scaling used to create the threshold levels, TH = 5 is 9 dB above the noise floor. With 71 dB attenuation, we estimate that the instrument has 9 to 12 dB of operational margin. The 3-dB granularity cannot be improved by these tests as detection measurements were taken at the system level using the 8-valued preset threshold scheme. Using analysis from the previous equations, the NLR is expected to operate satisfactorily ($P_d \geq 0.95$) given signal-to-noise power of 12 dB. Computations indicate that minimum signal-to-noise, SNR, at the required range, $R = 50$ km, is ≥ 18.5 dB.

5. Near Laser Rangefinder Data Products

We will analyze the fundamental topographic parameters (ellipsoidal axes, spherical harmonic coefficients) to characterize fundamental aspects of the shape over the wavelength spectrum (Zuber et al., 1997). This characterization will provide information on the extent to which these exogenic processes have controlled the surface evolution of Eros. With the first month of data we will produce a global reference surface that will likely supersede all current ellipsoids. If the shape of Eros turns out to be distinctly elliptical, then we will compute a best-fit ellipsoid. However, if Eros has a complex (e.g., jagged) shape, then we will choose a spherical harmonic surface to as many degrees and orders as are required to describe the global figure of the asteroid. There is a practical reason for choosing a reference surface that is more complex than an ellipsoid that is especially relevant to small, irregular planetary bodies: if the reference surface does not well approximate the shape of the body, it is difficult to map images onto it.

We will also produce a global hypsometric distribution of the topography of Eros, a topographic power spectrum, and a topographic error map and an error spectrum. The basis of our formal error estimation will be the variance-covariance matrix of the spherical harmonic topography solution. Our error analysis will provide a quantitative measure of the accuracy and integrity of the topographic field.

5.1. PROFILES

We will subtract initial orbits produced by the NLR Science Support Team from the range profiles to convert them to rough topographic profiles. These initial orbits will be available on a time scale of ~ 1 week and profiles with this correction will be created within a month. This data product is intended as a 'quick-release' (albeit imperfect) version of the data that will be easily accessible. The format of these data files will be of the form: latitude, longitude, height above preliminary reference surface.

When updated orbits and spacecraft state information become available later in the mission, we will compute more precise profiles, which we term Corrected Data Records (CDR). The CDRs will be profiles that use final precision orbits so these profiles will have the same horizontal resolution but higher vertical (and horizontal) integrity than the quick release data. The CDRs will be the final archived profile product.

5.2. GRIDS

After a month of continuous mapping, there will be sufficient global coverage to produce a global grid of topography with a grid resolution of 3 km or better at the equator. We will produce this grid, with heights initially referenced to the best-available ellipsoid. After 6 months we will produce a refined grid with a resolution

Figure 20. NLR integrated to instrument deck of NEAR spacecraft.

of $\sim$300 m referenced to an updated topographical surface that we will produce (see next section). Grids of intermediate resolution could be generated on an as-needed basis.

Our preferred algorithm creates a grid with points fit to a spherical cap. The algorithm then performs a bilinear interpolation to produce each grid point and for each point also provides a formal radial error estimate. The grid can be generated at any desired resolution but the algorithm can also be instructed to choose the best resolution that is allowed by the data in a statistical sense. We will release this grid when the data used in its generation are validated. As more data is collected we will produce densely sampled, updated grids, as well as higher resolution regional grids in areas of interest.

5.3. SPHERICAL HARMONIC TOPOGRAPHIC MODEL

We will also perform a spherical harmonic expansion of the topographic data. If the NLR operates for the equivalent of 6 months, the horizontal resolution will be of order 300 m that corresponds to a 120th degree and order model. For geophysical analyses, we will also produce a 15th degree and order model to be commensurate with the expected resolution of the gravity data.

5.4. HIGH RESOLUTION PROFILES OF SMALL-SCALE SURFACE FEATURES

The NLR, in conjunction with another NEAR instrument, the Multispectral Imager (MSI), will obtain important new information on small-scale topographic features such as craters or grooves (if present). NLR and MSI/NIS science teams will jointly produce a topographic map of one or more specific surface features with overlaid eight-color imagery plus 64-color data at lower spatial resolution. The density of the topographic grid will depend on the number of samples obtained, and the relative topographic accuracy will be about 5 m. In addition, we will produce a database containing NLR crater samples with simultaneous MSI images. High precision laser altimetry from NLR will combine effectively with high spatial resolution color imaging from MSI. The ranging data will provide unambiguous determinations of topography, whereas with imaging data alone, effects of topography (such as shadowing and reflectance changes with viewing geometry) can be difficult to distinguish from albedo changes. The imager, on the other hand, provides the geologic context in which to interpret the topographic profile obtained as NLR makes a crossing across a feature. The MSI spatial resolution is slightly higher than that of NLR. As described by Figure 2, NLR sampling consists of a series of laser spots on the asteroid surface that are close to contiguous, or which may overlap, given the current mission design.

Key to these experiments is the capability to measure the boresight co-alignment of MSI and NLR in-flight at Eros. MSI has been designed such that its 1050 nm filter passband includes the 1064 nm radiation from the NLR. MSI sensitivity is adequate to detect a NLR laser spot in a single image when the NLR operates in the 8 Hz mode over the darkside of the asteroid thus when the MSI obtains an image during an NLR sample, we will know precisely which imager pixels are illuminated by the laser. The NLR spot diameter at 235 μrad is larger than, but comparable to, the MSI pixel size of 95×161 μrad. During routine operations at Eros, when NLR is in its 1-Hz pulse repetition rate and MSI is imaging the dayside, NLR does not interfere with MSI observations in any way. This is because NLR fires its laser pulses at times when the MSI electronic shutter is normally closed (MSI CCD is not integrating the image). However, it is possible to operate the MSI such that its shutter is open when NLR fires in its 1-Hz mode, and in this case only, when MSI is in its 1050 nm filter, the laser pulse makes a small contribution to the detected brightness from the specific pixels illuminated by NLR. The magnitude of this contribution will be measured in-flight. The laser wavelength is outside the pass bands of the other seven MSI filters.

A second possible experiment will be carried out jointly by the NLR and Near-Infrared Spectrograph (NIS) science team. One or more individual surface features will be selected by the science team early in the mission for targeted observations by NLR, MSI, and the NSI. The goal will be to collect a large number of NLR samples with simultaneous MSI/NIS measurements, so as to be able to construct a 3-D topographic model of the feature and to overlay multispectral images and 64-color observations by NIS. From measurements at 20 km range, we will obtain a topographic grid from altimetry with 5-m spot sizes and with relative topographic accuracy of $\sim$5 m. These data will be combined with eight-color imaging at 1.9 m $\times$ 3.2 m per pixel and NIS 64-color data at 130 m $\times$ 260 m resolution. We will study the structure and morphology of the feature, relate these to the mineralogy, explore stratigraphic relations of the feature, and attempt to infer geologic processes.

6. Summary

Throughout testing, NLR performance was quite acceptable. We observed extremely quiet operation with noise levels on the order of 1-count bit. Altering threshold values produced the expected result of range walk (increase) with larger threshold values. Several scenarios were simulated with successful results through the versatility of the test support equipment.

Radiometric analyses and requirements were presented in Table II. The NLR performance met or exceeded all specifications. The inclusion of a 'free-air' lasing test provided significant data and confidence in the ability of the NLR to perform as required. Prior to this test, the NLR had been operated as a laser stimulating test equipment and a receiver being stimulated by test equipment. Issues relating to power levels, alignment, and other attributes associated with actual end-to-end testing were not addressed until this all-inclusive test was performed. Figure 20 is a photograph of the qualified NLR (enclosed by the circle) integrated to the NEAR spacecraft. Adjacent to the NLR are other NEAR instruments: on the left is the infrared imaging spectrograph (NIS) and on the right is the X-ray/gamma-ray spectrograph.

Although the NLR was a relatively simple laser radar compared to designs commonly in use by ladars today, the combined requirement of operating an instrument in deep space for a prolonged period under strong design constraints (weight, cost, schedule) contributed significantly to the complexity of this instrument. Fortunately, radiation effects are slight (10-kRad total dose), nevertheless, space exposure influenced our selection and design approach for the electronics and the optical components.

Acknowledgements

Acknowledgment is given to Dr Frederic Davidson (JHU) for the development of the APD output signal model used for NLR analysis, to Mr Ashruf El-Dinary for implementation of NLR software and testing, to Mr M. T. Boies for receiver optics design, to Mr R. A. Reiter for designing the analog signal section and power supply systems, and to Mr D. Rodriguez for development of the DPU and digital hardware.

References

Binzel, R., Gehrels, T., and Matthews, M. (eds.): 1989, *Asteroids II*, The University of Arizona Press, Tucson, AZ.

Boies, M. T., Cole, T. D., El-Dinary, A. S., and Reiter, R. A.: 1996, 'Optical System Development and Performance Testing of the Near Laser Rangefinder', *Photonics for Space Environments IV, SPIE*, Vol. 2811, pp. 169–184.

Burns, H. N.: 1994, *NEAR Laser Radar (NLR) Tilt/Decenter Analysis*, Technical Report to APL, Burns Engineering, Inc.

Cole, T. D. and Davidson, F. M.: 1996, 'Performance Evaluation of the Near-Earth Asteroid Rendezvous (NEAR) Laser Rangefinder', *Photonics for Space Environments IV, SPIE*, Vol. 2811, pp. 156–168.

Cole, T. D. et al.: 1996, 'Laser Rangefinder for the Near-Earth Asteroid Rendezvous (NEAR) Mission', *Lidar Techniques for Remote Sensing II, SPIE*, Vol. 2581, EUROPTO, Paris, France, pp. 2–26.

Culpepper, C., Kushina, M., Wiswall, C., and Cole, T.: 1995, 'Laser Transmitter for the Near-Earth Asteroid Rendezvous Spacecraft', *IEEE/OSA CLEO/QELS '95*, Baltimore.

El-Dinary, A. S.: 1995, *NEAR Laser Ranger Test Plan*, APL/JHU Technical Report No. 7361–3007.

El-Dinary, A. S., Cole, T. D., Boies, M. T., Reiter, R. A., and Rodriguez, D. E.: 1996, 'Pre-Launch and Post-Launch Testing of the Near-Earth Asteroid Rendezvous (NEAR) Laser Rangefinder', *Photonics for Space Environments IV, SPIE*, Vol. 2811, pp. 222–231.

Jelalian, A. V.: 1992, *Laser Radar Systems*, chapter 1.7, Artech House Publishers, p. 29.

Moore, R. C.: 1994, *NEAR NLR Flight Software Requirements Specification*, APL/JHU Report 7352–9069.

Penn, J.: 1994, *GaAs Design for Laser Rangefinder Timer*, APL Technical Memorandum No. S2R-94–050.

Reiter, A.: 1993, *Timing Precision of the NEAR Navigation Laser Rangefinder (NLR) Analog Electronics*, APL Technical Memorandum No. S2A-93–0201.

Rodriguez, D.: 1994, *NEAR Laser Rangefinder Digital Processor – Electrical Design Data Package*, APL Technical Memorandum No. S2F-94–0315.
Swenson, T.: 1994, *Specification of Microplasma 1064 nm Narrow Band Filter*, Application Note, Optical Corporation of America.
Zuber, M. T., Smith, D. E., Cheng, A. F., and Cole, T. D.: 1997, 'The NEAR Laser Ranging Experiment', *J. Geophys. Res. – Planets*, submitted.

NEAR MAGNETIC FIELD INVESTIGATION, INSTRUMENTATION, SPACECRAFT MAGNETICS AND DATA ACCESS

D. A. LOHR, L. J. ZANETTI, B. J. ANDERSON, T. A. POTEMRA, J. R. HAYES,
R. E. GOLD, R. M. HENSHAW, F. F. MOBLEY and D. B. HOLLAND
The Johns Hopkins University/Applied Physics Laboratory, Laurel, MD 20723–6099, U.S.A.

M. H. ACUÑA and J. L. SCHEIFELE
NASA/Goddard Space Flight Center (GSFC), U.S.A.

(Received 8 October, 1997)

Abstract. The primary objective of the investigation is the search for a body-wide magnetic field of the near Earth asteroid Eros. The Near Earth Asteroid Rendezvous (NEAR) 3-axis fluxgate magnetometer includes a sensor mounted on the high-gain antenna feed structure. The NEAR Magnetic Facility Instrument (MFI) is a joint hardware effort between GSFC and APL. The design and magnetics approach achieved by the NEAR MFI effort entailed low-cost, up-front attention to engineering solutions which did not impact the schedule. The goal of the magnetometer is reliable magnetic field measurements within 5 nT, which necessitates the use of an extensive spacecraft magnetic interference model but is achievable with the full year's orbital data set. Such a goal has been shown viable with recent in-flight calibration data and comparisons to the WIND magnetometer data. The NEAR MFI effort has succeeded in providing magnetic field measurements for the first flight in NASA's Discovery line.

1. Introduction and Objectives

The NEAR spacecraft will follow a 2-year solar system trajectory from launch in February of 1996 to the near Earth asteroid Eros 433, at which point the spacecraft will be inserted into orbits of progressively smaller radii, culminating in a one year, 35 km altitude orbit. The primary objective of the NEAR Magnetic Facility Instrument (MFI) is the search for coherent, body-wide, intrinsic magnetism of the asteroid Eros. If Eros was originally part of a differentiated parent body with an internal magnetic dynamo, much like the Earth, Mars, or Mercury, then it should have a coherent natural remanent magnetization (NRM). Further, if Eros' composition was metal rich during formation, the NRM could be large. In objects like the Sun, Earth, and Jupiter, currents flow in highly electrically conducting fluids and the magnetic field is maintained by the motions of the fluid. In a solid body such as Eros, the currents flow in the individual atoms such as iron and nickel. These currents are retained in the rock when it cools below the Curie temperature, if cooling was in the presence of a magnetic field. From studies of NRM on the moon, it is inferred that the moon once had a dynamo, although one does not presently exist.

Asteroids can be classified in two generic groups according to their optical properties (Chapman, 1975; Gaffey et al., 1993): The low albedo 'C' types with neutral colors in the visible, and the 'S-types', like Gaspra and Eros which have

© 1997 *Kluwer Academic Publishers. Printed in Belgium.*

high albedos and reddish color. The S-type asteroids are thus similar to meteorites that contain silicates and metallic iron and on this basis the existence of a detectable magnetic field at Eros is likely, particularly at the small radial distances planned for the NEAR mission orbits (approaching one asteroid radii, 1 Ra, 35 km). It is generally accepted that asteroids are the parent bodies for most meteorites reaching the Earth (Wasson and Wetherill, 1979; Feierberg et al., 1982). Thus the compositional differences that are observed in meteorites are expected to reflect that of asteroids. The two basic groups that characterize meteorites, undifferentiated and differentiated, reflect the thermal history of their parent bodies. However, no unambiguous match has been made between a meteorite and an asteroid (Gaffey et al., 1993). The remanent magnetization of the large population of meteorites considered to be from Mars, suggests that Mars once had a surface magnetic field comparable in magnitude to that of the Earth (Curtis and Ness, 1988). Thus the detection of a magnetic field at Eros would immediately establish it as an originally differentiated asteroid with iron rich concentrations of mass. The NEAR magnetic field measurements play an essential and fundamental role in the remote sensing of the composition and internal structure of Eros and have direct relevance to the understanding of the early processes of formation, evolution and differentiation of planetary bodies. NRM is expected from spectral analyses of similar S-type asteroids which indicate the presence of iron, at least as a surface material.

Published measurements of magnetic field disturbances from the Galileo flyby of the asteroid Gaspra (Kivelson et al., 1993) and subsequent solar wind interaction modeling (Baumgartel et al., 1994) have furthered evidence of NRM for S-class asteroids. Kivelson et al. (1993) extrapolated magnetic field measurements from 250 Ra back to Gaspra and set limits on the NRM of 10^{-3} to 3×10^{-2} Am^2 kg^{-1}; the higher limit gives a surface field actually equivalent to that of the Earth. Nevertheless, the lower limit of the Gaspra inferences would indicate an approximately 50 nanoTesla (nT) field at the 1 Ra altitude orbit, and NEAR will remain within the Eros magnetosphere for extended periods. If Eros has even smaller NRM, the magnetic sphere of influence will be smaller than NEAR's 35-km altitude orbit (perpendicular to the sun–asteroid line), and the strength of the moment must be assessed indirectly by studying the magnetic perturbations associated with the Eros bow wave. Special bandpassed channels have been designed into the NEAR MFI, channels which are based on previous boundary identification from Earth-orbiting satellites (Anderson et al., 1990, 1992) and which have been suggested as generalized low-altitude as well as magnetopause boundary identifiers (Zanetti, et al., 1995). For intermediate NRM, the spacecraft may intermittently cross the Eros magnetopause, and magnetic fluctuation measurements will be used to assess these events.

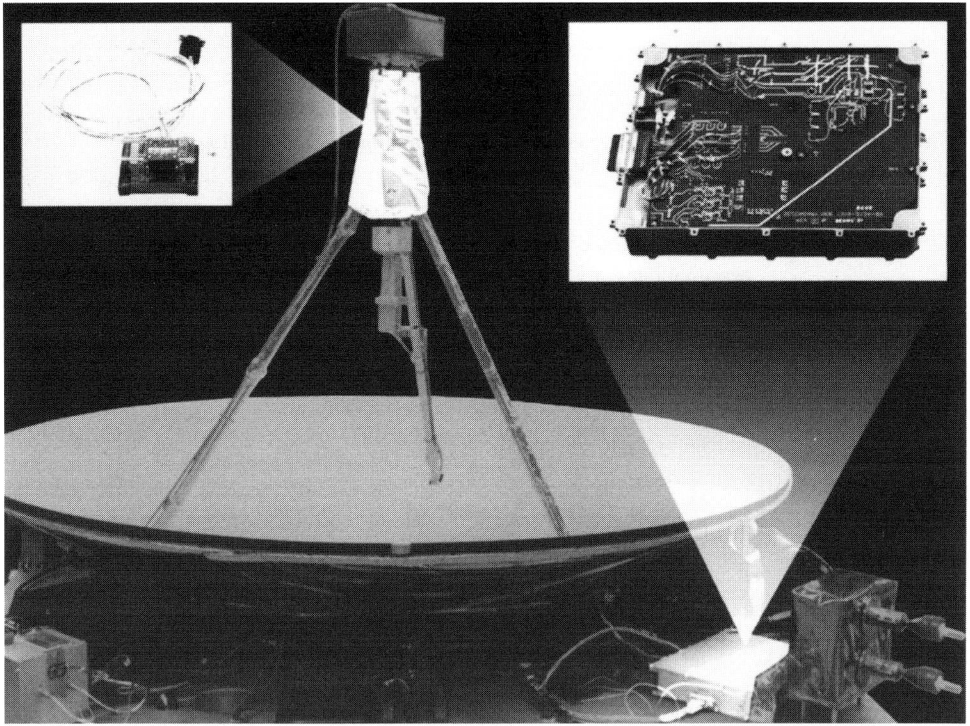

Figure 1. Schematic of NEAR MFI spacecraft installation, specifications.

2. Instrument Description

The heritage of the NEAR 3-axis fluxgate magnetometer includes GSFC (Acuña et al., 1974, 1992) and APL (Zanetti et al., 1994) instruments aboard Voyager, Freja, WIND, Mars Observer, Geotail, Giotto, Magsat, Viking, Pioneer, POGS, AMPTE/CCE, UARS, and DMSP. The NEAR MFI instrument is a joint hardware effort between GSFC and APL. The magnetometer sensor mounting on the high-gain antenna feed structure is similar to that used on the successful Giotto mission to the Comet Halley. Figure 1 shows the sensor configuration and details of the MFI electronics package which is mounted on the top deck. The analog and digital electronics package used a recent chassis (one-half of the Swedish Freja magnetometer) design in order to develop a fast mechanical interface definition. The chassis and GSFC and APL electronics boards were designed, fabricated, integrated, tested and delivered within one and a half years' time without a full time dedicated effort. The fabrication philosophy was tailored to the fast development schedule of NEAR (24 month spacecraft plus instrument delivery) and necessitated a parallel approach in design and implementation. Thus the magnetometer package is not self contained as with past projects but rather the MFI shares a digital processing unit (DPU) with the NEAR Infrared Spectrograph (NIS).

Table I

Specification for NEAR magnetometer

- Three-axis fluxgate sensor, antenna feed mounted

- Electronics: analog sensor circuitry, ranging, A/D, filters; logic and gate array, and DC/DC convertor

- Output samples: 80 bits
    Rate: 0.01–20 Hz, commandable
    Nominal rates: 1 s^{-1} (DC), 0.0–1 s^{-1} digital field
    Band-passed: (AC) 1.0–10.0 Hz, r.m.s. amplitude

- Eight ranges:

Range	Maximum field	Magnetic resolution
0	±4 nT	*
1	±16 nT	*
2	±64 nT	0.002 nT
3	±256 nT	0.008 nT
4	±1024 nT	0.031 nT
5	±4096 nT	0.125 nT
6	±16384 nT	0.500 nT
7	±65536 nT	2.00 nT

*Resolution limited by instrument noise.

- Software: range selection, digital filtering, anti-aliasing, band-pass, noise rejection, command, control, telemetry formatting.

- Internal sample rate: 20 s^{-1} 20 bits resolution

- Resources: 1.5 kg, 1.5 W, 80 bits s^{-1} (nominal)

The NEAR MFI 50 nT measurement requirement (from the lower limit of the Kivelson et al. (1993)), was the basis for the accuracy level imposed on the entire system which consisted not only of the instrumentation but also of, in fact dominated by, the spacecraft itself. This requirement is further complicated in that the sensor is body mounted rather than on a boom. Up-front engineering was employed to reduce magnetic noise as much as possible without impacting schedule; in fact neither schedule nor funding permitted calibration in a magnetics facility beyond the package level. The magnetics procedure concentrated on dynamic interference signals; static fields were assessed, and compensated for, to many tens of nT in the presence of Earth's field. The philosophy adopted was to insure reasonable cleanliness such that a suitable, full dynamic range could be maintained that had sufficient digitization (16 bits) to retain resolution. The ±256 nT and ±1024 nT ranges have been maintained in flight, which have digitization levels of 8 and 31 picoTesla.

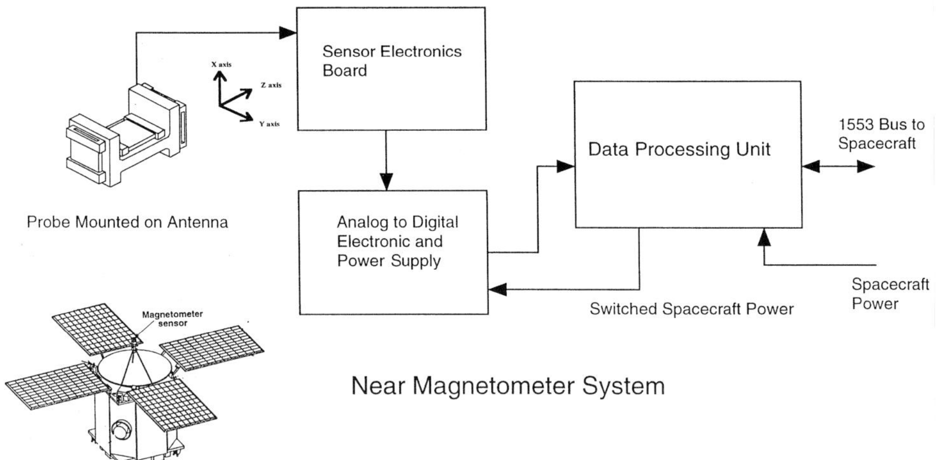

Figure 2. Block diagram of the NEAR Magnetic Field Investigation. Due to spacecraft resource limitations the triaxial sensor assembly was mounted inside the antenna feed structure as shown.

Table I includes a summary of the specifications which include commandable data rates that will accommodate on-board storage capacity in order to minimize ground operations. Three axis, full-range data are accumulated along with a band-passed channel, selectable to any one axis, to observe fluctuation information of solar wind boundaries and possible interactions with the asteroid. The eight ranges permit high-resolution measurements on all but the two most sensitive ranges as detailed in the table. The current spacecraft residual field prevents operation in ranges 0, 1, or 2 due to electronics saturation. The electronics samples the sensor coils internally at 20 Hz and the A/D converter has 20 bit resolution; 16 bits are transmitted for 3-axis data and the additional bits are used for the band-passed channel (labeled AC). Amplification was not needed on this channel as is usually done because the fluctuations in solar wind structures and body-solar-wind interaction disturbances are similar to the background field. The anti-aliasing and band-passed filtering, as well as noise spike rejection, are performed on an RTX2010 microprocessor. Filtering coefficients and frequency ranges, offsets and calibration factors are included in the on-board processing and can be uploaded during flight.

Figure 2 is a functional block diagram of the NEAR MFI instrument. The hardware used to implement the magnetometer was housed in two separate hardware assemblies. The hardware used to transform the magnetic signals to digital values was located on the upper spacecraft deck and consists of a small magnesium chassis containing two printed circuit assemblies, one from GSFC and one from APL. The sensor electronics is self powered to minimize interference leaking in from the spacecraft power system and was necessitated by the high-resolution and low noise requirements (Lohr et al., 1992). The processing part of the magnetometer was located on the lower deck and was shared with the NEAR Infrared Spectrograph

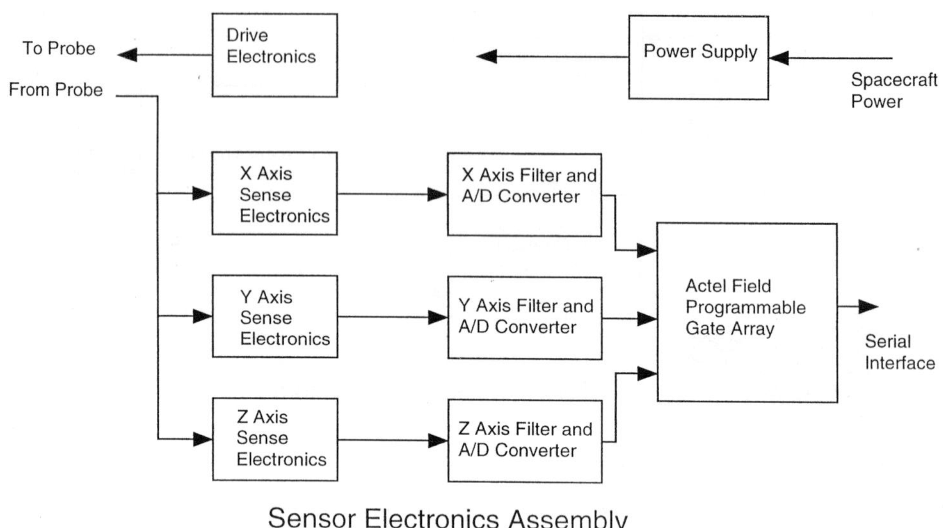

Figure 3. Schematic illustration of the NEAR magnetometer internal functional blocks. In addition to the three axis of magnetic signal processing, digital filters are included to detect higher frequency fluctuations in the ambient field.

(NIS) spectrophotometer. This NIS/MFI data processing unit (DPU) consists of six printed circuit assemblies forming a processing unit which collects digital data from the sensor electronics, processes and stores the data from the instruments, controls the experiments, and transmits the data via the MIL-STD-1553 interface to the spacecraft for transmission to Earth.

Figures 3 and 4 show the details of the sensor electronics unit and the DPU respectively. The sensor electronics contains two printed circuit assemblies. The first assembly contains a common circuit to excite all three axes of the probe and three identical circuits to null the field in their respective axes. The magnitude of the magnetic field used to null the externally sensed field is a parametric indication of the true field at the sensor. The output voltage is thus proportional to the magnetic field at the sensor. This parametric voltage is then passed to the second assembly where it is filtered with a single pole, low-pass filter and then converted to a digital value. The unit also contains electronics to condition the probe temperature sensor to analog voltage required by the unique low-magnetic-signature temperature sensor used. The unit contains circuitry to cause a $\frac{1}{4}$ full scale offset in the electronics to permit calibration of the electronic gain present in the system.

The second assembly (right half Figure 3) contains the power converters for both units, the A/D converters and the digital logic to collect data and control the sensor electronics. The power conversion is handled by a hybrid module converter providing $+15$ V and -15 V. The $+5$ V required by the A/D converter is obtain by a linear regulator. The A/D conversion utilizes a sigma delta technique to achieve its high resolution and inherent high-frequency rejection. The A/D converter also

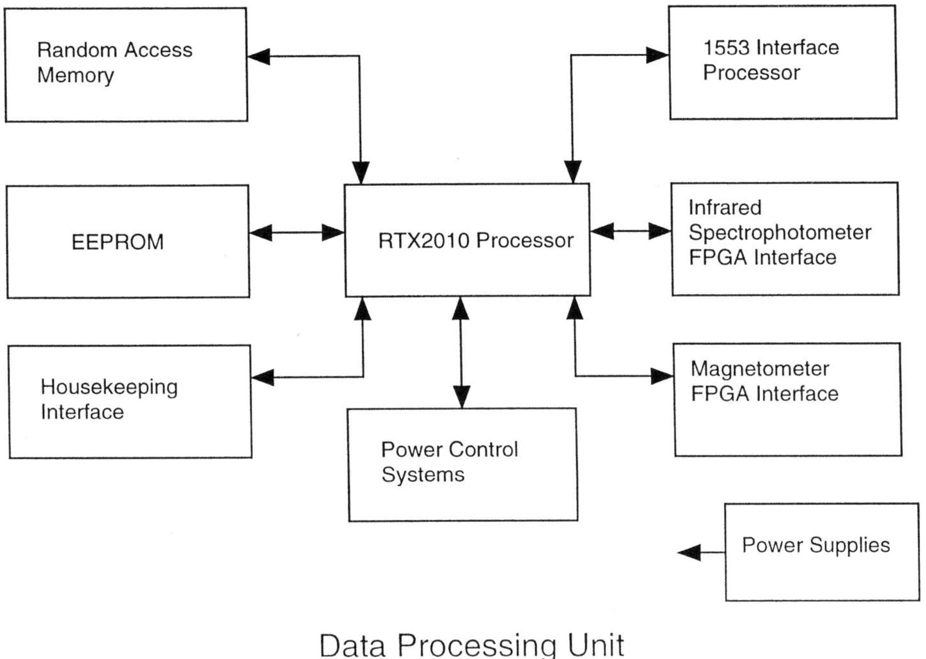

Figure 4. Schematic illustration of the NEAR DPU internal functional blocks, FPGA is for the field programmable gate arrays by ACTEL.

contains circuitry to perform internal self calibration to maintain its high accuracy after radiation induced component parametric changes. The A/D converters' ultra-low power requirements permit linear regulation to power the unit affording additional power-supply interference rejection. The digital signal processing utilized by the A/D converter appears very much like a low-pass filter on the analog data without adversely affecting the converted digital data in the pass-band. The data from the A/D converters is serially sent to the field programmable gate array (ACTEL, ACT II family). This gate array contains circuitry to control the range setting of the analog electronics, the operation of the internal calibration circuitry, the internal calibration of the A/D converter, the control of the A/D converters and the collection of the digital data for transmission to the DPU. The gate array communicates with the DPU via two serial data lines with a common clock. This serial interface minimizes connection to the sensor electronics box to minimize potential interference coupling into the high-resolution, low-noise circuitry. The entire sensor electronics assembly is isolated from its chassis to minimize ground-induced errors and currents into the assembly. This approach has yielded resolution much greater than possible, given normal spacecraft-induced interference. This extra resolution has insured that the spacecraft will be the ultimate limit on the resolution of the magnetometer system.

The block diagram of the DPU is shown in Figure 4. The DPU contains a general purpose processor, a special purpose 1553 processor, memory, and interfaces to complete the magnetometer system. The unit utilizes a radiation hardened RTX2010 for general instrument control and for data processing. The computer as well as the FORTH operating system (Hayes et al., 1987) are well suited for these functions. The program for controlling the processor is contained in electrically erasable programmable read-only memory (EEPROM) which is copied to random access memory (RAM) for execution. The use of EEPROM permits changes in the programming after launch should the need arise from unforeseen requirements. The DPU also contains a specialized processor to handle the interface with the 1553 bus utilized to transmit the data to the spacecraft. This interface is bidirectional and is used to receive commands sent to the experiment. The ACTEL interfaces to both the NEAR MFI and NIS experiments are contained in the DPU.

3. Flight Software

The common NIS/MFI DPU controls and monitors the NIS and MFI instruments, as well as their data and command streams. The DPU implements the interfaces between these instruments and the redundant Command and Telemetry Processors (CTPs), which provide command routing, data storing, and downlink services to all of the common DPUs. The DPU and the CTPs communicate with each other via a MIL-STD-1553 bus, with the active CTP acting as bus controller and each of the DPUs acting as a 1553 remote terminal. There are also 2 Mbit s^{-1} serial links from the NEAR Multi-Spectral Imager DPU through the redundant CTPs to the redundant solid state recorders for high-speed image data transfer. Details of the NIS/MFI DPU software which is common to all the NEAR DPUs is available (Hayes, 1995a).

The DPU receives vector magnetic field samples, 20 bits per axis, from the magnetometer electronics at 20 Hz. It filters and subsamples this 20 Hz data to generate the following output rates per second: 0.01, 0.02, 0.05, 0.1, 0.2, 0.5, 1, 2, 5, 10, or 20. Since all data are discretely sampled at 20 times a second the possibility for aliasing of higher frequencies to lower frequencies exists if we simply throw away data samples in order to reduce the data actually telemetered. The simple analog filters would still permit extensive aliasing when mission requirements limit data rates to much lower sampling rates. Nyquist sampling criteria require that a continuous frequency domain signal be sampled at a rate greater than twice the frequency of interest in order to recover the information from the discrete digital samples. It is possible to correctly reconstruct continuous data from discrete sampled data as long as the sample rate is greater than twice the sample signal. In order to minimize Nyquist sampling limitation errors we employed digital filters to limit the bandwidth of the signals when in reduced data rates.

The DPU continuously builds telemetry packets with this processed magnetometer data (processed by the DPU). It starts each packet with a housekeeping header, sampled at the time indicated in the secondary packet header. It fills the remainder of the packet with 44 samples of magnetometer data, using the 16 most significant bits of each processed DC field strength value. When it completes a full telemetry packet, the DPU sends it to the CTP for downlink, and starts preparing the next telemetry packet.

The DPU controls power to the magnetometer electronics via the magnetometer on/off discrete output. This discrete output controls a solid-state relay that turns the $+28$ V primary power to the DC/DC converter in the magnetometer electronics (Figure 3) on and off. The MFI converter provides $+15$ V, and -15 V to the sensor electronics. The DPU controls three discrete lines which set the gain of the magnetometer detector's front-end electronics. The DPU sets these bits as determined by the 'range automatic/manual' command. In manual mode, the DPU sets the bits to the value in the most recent 'manual range setting' command. In automatic mode, the DPU implements an auto-ranging function by examining each X, Y, and Z magnetic field strength sample, and determining how to set the range bits for the next sample with the following logic:

(1) find the component (X, Y, or Z) with the largest absolute value, V; (2) if ($V > 0.875 \times$ full scale of current range) and (current range <7) then increment range; (3) if ($V < 0.19 \times$ full scale of current range) and (current range >1) then decrement range.

Note that the DPU output data includes the range setting for each vector sample, so the absolute magnetic field strengths may be determined.

The magnetometer contains a calibration circuit which can inject a known bias current into the front-end detector's electronics, simulating a magnetic field offset of approximately one-half of full scale in each axis. The DPU controls this calibration circuit by means of the 'calibrate on/off' command. The A/D converter in the magnetometer electronics has a calibration mode which is exercised periodically by command. When the DPU receives the 'calibrate A/D' command, it finishes building the current telemetry packet and queues it for transmission, then sets the 'calibrate A/D' bit in its interface to the magnetometer. For the next two 20 Hz magnetometer cycles the DPU discards the serial data which the magnetometer sends. At the end of these 2 cycles, it resets the 'calibrate A/D bit' and then resumes normal data collection from the magnetometer on the following cycle.

The processing which the DPU performs depends on the commanded state of the instrument: output sample rate, noise-rejection on or off, software anti-alias filters on or off, and AC axis selected. If noise rejection is on, the DPU first performs a noise-rejection function on the 20 samples/second data from all three input channels to eliminate large spikes (see Appendix for details on the rejection algorithm). The output sample rate (0.01, 0.02, 0.05, 0.1, 0.2, 0.5, 1, 2, 5, 10, or 20 Hz) determines how the DPU filters (anti-aliasing) and the 20 Hz magnetometer data fit within the allotted telemetry bandwidth. If software anti-alias filters are on, the DPU performs a digital low-pass filter on the 20 Hz output samples from the noise-rejection logic, with a cutoff at one half the output sample frequency. The DPU subsamples the resulting data to match the output sample rate and builds telemetry packets with the samples selected. Software filtering is not necessary at the 20 Hz output frequency, because the analog signals are already filtered in hardware with a 10 Hz cutoff (3 dB point with 6 dB/octave) and the sigma delta A/D convertor adds an additional 6 dB per octave from its 3dB point of 15 Hz. In addition, for all output rates of 1 Hz or less, the DPU uses the low-pass filter with 0.5 Hz cutoff (3 dB point with 66 dB/octave) and subsamples the results as necessary to produce the desired output rate. This approach allows some aliasing, but prevents excessive smearing of events in the time domain if additional digital filtering were employed. Note that to process the data, both for noise rejection and anti-aliasing, the DPU must normalize each sample based on the range in effect when the sample was taken. By doing so, transitions across range changes are smooth. After anti-alias filtering, the DPU rescales each output sample back to the range originally indicated for the corresponding input sample from the magnetometer.

In addition to the vector field values, the DPU also computes a 9-bit AC axis value for each sample. It computes this value by filtering the X, Y, or Z axis field values (after noise rejection), as indicated by the 'AC axis select' command, with a 1–10 Hz passband digital filter. To produce an output value to include with each subsampled magnetic field vector, the DPU computes the root mean square amplitude of this filtered signal over the subsampling interval. It then codes the resulting value into a 9-bit data item with a pseudo-logarithmic representation. The DPU generates the magnetometer's real-time housekeeping data, which consists of: output sample rate, AC axis selects, software anti-alias filters on/off, range manual/automatic, range, noise rejection on/off, calibration on/off, calibration normal/flip, calibrate A/D, spare, +28 V current, probe temperature, magnetometer electronics temperature, number of noise rejected points in most recent science record, Bx, By, Bz. The time tag indicates the time of the first magnetometer data sample in the science record, and provides more accuracy than the time tag in the packet header. The time tag is the delta time, in 0.05 s increments, of first sample from the 1 s time tag in the packet header. The other housekeeping data items reflect the commanded state of the instrument.

4. Calibration

The NEAR MFI was calibrated at the package level at the NASA/GSFC 40-ft magnetics facility. The objective of this calibration was to test the various ranges to as sensitive a level as possible. The most important aspect of this test was to evaluate the linearity, orthogonality, and cross talk of the sensor block for each axis and for each range. As discussed above the instrument has dynamic ranges (7 to 0) of ±65536 nanoTesla (nT) to ±4 nT with 20 bit internal data, all of which were recorded during the calibration; 16 bit data are downlinked during flight. The linearity was calibrated down to range 3 (± 256 nT), at which point the GSFC facility was at a comparable digitization noise level; the linearity procedure was continued at APL in concentric permalloy shield cans (zeroed to less than 1 nT) to exclude the ambient field as well as external noise. The linearity tests in the shield cans were performed with a finely varied, calibrated current source exciting a solenoid surrounding the sensor; Range 6, selected for maximum linearity, was used to normalize the transition from the GSFC test data to the more sensitive range tests performed in the shield cans Appendix Tables A1 and A2 list the input calibration data as the field was changed in each direction parallel to the X, Y, and Z sensor block axes, for ranges 7 to 3. For the sensor alignment calibration, a constant 15 000 nT field was applied in all six directions to assess the leakage or cross-talk field in orthogonal axes. For example, Appendix Table A1 (bottom left) indicates that the 15K E from the facility imposed field gave the dominant field in Z and an approximately 0.6% leakage field in both X and Y.

Figure 5 illustrates the procedure used to assess the linearity of the different ranges and axes as well as the quality of the test data. Figure 5(a) is a plot of counts vs applied field (nT) for the y-axis sensor in range 3 (± 256 nT). The bottom panel is the residual counts as a function of the applied field which is used to determine the goodness of fit. This example was shown to illustrate the analysis process and the errors encountered, e.g., the range 3, y-axis, -100 nT applied field test point (APL shield cans) was in obvious error; with this point removed, the fit parameters improved to acceptable quality. Figure 5(b) shows the linearity fit for range 4, y-axis data, using GSFC's magnetics facility test data, along with residual errors at nominal levels. Table II is a list of the linearity test results performed at the GSFC's magnetics facility for ranges 7 through 3. The APL data was collected using permalloy shield cans with less than 1 nT residual field to obtain calibration coefficients for ranges 0, 1, and 2. The coefficients for the two different calibration techniques are remarkably similar (% Dev. column) given the different input test data sets. One should note that the numbers presented in Table II are given in raw counts of 20 bit values as displayed on ground-support equipment without any scaling or offsets.

The offsets from the prelaunch tests, inferred from Figure 5 for zero applied field, are not meaningful and are being further assessed with in-flight measurements of the solar wind and intercomparisons with the NASA WIND spacecraft data. Following the conversion of the range 6 test data to magnetic field values, the sensor alignment data (Appendix A, Table A2) were used to determine an orthogonality, cross-talk matrix which will convert the registered counts to ideal, sensor-block-oriented magnetic-field measurements. Further rotations will be used to transform the data to spacecraft coordinates and then to physical coordinates. The preliminary alignment matrix, based on the prelaunch linearity testing, is listed at the bottom of Table II and is applied to the data after the scaling coefficients of the appropriate range have been used to get the measurements into magnetic field units (nT).

D. A. LOHR ET AL.

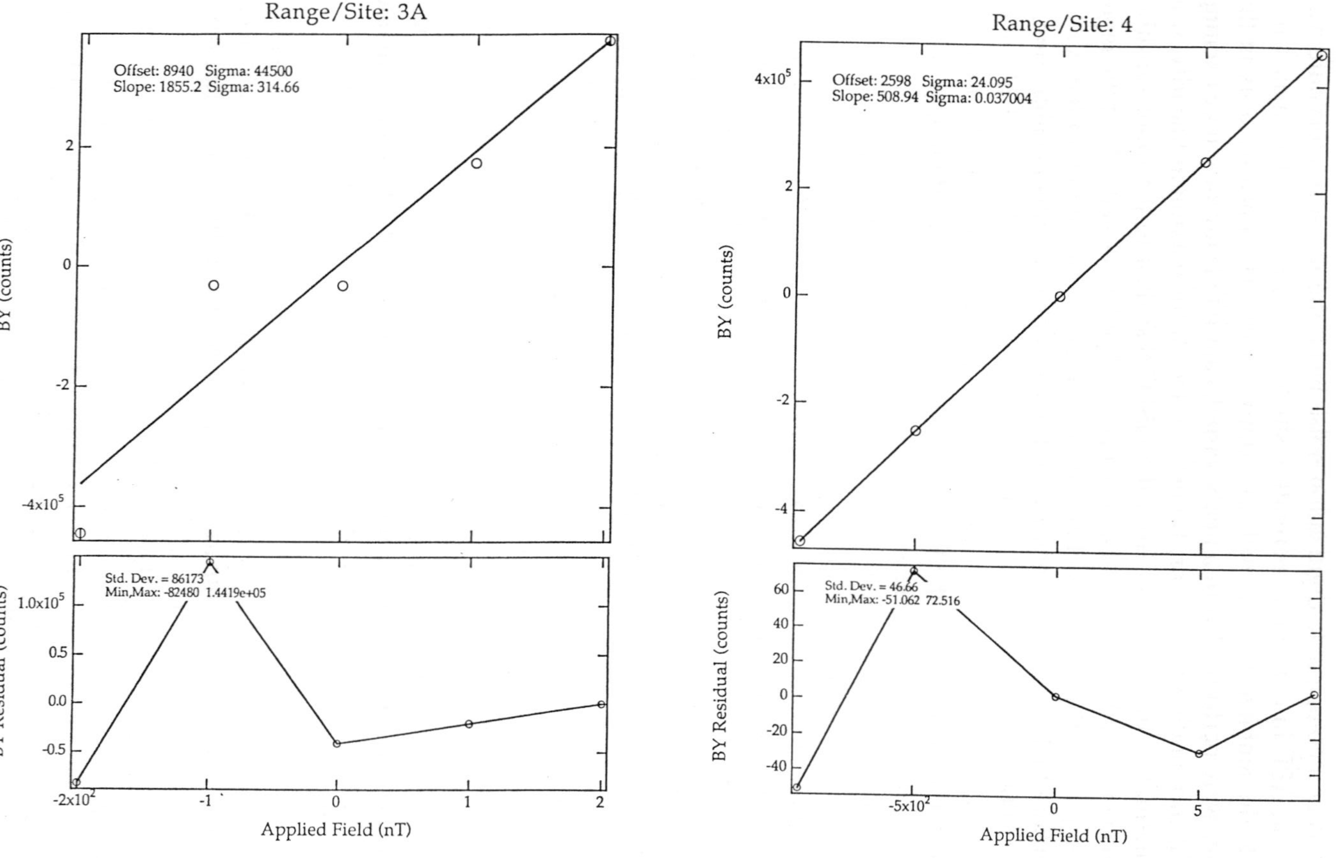

Figure 5. Graphs of linearity results, response *vs* applied field, for Ranges 3 and 4, ±256 nT and ±1024 nT, respectively.

Table II

NEAR magnetometer calibration: counts nT^{-1} (20 bit values)

Range	Component	GSFC result	APL result	% Dev.
0	X		124850 ± 1300	
	Y		126700 ± 920	
	Z		125700 ± 1400	
1	X		32320 ± 55	
	Y		32316 ± 66	
	Z		32596 ± 24	
2	X		7927.5 ± 6.3	
	Y		7981.2 ± 6.7	
	Z		8015.5 ± 2.9	
3	X	2019.3	2020.0	−0.035
	Y	2027.1	2027.5	−0.020
	Z	2038.8	2040.3	−0.074
4	X	509.90	510.06	−0.031
	Y	508.94	508.85	0.018
	Z	513.12	513.61	−0.095
5	X	129.75	129.83	−0.062
	Y	129.34	129.32	0.015
	Z	130.51	130.66	−0.115
6	X	31.867	31.872	−0.016
	Y	31.869	31.871	−0.006
	Z	32.087	32.106	−0.059
7	X	8.1103	8.1143	−0.049
	Y	8.1010	8.1030	−0.025
	Z	8.1616	8.1717	−0.124

Inverted correction matrix

1.0000182	0.0049797	0.0055376
−0.0014784	0.9999819	−0.0055371
0.0045922	0.0019468	1.0000148

5. Magnetics Program

The design approach achieved by the NEAR magnetometer effort entailed low-cost, up-front attention to engineering solutions that did not impact schedule. The goal of the magnetometer team is to achieve reliable measurements within 5–10 nT; recall the more rigid requirement to assess a 50 nT intrinsic Eros magnetic field. The focus of the magnetics effort was on dynamics, especially dynamics that are not traceable, such as the autonomous heater circuits. Internal wiring loops on the spacecraft terminal connection board for the heater circuits were individually compensated with current reverse loops. Single-point ground was implemented from the beginning to reduce noise from unintended ground loops. The engineering solutions to divulged problems were not comprehensive but limited to those that could not be dealt with subsequently by processing or spacecraft modeling. For example, magnetic fields created by commanded, traceable subsystem loads are being modeled and removed during analysis using housekeeping monitors of subsystem currents. Thermal-vacuum testing at GSFC was performed with small 'Helmholtz' coils rigged around the high-gain antenna tube and driven by constant current supplies. This arrangement allowed lower dynamic ranges to be used and subsystem noise levels to be better assessed.

Figure 6 summarizes a few of the major systems that had design input from the magnetometer team. Design effort for low-magnetic signal was applied to the power-system electronics as well as the solar-panel circuit layout. Interference from the solar arrays was further minimized by taking advantage of the symmetry of the four panels that surround the center-axis location of the magnetometer sensor on the high gain antenna feed structure. To maintain the symmetry the shunting sequences were balanced from one side to the other. The solar cell strings were not individually back wired but levels below 1 nT were determined from balanced cancellation (see Figure 6, graph of Solar Panel Contamination Magnetic Field, the NEAR MFI sensors are approximately 1.5 m from the panel centroids). A custom-wired-model panel was suggested to minimize the field at the specific sensor location, but was not implemented.

Generally static fields were individually checked and most subsystems would not interfere with the measurement objectives. The major exception to residual magnetization levels, noted in Figure 6, consisted of the latch valves (total of 9 permanent magnets) from the propulsion system. An initial survey of latch valve assemblies was performed at Aerojet with support from UCLA. The engineering model of the Vacco latch valves was independently at about 800 nT at 1 m distance (also confirmed by similar UCLA work on Cassini). It was also determined that the valve position, open or closed, had little effect on this leakage field. Metglass® shielding was installed in seven levels but failed to contain the magnetic flux. The performance of the valves was tested and was unaffected by the shielding (possibly improved). Metglass shielding has been successfully applied at a stand-off distance from smaller moments for latch valves on the NASA/Advanced Composition Explorer spacecraft. Due to the high fields involved at the latch valves the Metglass saturated and was ineffective at reducing the external field. Approximately 800–900 nT of static field from the NEAR propulsion system latch valves was nulled with the careful placement of compensation magnets onto the spacecraft which countered other spurious fields as well. These magnets and their placement are noted in the top deck schematic at the bottom of Figure 6. The procedure for doing this compensation is as follows. Before the antenna and flight sensor were installed, a station magnetometer was positioned, during quiet times but in Earth's field, at the flight-sensor position. The spacecraft was moved into and out of this position until repeatable background levels fell to within about 10s of nT.

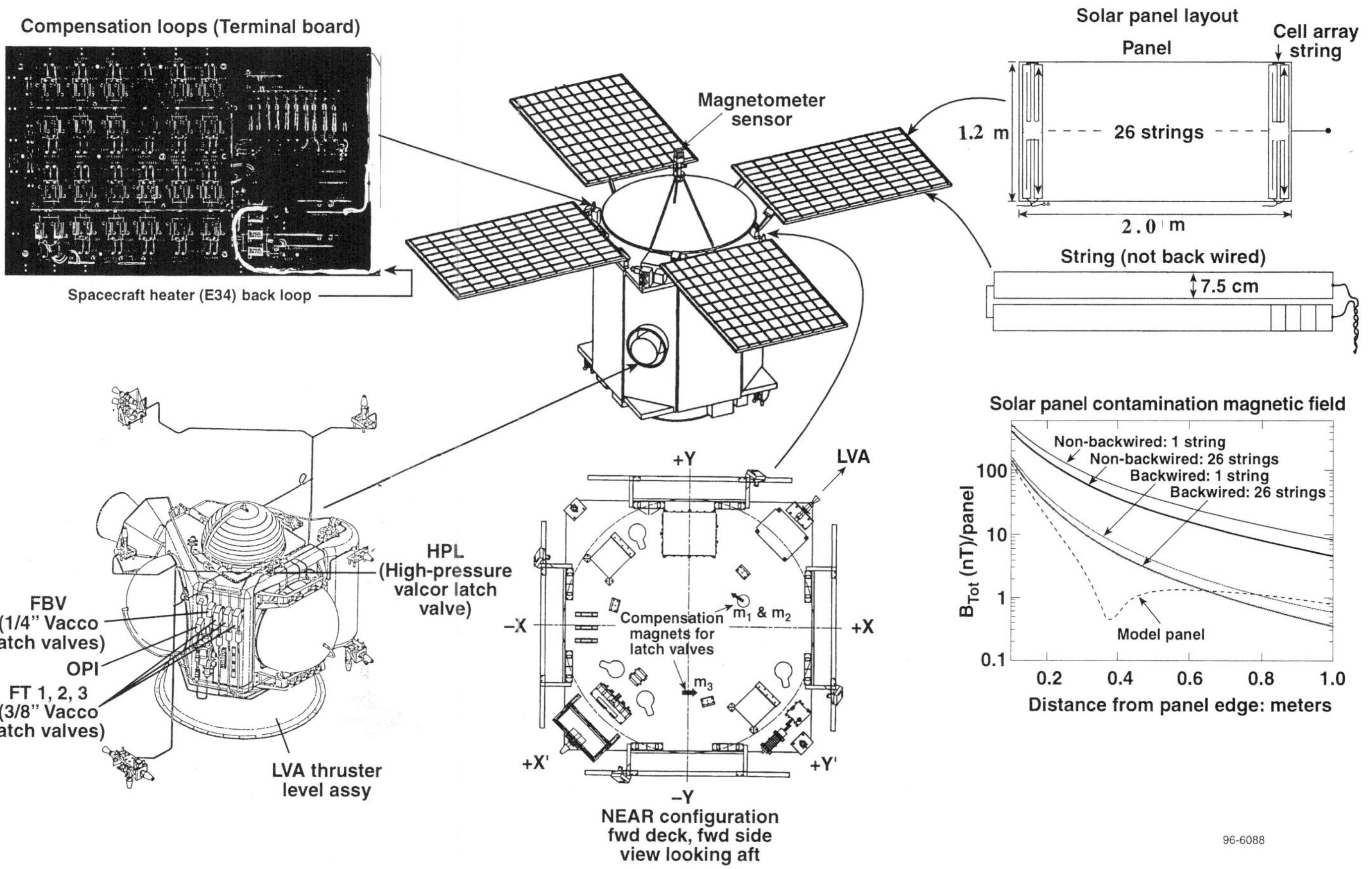

Figure 6. Magnetics summary for the NEAR MFI program.

Table A3 is a detailed list of the individual systems or problems that were addressed and actions taken with regard to magnetics. For example, custom RF connectors were fabricated of nonmagnetic beryllium copper to minimize magnetic fields local to the sensor. The goal of NEAR's magnetics program was to maintain the bulk of the measurements distant from the asteroid within the 256 to 1024 nT full scale ranges in order to provide sufficient resolution. The objective goals of an Eros magnetic field measurement of 50 nT as a lower limit should be easily realizable, especially with the advantage of the integration time of one year's orbit.

6. First NEAR Magnetometer Data

The nominal telemetry data rate for NEAR at the Eros orbit is about 4 kbps at an approximately one-third duty cycle. Given the present existence of large capacity magnetic disks, data will be archived as on-line files. Given the present network capability through the World Wide Web (WWW), some data as well as file address access is through the NEAR Home Page (http://sd-www.jhuapl.edu/NEAR). The Hierarchical Data Format (HDF) standard is in use for the MFI flight data, as well as the spacecraft housekeeping, presently totaling about three months of 1 minute data, centered around the sun–Earth heliocentric longitude location of NEAR (1.2 to 1.7 AU) from April 27 to August 8, 1996.

Pre-launch data are archived from the ground support equipment (GSE) during operational testing, GSFC thermal vacuum, and other field testing into a HEX format with reading and IDL plotting routines available from the above Home Page address. The access procedure is to: (1) go to Science Data Center; (2) go to Pre-Launch; (3) go to MAG file listings; (4) list README.TXT. Housekeeping files are archived as HDF files as well, which contain stated information on the circuits powering various subsystems. In-flight test data have been used to correlate magnetic field shifts with the NEAR X-ray/gamma-ray instrument and an associated thermo-electric cooler loading of the power system which creates a predetermined field from the current loops on the terminal boards. Also correlated with field shifts are digital shunt circuits controlling the individual circuits of each panel and thus unbalancing the signature cancellation due to the symmetry of the panel circuits as discussed above. The housekeeping current is being used to model the dynamic spacecraft magnetic signature, which will then be subtracted from the instrument data. Demonstation of this spacecraft preliminary model applied to in-flight calibration data is shown in the companion science paper (Acuña et al., 1997).

Figure 7 summarizes access to and display of pre-launch calibration data; some post-launch data are available in a similar manner. The figure is a screen dump of WWW access to GSE archived files labeled with local time and day-of-year extension. In the bottom section is a listing of an ASCII data file from November 17, 1995 (DOY 321) ordered by: mission elapsed time (MET) in seconds, Bx, By,

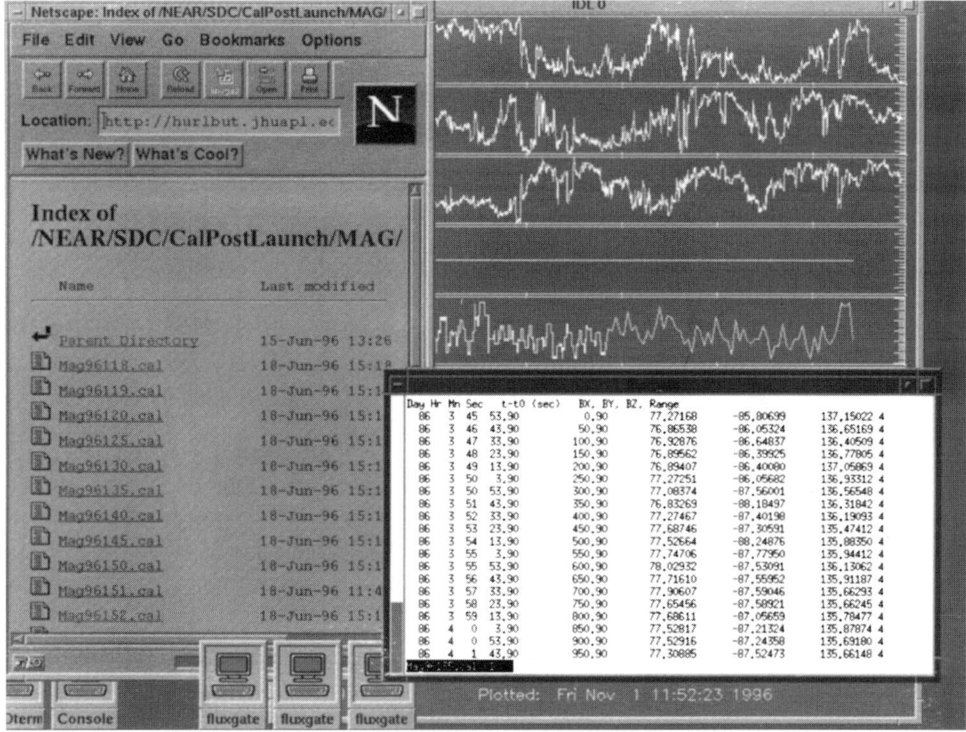

Figure 7. WWW access for NEAR MFI data: http://sd-www.jhuapl.edu/NEAR.

and Bz in sensor coordinates, AC channel, noise-spike counter, range, counts x, counts y, and counts z. These data were collected during thermal vacuum testing at GSFC. In the middle right area is an IDL plot (using rpljz_near.pro) of the test data showing a 10 nT positive level shift in the Bx measurement that was correlated with a switch in antenna and telemetry power units.

The data at URL (http://sd-www.jhuapl.edu/NEAR/SDC/SDC_HDF/MAG) will be archived by year and day of year. The data files are in Hierarchical Data Format. HDF is a standard for science-data files developed by the National Center for Supercomputing Applications to provide machine independent, self-documenting files capable of containing complex data types. The HDF files organization, is based on the experiment data record defined in the instrument DPU requirements document (Hayes, 1995a, b). Each record (described in detail in Appendix Table A4), contains time-tag information, housekeeping values and sets of 44 samples of magnetic-field values, AC-field values and noise flags. Examples of IDL-read routines are also available through the WWW project areas that are controlled by the individual science teams.

Figure 8 shows the relative locations of the WIND and NEAR spacecraft near the April 26–28, 1996, conjunction date. Figure 9 is a composite of NEAR MFI first data on the top panel (Bx, By, and Bz spacecraft coordinates in nT) to be

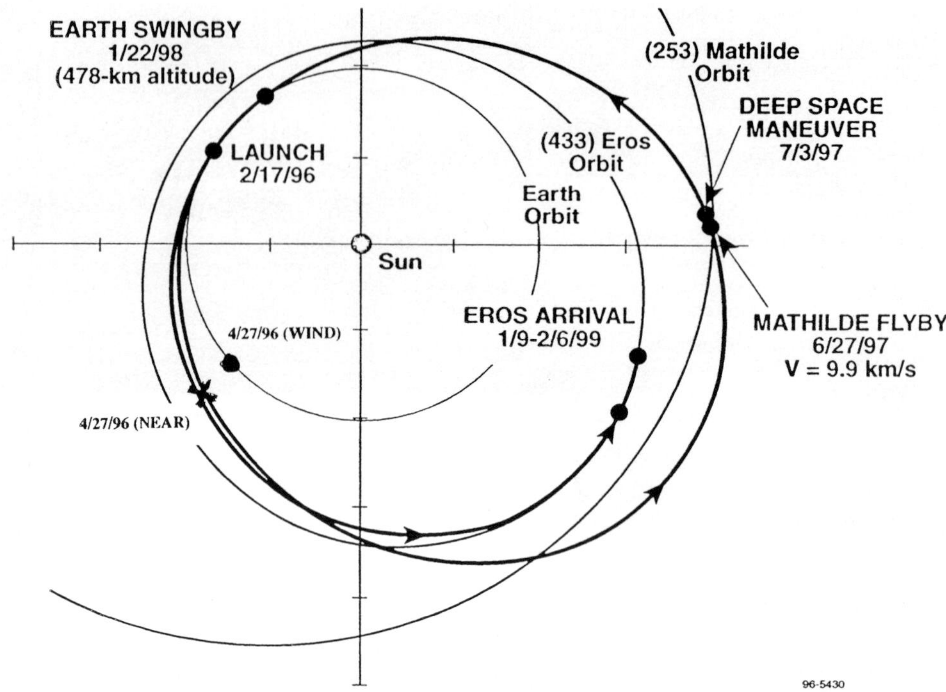

Figure 8. Relative locations of the NEAR and WIND spacecraft, April 27, 1996.

calibrated against the WIND Magnetic Field Investigation measurements on the bottom panel (Bx, By, and Bz Geocentric Solar Ecliptic coordinates; on WWW from R. Lepping, WIND Principal Investigator; these WIND data were available with a few days delay through the WWW WIND Home Page and are for survey purposes only). These one-minute data cover a few days from April 26–29, the first recording period for the NEAR MFI instrument. The opportunity exploited here is the sun–Earth line conjunction of NEAR discussed above to compare with WIND data; the time lines of Figure 9 have been shifted by about one day, the approximate and average shift required for the simple travel time from 1.0 AU (WIND) to 1.25 AU (NEAR).

No obvious correlation can be seen (BXGSE for WIND is approximately equal to BZ spacecraft for NEAR, thus BX and BY are approximately transverse to BXGSE) for the hour time scales, longer integration periods will be analyzed (although interesting similarities appear in transverse fields on the first day in each data period). Be advised that these data and quoted results are at an extremely preliminary analysis level; nevertheless, the NEAR MFI first data are shown in Figure 9 with baselines removed of: $BX_0 = 83$ nT, $BY_0 = 73$ nT, and $BZ_0 = 133$ nT. These are stable offsets combined with a sum of expected powered subsystem circuit interference; this sum will be dynamically removed by the magnetic model of the spacecraft, driven by the housekeeping current monitors. Many features still

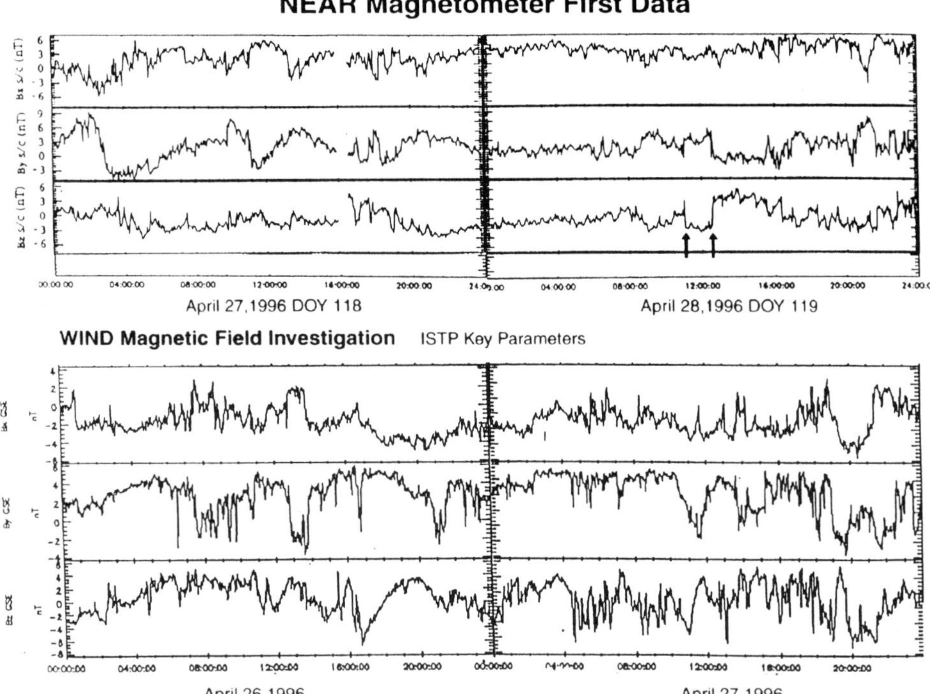

Figure 9. First data for the NEAR MFI (spacecraft coordinates), WWW ISTP Key Parameter preliminary data for WIND (courtesy, R. Lepping).

need to be removed from the NEAR MFI data, such as those indicated by arrows at about 12 UT on April 28, 1996. In addition, it is very encouraging to see the similarity in the amplitude of the small time-scale features between both data sets, indicating that the NEAR MFI is truly observing solar wind features; in fact, the RMS noise level appears to be well below the 1 nT level. These are similar data as will be seen on approach to Eros, but reduced in amplitude by about a factor of two. As a further analysis, the preliminary spacecraft magnetics model has been applied to an exceptionally quiet (quiet in both solar wind and spacecraft activity), six-day period from June 18–23, 1996. The purpose was to verify the above mentioned discussion with regard to the design intending to be spacecraft noise limited, not instrument noise limited, and to test the ability of the MFI to observe a natural solar wind spectrum. Figure 10 is an FFT power spectrum of log power $(nT^{-2}\ Hz^{-1})$ vs log frequency (milliHz) of 10 min data from the x sensor for June 18–23. The spectrum appears to be a decreasing power law function with possible event peaks at about 0.02 and 0.05 mHz (14 and 5 hour periods). In addition the high-frequency part of the NEAR MFI spectrum (Figure 10) compares favorably to one having a $(frequency)^{-(5/3)}$ dependence which is the representative spectrum

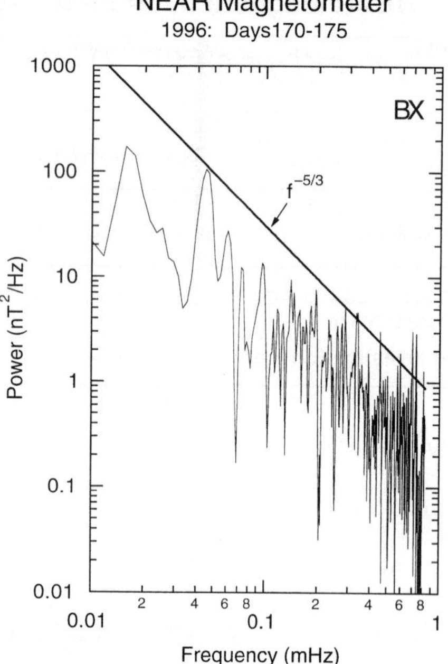

Figure 10. FFT of first data collected.

for the Kolmogorov theory of turbulence (1941), and which is typically found in the solar wind.

7. Summary

In conclusion, the NEAR MFI effort has succeeded in providing magnetic field measurements for the first flight in NASA's Discovery line. The primary objective of the investigation is the search for body-wide magnetic fields of the asteroid Eros, and therefore magnetic requirements were not as stringent as in traditional interplanetary missions. This objective, coupled with a year's worth of orbital data around Eros, allowed the low-cost approach to be relatively risk-free. The success, however, has been a few orders of magnitude better than the requirements (although close to the optimistic goals) due to common sense, but diligent, cleanliness and interplanetary quality instrumentation. NEAR will rely on in-flight calibration to produce averaged solar wind measurements that may approach the averaged solar wind measurements of the Giotto and WIND magnetometers.

The magnetics program was conducted in a noninterference manner, no observatory level magnetics facility testing was available in the tight 24-month delivery schedule. Design effort to minimize magnetic interference was inserted to the extent possible and engineering solutions were exercised when problems arose.

Subsystems were tested or examined individually and the spacecraft was surveyed and compensated for near the end of integration, albeit in the Earth's field.

The NEAR MFI is a premier example of the new Discovery approach. This investigation was achieved with approximately $\frac{1}{25}$th of the resources allotted to the instrument suite which included the modest magnetics program. The magnetometer instrument sensor and electronics itself is typical of previous interplanetary GSFC magnetometers, approaching picoTelsa noise levels and nT offset drifts over mission lifetimes; the instrument itself is not the limiting factor in the overall measurement noise level. The NEAR MFI is working better than expected and there is full confidence in achievement beyond mission goals. Contracted schedules and limited resources do not necessarily reduce the quality of the experimental data.

Acknowledgements

The UCLA magnetics group is acknowledged, in particular, Robert Snare, for supporting the Aerojet propulsion magnetics testing. The authors and NEAR Science Team wish to acknowledge the excessive contributions of the NEAR engineering teams, particularly given the unprecedented schedule of 26 months from project go-ahead to launch in February, 1996. We wish to thank the NASA/Discovery management of Headquarters for the proper and on-time support and freedom to achieve this mission. We acknowledge the dedicated Mission Operations for the data already delivered and those to come.

Appendix A

NEAR MFI Noise Rejection Algorithm

The MFI noise-rejection algorithm replaces samples which are too far out of line with the average of the previous and following samples (i.e., a straight line interpolation). The processing computes a median first difference function to determine whether each sample is 'too far out of line'. Given a time series sample xi, it computes the first difference series 'd1' of the 21 samples centered on xi:

$$d1i-9 = (xi-9) - (xi-10)$$
$$through$$
$$d1i+10 = (xi+10) - (xi+9)$$

It then sorts the absolute values of these differences and computes 'd1m', the average of the 5 values which fall in the middle of the sorted order (positions 8 through 12). Multiplying d1m by the commanded rejection threshold factor 'rtf' produces 'rt', the rejection threshold. The following three conditions must hold to reject xi:

$$|\,d1i\,| > rt;$$
$$|\,d1i{+}1\,| > rt; \text{ and}$$
$$d1i \times d1i{+}1 < 0$$

In other words, the magnitudes of the differences between xi and the samples immediately before and after xi must both be greater than the rejection threshold, and these two differences must have opposite signs. Normally, the noise-rejection software's output is the same as its input. However, it replaces each sample xi which meets all the criteria above with the average of the preceding and following samples:

$$(xi{-}1 + xi{+}1)/2$$

Notice that the software must retain the original values, since they may be needed in the median filter computation for later samples. The software also maintains a flag to indicate whether any 20 Hz raw X, Y, or Z axis samples have been rejected in each filtered, subsampled output magnetic field vector. This 'noise rejection applied' flag is part of each downlinked magnetic-field data sample.

Table A1

Raw calibration data values collected at GSFC Calibration Range

GSFC NEAR magnetometer package-level magnetics, April 28, 1995
Sensor block: X axis – UP; Y axis – SOUTH; Z axis – EAST

Counts (20 bits)	UP (X)	South (Y)	East (Z)	nT
Range 7: ±64 K	405610	405352	408310	50000
	202828	202740	204265	25000
	67	250	214	0
	−202675	−202220	−203825	−25000
	−405422	−404790	−407848	−50000
			(−367044)	(−45000)
Range 6: ±16 K	478705	479380	482710	15000
	239710	240355	242057	7500
	720	1335	1400	0
	−238320	−237677	−239250	−7500
	−477300	−492644**	−479905	−15000
				**bad data
Range 5: ±4K	454650*,[1]	427400**	456890	3500
	260140	259260	261140	2000
	620	600	120	0
	−258900	−258090	−260900	−2000
	−453600*,[2]	−452089	−456680	−3500
		* 3.5/4 times 1) 519600, 2) 518400		
Range 4: ±1 K	461770	460650	462780	900
	257650	257040	257650	500
	2800	2600	990	0
	−252220	−251880	−255500	−500
	−456060	−455500	−460820	−900

Range 3: ±256: data stored as files
too noisy
Sensor alignment

nT	X	Y	Z
50 K E	−2188	2374	408310
zero	69	252	220
15 K E	−610	888	122650
15 K W	744	−385	−122208
15 K N	675	−121227	452
15 K S	−534	121738	−19
15 K U	121718	425	−345
15 K D	−121580	72	780
zero	69	250	219

Inflight Cal: on/off Ranges 7-0

X	Y	Z	
142085	145235	142340	R7
160	205	227	
140205	143528	140865	R6
1066	1137	1470	
144100	14521x	142388	R5
2160	168	270	
148150	14337x	141020	R4
9000	99x	1630	
174800	1444xx	142400	R3
32880	−5xx	2300	
2683xx	1406xx	144xxx	R2
129xxx	−155x	1500	

Table A2

Raw calibration data values collected in magnetic shield cans

In-shield-cans calibration to extend ranges downward; X, Y, Z performed separately, zeros repeated; APL May 4, 1995; repeat of GSFC on May 2, 1995

	nT	X	Y	Z
R7	50000	405975	405285	408900
	25000	203023	202552	204564
	0	206 (154, 292)	34 (102, 10)	288(381)
	−25000	−202656	−202430	−203923
	−50000	−405472	−405098	−408319
R6	15000	479299	478617	483158
	7500	240306	239622	242496
	0	1295 (1236)	597(604)	1786 (1787)
	−7500	−237770	−238425	−239129
	−15000	−476848	−477520	−480020
R5	3000	392854	385022	392799
	1500	198115	190995	196785
	0	3388 (3388, 3300)	−3009 (−2948)	789 (830, 817)
	−1500	−191377	−196928	−195145
	−3000	−386154	−390913	−391165
R4	800	42148x	39584x	414660
	400	21746x	19227x	20921x
	0	1343x (1343x)	−11249 (−1123x)	3779 (379x)
	−400	−19058x	−21474x	−201649
	−800	−39460x	−41835x	−407129
R3	200	4541xx	35654x	41700x
	100	2522xx	15383x	21297x
	0	502xx (501xx)	−4888x (−4889x)	891x (90xx)
	−100	−1518xx	−25179x	−19515x
	−200	−3737xx**	−45437x	−39909x
R2	40	3152xx	3255xx	3561xx
	20	1575xx	16609x	1960xx
	0	−12xx (−12xx)	68xx(68xx)	359xx (358xx)
	−20	−1601xx	−1537xx	−1248xx
	−40	−3186xx	−3126xx	−2849xx
R1	10	3158xx	3488xx	3049xx
	5	1548xx	1882xx	1415xx
	0	−77xx (−79xx)	268xx (279xx)	−201xx (−218xx)
	−5	−1699xx	−1367xx	−1846xx
	−10	−3297xx	−2965xx	−3468xx
R0	4	478xxx	4887xx	420xxx
	2	220xxx	228xxx	170xxx
	0	−27xxx (−32xxx)	−19xxx (−15xxx)	−90xxx
	−2	−270xxx	−2739xx	−330xxx
	−4	saturated	$\longrightarrow$	

Table A3

Specific actions with respect to spacecraft's magnetics program

---

NEAR Spacecraft Magnetics: Compensation, sniffing

Concentration on dynamic fields, low impact/low cost

Goal: <10 nT dynamic at magnetometer probe

- **Single point ground, twisted pair distribution**

- **Power system**

    a) Solar panel/control circuit layouts modified for minimal magnetic moments

    b) 4—panel symmetry, shunt sequence to retain balance

    c) Battery/cell wiring for minimal magnetic moment

    d) Power distribution unit/relays static-tested cable routing for minimum loop

- **Instrument motors/reaction wheels tested** (<0.2 nT at the probe)

- **Nonmagnetic hardware in sensor close vicinity**

    High-gain antenna cables and stainless steel connectors replaced

- **Replaced 1553 stainless steel couplers**

- **Autonomous heater circuits, terminal board**

    a) Cancellation loops installed on terminal board compensating for three 0.04 amp—meter^2 current loops

    b) Remaining circuits loops tested, modeled with in-flight data

    c) Terminal board circuit loops simulated and measured

- **Propulsion system/Attitude**

    a) Propulsion latch valves assessed, Cassini results matched, 900 nT at probe

    b) Metglas$^{\circledR}$ shielding failed to contain high-magnetic flux

    c) Spacecraft magnetics assessment, APL high-bay area
       compensation magnets installed on top deck:
       2-2 inch Alnico magnets @ 1880 pole-cm each on $+x$ s/c, north poles $+z$, slightly inward tilted
       1-1 inch @ 800 pole-cm horizontal on $-y$, north pointed $+x$

    d) Reaction wheels tested, <0.2 nT at sensor

- **Instruments**

    a) Stepper motors tested, <0.1 nT at sensor

    b) Package-level testing

    c) Grounding per power system

- **Magnetometer operation during I&T, s/c thermal vacuum testing**

    Check subsystem interference, model magnetics

- **Cruise**

    a) Calibration of subsystems' magnetic effects

    b) Roll assessment of static field

    c) Solar wind average for total field

    d) Correlation with WIND MFI

Table A4
HDF telemetry format definition

HDF data type and name	Field name or science data set size	Field data type
VDATA	MET	Unsigned32 bit integer
MAG_HEADER_VA	UTC	64 bit floating point
RS	DQI	Unsigned 32 bit integer
	MAG_ON_OFF	Unsigned 8 bit integer
	CAL_ON_OFF	Unsigned 8 bit integer
	CAL_A_D	Unsigned 8 bit integer
	RANGE_MAN_AUTO	Unsigned 8 bit integer
	NOISE_REJ_ON_OFF	Unsigned 8 bit integer
	SW_ANTI_ALIAS_FILT	Unsigned 8 bit integer
VDATA	MANUAL_RANGE	Unsigned 8 bit integer
MAG_HEADER_VARS	ACTUAL_RANGE	Unsigned 8 bit integer
	AC_AXIS_SELECT	Unsigned 8 bit integer
	RNG_CHNG_MORE_SENS	Unsigned 8 bit integer
	RNG_CHNG_LESS_SENS	Unsigned 8 bit integer
	NOISE_SPIKE_COUNT	Unsigned 8 bit integer
	SAMPLE_RATE	Unsigned 8 bit integer
	MAG_DC_DC_CURR	Unsigned 8 bit integer
	PROBE_TEMP	Unsigned 8 bit integer
	MAG_ELECT_TEMP	Unsigned 8 bit integer
	DPU_DC_DC_TEMP	Unsigned 8 bit integer
	RECORD_ABORTED	Unsigned 8 bit integer
	TIME_TAG	Unsigned 8 bit integer
SDS X_DATA	44 Samples	Unsigned 16 bit integer
SDS Y_DATA	44 Samples	Unsigned 16 bit integer
SDS Z_DATA	44 Samples	Unsigned 16 bit integer
SDS AC_DATA	44 Samples	Unsigned 8 bit integer
SDS NOISE_DATA	44 Samples	Unsigned 8 bit integer

References

Acuña, M. H.: 1974, 'Fluxgate Magnetometers for Outer Planets Exploration', *IEEE Trans. Magnetics* **Mag-10**, 519.

Acuña, M. H., Connerney, J. E. P., Wasilewski, P., Lin, R. P., Anderson, K. A., Carlson, C. W., McFadden, J., Curtis, D. W., Reme, H., Cros, A., Medale, J. L., Sauvaud, J. A., D'Uston, C., Bauer, S. J., Cloutier, P., Mayhew, M., and Ness, N. F.: 1992, 'Mars Observer Magnetic Fields Investigation', *J. Geophys. Res.* **97**, 7781–7798.

Acuña, M. H., Zanetti, L. J., Russell, C. T., and Anderson, B. J.: 1997, 'The NEAR Magnetic Field Investigation – Science Objectives at Asteroid Eros 433 and Experimental Approach', *J. Geophys. Res.*, in press.

Anderson, B. J., Takahashi, K., Erlandson, R. E., and Zanetti, L. J.: 1990, 'Pc1 Pulsations Observed by AMPTE/CCE in the Earth's Outer Magnetosphere', *Geophys. Res. Lett.* **17**, 1853.

Anderson, B. J., Potemra, T. A., Bythrow, P. F., Zanetti, L. J., Holland, D. B., and Winningham, J. D.: 1993, 'Auroral Currents during the Magnetic Storm of November 8 and 9, 1991: Observations from the Upper Atmosphere Research Satellite Particle Environment Monitor', *Geophys. Res. Lett.* **20**, 1327.

Baumgartel, K., Sauer, K., and Bogdanov, A.: 1994, 'A Magnetohydrodynamic Model of Solar Wind Interaction with Asteroid Gaspra', *Science* **263**, 653–655.

Chapman, C. R.: 1975, 'The Nature of Asteroids', *Sci. Am.* **232** (1), 24–33.

Curtis, S. A. and Ness, N. F.: 1988, 'Remanent Magnetism at Mars', *Geophys. Res. Lett.* **15** (8), 737–739.

Feierberg, M. A., Larson, H. P., and Clark, C. R.: 1982, 'Spectroscopic Evidence for Undifferentiated S-Type Asteroids', *Astrophys. J.* **257**, 361–372.

Gaffey, M. J., Burbine, T. H., and Binzel, R. P.: 1993, 'Asteroid Spectroscopy: Progress and Perspectives', *Meteoritics* **28**, 161–187.

Hayes, J. R.: 1995a, *NEAR Instrument DPU Common Software Requirements Specification*, JHU/APL 7358–9001.

Hayes, J. R.: 1995b, *NEAR Infrared Spectrograph/Magnetometer (NIS/MFI) DPU Software Requirements Specification*, JHU/APL 7357–9001.

Hayes, J. R., Fraeman, M. E., Williams, R. L., and Zaremba, T.: 1987, 'An Architecture for the Direct Execution of the Forth Programming Language', in *Proceedings of the Second International Conference on Architectural Support for Programming Languages and Operating System*, The Compute Society of the IEEE, pp. 42–49.

Kivelson, M. G., Bargatze, L. F., Khurana, K. K., Southwood, D. J., Walker, R. J., and Coleman, P. J., Jr.: 1993, 'Magnetic Field Signatures near Galileo's Closest Approach to Gaspra', *Science* **271**, 331–334.

Kolmogorov, A. N.: 1941, 'The Local Structure of Turbulence in Incompressible Viscous Fluids for Very High Reynolds Numbers', *Dokl. Akad. Nauk SSSR* **30**, 301.

Lohr, D. A.: 1992, 'Design of a Very Low Noise Magnetometer, in TEO Transactions TEO92-036', *JHU/APL Technical Report*, pp. 67–69.

Parancias, C. and Cheng, A. F.: 1996, 'Predictions of Magnetic Field Signatures of 433 Eros', *Geophys. Res. Lett.*, submitted.

Wasson, J. T. and Wetherill, G. W.: 1979, 'Dynamical, Chemical, and Isotopic Evidence Regarding the Formation Locations of Asteroids and Meteorites', *Asteroids*, pp. 926–974.

Zanetti, L. , Potemra, T., Erlandson, R., Bythrow, P., Anderson, B., Lui, A., Ohtani, S., Fountain, G., Henshaw, R., Ballard, B., Lohr, D., Hayes, J., Holland, D.; Acuña, M., Fairheld, D., Slavin, J., Baumjohann, W., Engebretson, M., Glassmeier, K., Gustafsson, G., Iijima, T., Luehr, H., and Primdahl, F.: 1994, 'Magnetic Field Experiment for the Freja Satellite', *Space Sci. Rev.* **70**, 465–482.

Zanetti, L. J., Potemra, T. A., and Anderson, B. J.: 1995, 'Boundary Determinations Based on Low Frequency Magnetic Field Measurements, Proceedings of the International Conference on Comparative Planetology', *Earth, Moon and Planets* **67**, 175–178.

THE NEAR SCIENCE DATA CENTER

K. J. HEERES, D. B. HOLLAND and A. F. CHENG
Johns Hopkins University, Applied Physics Laboratory, Laurel, MD 20723, U.S.A.

(Received August, 1996)

Abstract. The NEAR (Near Earth Asteroid Rendezvous) Science Data Center (SDC) serves as the central site for common data processing activities needed by the NEAR science teams in particular and the scientific community in general. The SDC provides instrument and spacecraft data to the science teams from around the world and redistributes science products produced by those teams, all the science teams to focus on analysis. This data and the accompanying documentation are available at 'http://sd-www.jhuapl.edu/NEAR/'. In addition the SDC is responsible for archiving spacecraft, instrument, and science data to the Planetary Data System (PDS).

1. Introduction

This paper describes the SDC design, its rationale, and its implementation. The description focuses on the following areas, design considerations and decisions that determined the implementation, as well as a functional view of the tasks performed and the services provided by the SDC. A glossary of acronyms used throughout this paper and a list of tables and figures can be found at the end of the paper.

The major sections of the SDC processing, Telemetry Server Processing (TSP), Science Archive Processing (SAP), and Product Server Processing (PSP) are discussed, covering the systems in the sequence of the data flow (see Figure 1). The TSP handles the raw telemetry data and produces the Telemetry Archive. The SAP uses the Telemetry Archive and ancillary information to produce the Science Archive (SA) and the Catalog. The PSP uses the Science Archive and the Catalog to produce products for the Science Teams and to archive data to the Planetary Data System (PDS).

Because the SDC is being developed during the NEAR cruise period, between launch and encounter, parts of the SDC are at varying stages of development. The development and the function of each phase are discussed. A physical description of SDC, including computer, display, and networking facilities is also provided. Unique mission design concepts and the following design considerations have driven the project.

2. Mission and Technology Aspects

The design of the NEAR SDC takes advantage of certain features of the project and advances in technology. The SDC can be smaller and less expensive than similar facilities that have been previously implemented and therefore more cost-effective.

Space Science Reviews **82:** 283–308, 1997.
© 1997 *Kluwer Academic Publishers. Printed in Belgium.*

Foremost, the amount of data that will be telemetered each day from the spacecraft is small when compared to recent design experience with Earth orbiting missions (for NEAR 95 MB/day of telemetered data). The total amount of science data that will be collected over the mission is small (for NEAR 30 GB of telemetered data during rendezvous*). The NEAR telemetry stream is simple. It consists of fixed length Consultative Committee for Space Data Systems (CCSDS) transfer frames that are assembled by the onboard Command and Data Handling (C&DH) system. These transfer frames are primarily composed of fixed length instrument packets formed by the instrument Data Processing Units (DPUs). A second simplification for the SDC is that the packaging of the telemetry stream necessary to transmit the data to earth has been removed by the Mission Operations Center (MOC). The data appear to the SDC as they were formed by the C&DH system.

The instruments package their science records into packets for collection by the C&DH system. The instrument science records were designed to contain all of the instrument state information that is necessary to interpret the instrument data. This makes it easy to correlate measurements with state information. An exception to this is the magnetometer, which will need currents from throughout the spacecraft to correctly interpret the data.

The project has a long cruise period (approximately 3 years from launch to rendezvous with Eros) during which no scheduled science activities will occur. As a result, the SDC will not need to be fully operational until just prior to rendezvous. In the meantime, the science teams are encouraged to use the instrument Ground Support Equipment (GSE) software for Quick-Look analysis and calibration during the cruise.

Advances in technology are an element of design. Because of the growth in on-line storage capacity, commonly available magnetic media has a capacity of $\sim$9 GB per drive. By the time of rendezvous, it is expected to be $\sim$50 GB. This means that it will be easier to store all of the data related to the mission on-line and the design of all processing and analysis software can be simplified. The need for operator staff during rendezvous operations can be reduced because staff is not needed to archive and restore off-line media and software does not have to be developed to support this activity. The available on-line storage makes access more immediate and helps with correlative studies.

The phenomenal increase in the computing power of desktop systems, both in PCs and in workstations, allows us to defer some of the data processing to data access. Preprocessing of data is therefore reduced and data can be stored in a very raw format. Most, if not all, analysis can be deferred to the scientist's desktop.

The SDC assumes that all members of the NEAR science teams have access to the global Internet. With universal networking and low data rate, we can deliver products through the Internet to the science teams in electronic form.

---

* These values use conservative estimates of antenna coverage. If the project receives more coverage, these values will increase. MSX downlinks between 2 and 7 GB per day.

3. SDC Design Considerations

The SDC design is based on the previously described mission aspects and technology advances leading to the following top-level design considerations.

The SDC performs common data processing tasks and makes spacecraft and navigation data available to the science teams and its members. Data is archived and retrieved in a format that has been developed by the SDC and the science teams and through the SDC, the science teams will have access to all the science team's data. Correction and calibration algorithms developed by the science teams will be incorporated into SDC processing and make validated science data generally available. The science teams are responsible for supplying final calibration algorithms for archival to the Planetary Data System (PDS) at the end of the mission.

The SDC will make extensive use of the Internet. All outputs from the SDC are delivered over the network, and in most cases, only over the network. The SDC will not make tapes or hard copies of plots. Since the SDC is delivering data to the desktops of scientists, not software or applications, data will be delivered in common formats such as FITS (Flexible Image Transport System) and HDF (Hierarchical Data Format). Data in these formats can be easily read by commercial and public domain software. The science teams using libraries that can read these formats will also develop analysis software.

4. Data and Processes

The major sources of data for the SDC are Telemetry downlink; commands executed on the spacecraft, navigation data, and validated scientific data. The first two are delivered to the SDC by the Mission Operations Center (MOC) at JHU/APL. Mission Operations is responsible for command and control of the spacecraft and instruments and for the retrieval of telemetry from the Deep Space Network (DSN). The Jet Propulsion Laboratory (JPL) Navigation Team provides navigation data and the science teams provide validated science. SDC can then distribute the data to other science teams, the general scientific community, and the PDS.

5. Telemetry Server Process

The TSP removes and logs errors in the telemetry stream, sorts the data according to time, breaks the telemetry stream into packets, and removes duplicate packets. In close to real-time, TSP outputs a set of minimally cleaned and sorted files for use by GSE Quick-Look displays. It also provides a merged telemetry data archive that becomes the basis for science archive processing.

The telemetry data are delivered to the SDC in the form of CCSDS (see Reference [1]) VC0 (recorder), VC3 (real-time), and VC2 (image) transfer frames with a data quality header attached by the MOC. The transfer frames delivered to the SDC

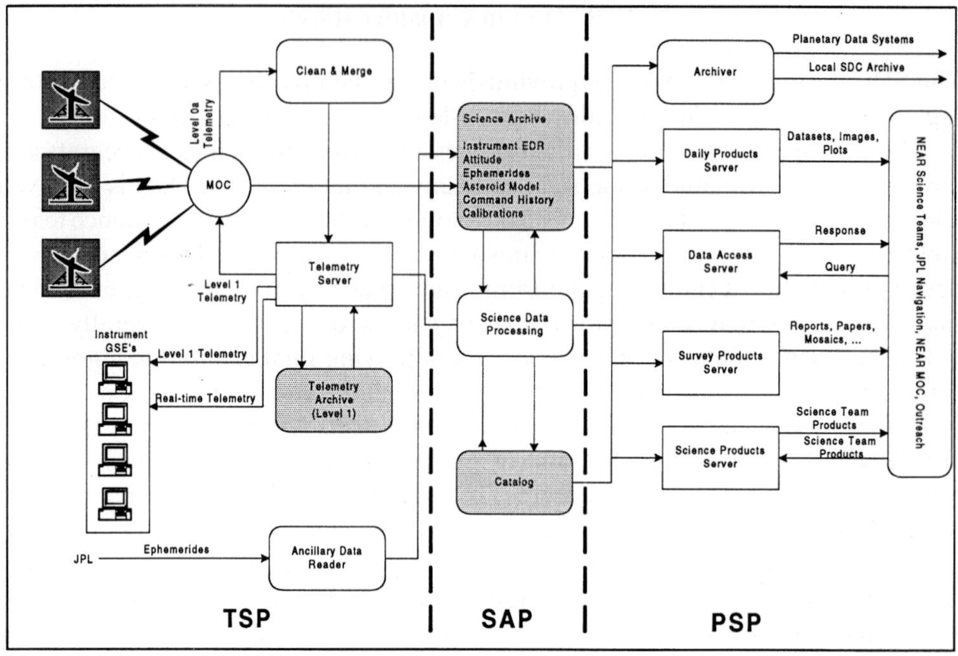

Figure 1. NEAR SDC data flow.

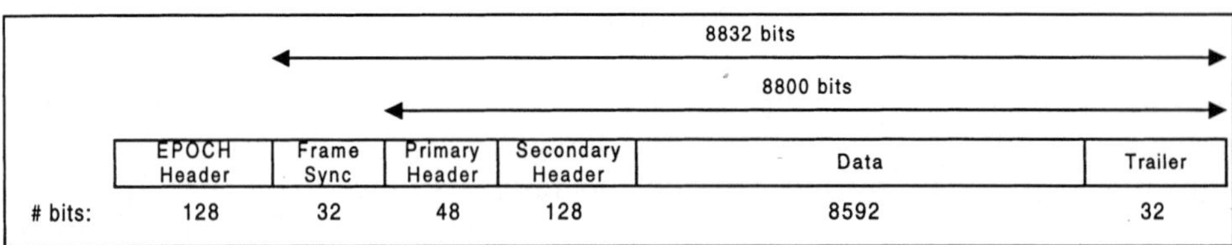

Figure 2. CCSDS transfer frame.

consist of an EPOCH Data Quality Header, Frame Sync, Primary and Secondary Headers, a fixed size data portion and a Trailer as shown in Figure 2.

A description of the EPOCH header is found in Table A1 of Appendix A. The EPOCH Data Quality Header is also found in Appendix A. A description of each of the fields within the transfer frame primary and secondary headers is found in Table A2 and Table A3, respectively.

The data portion of the VC0 and VC3 transfer frames are composed of three fixed-size packets. The packets consist of Primary and Secondary headers followed by a fixed-size data portion as shown in Figure 3.

A description of each of the fields within the packet primary and secondary headers is found in Tables A4 and A5, respectively, of Appendix A. The content of the data portion of the packet depends on the subsystem and instrument producing

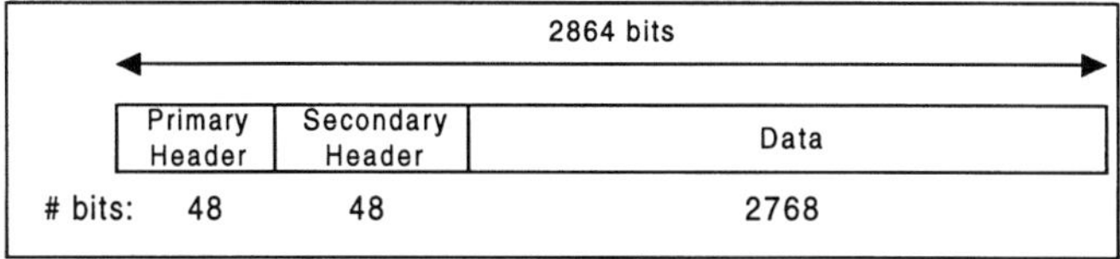

Figure 3. CCSDS packet.

the packet. Table I shows the contents by instrument. A more detailed discussion of the layout is found in References [1]–[6].

The telemetry data are written to a series of files on a front-end* machine within the MOC. These files are read by a process running on the MOC front-end machine and passed, via a TCP/IP socket, to a process running on a machine in the SDC. This design means the processing in the SDC can be asynchronous with processes running in the MOC. There is no need for elaborate control and signal passing between the MOC and SDC. The SDC can incur downtime without affecting critical functions in the MOC. If the SDC loses contact with the MOC, no data are lost. The TSP resumes processing when contact is reestablished. The MOC can request that the DSN resend the data if there is a problem between the MOC and DSN during a pass.

5.1. REAL-TIME PROCESSING FOR GROUND SUPPORT EQUIPMENT (GSE) QUICK LOOK

The NEAR GSEs are PCs with software to provide a quick look at the NEAR science telemetry data. The GSEs were developed for instrument checkout. They were designed to minimize expense by providing a view of the science data while the SDC processing and science team displays are being developed. The Telemetry Server provides data to all GSEs, local and remote, through the Network File System (NFS) or File Transfer Protocol (FTP). The GSE displays are being augmented by Visual Basic programs to display real time Command and Telemetry Processor (CTP), Attitude Interface Unit (AIU) and Flight Computer (FC) packets. The directory layout of the real-time telemetry processing is shown in Table II.

The SDC real-time telemetry processing does a quick sort of the incoming telemetry transfer frames. These frames are sorted by Mission Elapsed Time (MET) and virtual channel frame count using a short in-memory sorted cache of the transfer frames. Any duplicates that appear within the cache are removed and their occurrence is logged. When the cache is full or a timer has expired, the transfer frames are written to files for further processing.

The VC0 and VC3 transfer frames are broken into packets and written to files for use by the instrument GSEs. The VC2 transfer frames are kept intact. The

---

* The front-end machine performs the real-time capture of the telemetry downlink. The data is delivered to S/C displays within the MOC and processes in the SDC.

Table I

Instrument packet data description

Instru-ment	Packet type (file name)	Science data description
MSI	Full image (imager.pkt, vx2.xfr)	Full image contains 244 lines of 537 pixels with header and housekeeping. A full image requires up to 183 VC2 transfer frames or 569 VC0/VC3 packets, depending on compression and 8 or 12-bit pixel size. The header contains filter position, exposure time and compression mode.
MSI	Summary image (imager.pkt)	Each packet contains a summary image consisting of 22 lines of 26 bit superpixels with header and housekeeping.
XGRS	Gamma ray (xg-spec.pkt)	A gamma-ray science record requires 31 packets. A science record contains housekeeping, and these spectra (incremented when) (number of 16 bit bins): Nal integral (any Nal event)(1024), BGO integral (any BGO event)(1024), Nal anti-coincidence (Nal event, no coinc.)(1024), Nal single escape shifted (coinc., BGO in single esc. window) (1024), BGO single escape peak (coinc., BGO in single esc. window) (42), Nal double escape shifted (coinc., BGO in double esc. window) (1024), BGO double escape peak (coinc., BGO in double esc. window) (42).
XGRS	X-ray (xg-spec.pkt)	An X-ray science record requires 7 packets. A science record contains housekeeping and 4 types of spectrum, unfiltered, Mg, Al and active solar X-ray. Each spectrum has 256 bins, with 16 bits per bin, of valid events.
XGRS	X-ray, gamma-ray Summary (xg-spec.pkt)	One packet with housekeeping and spectra for: Unfiltered X-ray, Mg filtered X-ray, Al filtered X-ray, Active solar monitor, Integral Nal, and Integral BGO. All spectra have 16 bins with compressed 8 bit values.
NIS	Streamed science records (nis-mag.pkt)	Each packet can contain approximately 2.3 NIS science records. Science records immediately follow the previous record. Each science record contains housekeeping, a Ge spectrum $(0.8\mu–1.5\mu)$ and a InGaAs spectrum $(1.3\mu–2.6\mu)$, both with 32×16 bit bins. The housekeeping values contain gain state and slit position.
MAG	Science record (nis-mag.pkt)	Each packet contains a single science record with housekeeping and 44 16 bit samples of X, Y and Z components of the DC field and 1 AC sample.
NLR	Normal rate (lidar.pkt)	Each packet contains a 60 bit status section and 56×48 bit data sets. A data set includes the Mission Elapsed Time, threshold setting, calibration range value, data quality flags and a 21 bit range value.
NLR	High rate (lidar.pkt)	Each packet contains a 60 bit status section and 112×24 bit data sets. A data set includes data quality flags and a 21-bit range value.

Table II
Real-time telemetry directory structure

Directory structure	Purpose
Telemetry (dir)	Contains all of the telemetry files
realtime (dir)	Contains current telemetry files
vc2.xtr	Contains current VC2 transfer frames with EPOCH ceaders
vc2.idx	Index file for use by instrument GSEs
vc0(dir)	Contains the current down linked recorder data
vc0.xfr	Contains current VC3 transfer frames with EPOCH headers
aiu1.pkt	Contains AIU packets with APID 2
aiu2.pkt	Contains AIU packets with APID 3
ctp1.pkt	Contains C/TP packets with APID 0
ctp2.pkt	Contains C/TP packets with APID 1
fc1.pkt	Contains flight computer packets with APID 10
fc2.pkt	Contains flight computer packets with APID 11
imager.pkt	Contains MSI packets with APID 4
lidar.pkt	Contains NLR packets with APID 7
nis-mag.pkt	Contains NIS/MAG packets with APID 6
xg-spec.pkt	Contains XGRS packets with APID 5
xxx.idx	Index files for each of the packets files, for use by GSEs
vc3	Contains all of the down linked realtime data
same as vc0	
contact (dir)	Contains current transfer frame files for use by Sort and Merge
vc0.xfr	
vc2.xfr	
vc3.xfr	
yydddhhmmsss.psu	
yydddhhmmsss.err	
process (dir)	
vc0.xfr	
vc2.xfr	
vc3.xfr	
error-epoch (dir)	Contains transfer frames that were found to be in error
yyyy (dir)	Year that the EPOCH file was created
ddd (dir)	Day that the EPOCH file was created
yydddhhmmss.er	Bad transfer frames from EPOCH
yydddhhmmss.log	Log of EPOCH processing errors

real-time files are fixed length circular queues of instrument data for the GSEs and a paired index file reflecting the current position being written. The MOC files that the process reads are approximately 1 MB in size. After this size is reached or if approximately 1 h has elapsed, the MOC will create a new file. When the MOC starts a new file, the Telemetry Server starts a merge, and the data from the file are added to the full day telemetry archive file.

5.2. Telemetry Archive/Clean and Merge

The process of adding the data is called 'Clean and Merge'. The telemetry reading process is momentarily halted. The transfer frames files that were being written are moved from the contact directory to the process directory (see Table II), and the telemetry reading process continues. In the case of VC0 and VC3, the moved files are sorted by Mission Elapsed Time (MET). In the case of VC2, they are sorted by MET and count and merged with data from the recent days. Transfer frames that are not time tagged within 7 days of the current date are logged as bad and placed in an error file. Transfer frames that have duplicate indexes but different data, are also logged as bad and placed in an error file. The most recent retrieved data are used. Transfer frames that have duplicate indexes and duplicate data are logged. The VC0 and VC3 transfer frames are broken out into packet files. These packet files are sorted, duplicates logged, and removed. The process creates files of Multi Spectral Imager (MSI), X-Ray/Gamma-Ray Spectrometer (XGRS), NEAR Infrared Spectrograph/Magnetometer (NIS/MAG), CTP, AIU, and FC packet types. Detailed descriptions of the data layout of the instrument files are found in the instrument software requirement specifications (References [3]–[6]).

5.3. Telemetry Archive

The telemetry archive is organized by Universal Time (UT) year, and UT day within UT year (Table III). The UT time is the time that the transfer frame was formed on the spacecraft expressed in UT. These data are made available to the science teams, engineering teams, and the MOC. The data are available locally using NFS, AppleShare, and Microsoft Networking. The data are also available remotely via World Wide Web (WWW) or FTP.

5.4. Command history

A history of commands, as executed on the spacecraft, is formed by the MOC. The purpose of these data is to provide the science teams with the capability to verify how their instrument was commanded and to correlate the information to observed measurements. The exact format and delivery method for command histories will be developed during the cruise portion.

5.5. Navigation data

The navigation and pointing information for the mission will be distributed to the science teams via SPICE* kernels. The SPICE kernels are comprised of spacecraft and asteroid ephmemerides produced by the JPL Navigation Team, clock and

---

* SPICE is an information facility for providing ancillary observation geometry data. The SPICE acronym stands for spacecraft (S), planet (P), instrument (I), 'C-matrix' (C), events (E).

Table III
Telemetry archive file organization

Directory structure	Purpose
Telemetry (dir)	Contains all of the telemetry files
data (dir)	Contains telemetry archive
yyyy (dir)	Year of transfer frame formation
ddd (dir)	Day of transfer frame formation
vc2.xfr	Contains VC2 transfer frames with EPOCH headers
vc2.idx	Index file for use by instrument GSEs
vc0 (dir)	
vc0.xfr	Contains VC3 transfer frames with EPOCH headers
aiu1.pky	Contains AIU packets with APID 2
aiu2.pkt	Contains AIU packets with APID 3
ctp1.pkt	Contains C/TP packets with APID 0
ctp2.pkt	Contains C/TP packets with APID 1
fc1.pkt	Contains flight computer packets with APID 10
fc2.pkt	Contains flight computer packets with APID 11
imager.pkt	Contains MSI packets with APID 4
lidar.pkt	Contains NLR packets with APID 7
nis-mag.pkt	Contains NIS/MAG packets with APID 6
xg-spec.pkt	Contains XGRS packets with APID 5
xxx.idx	Index files for each of the packet files, for use by GSEs
vc3 (dir)	
save as vc0	
tar (dir)	GNU zip compressed tar files for older files
error-cm (dir)	Contains transfer frames that were found in error
yyyy (dir)	Year of transfer frame formation
ddd (dir)	Day of transfer frame formation
yyddd.er	
yyddd.log	

attitude kernels produced by the SDC, and binary planet kernels produced by the NAIF (Navigation Ancillary Information Facility). The binary planet kernel is an extension to SPICE, which will encapsulate the dynamics of the asteroid. A second extension to SPICE, built explicitly for NEAR, will allow the use of a plate model of the asteroid. The SDC maintains an on-line archive of SPICE files organized by mission phase and data source (predictive or determined).

6. Science Data Processing

The telemetry data are processed into science Experimental Data Records (EDR) and housekeeping data, then stored within the Science Archive. Each instrument has defined an EDR in an instrument requirements document. The EDR is one unit

Table IV

HDF file organization

Directory structure	Purpose
SDC_HDF (dir)	Contains all science archive HDF files
MSI (dir)	One directory per instrument
yyyy (dir)	Year of image aquisition
saiyydd.hdf	All images for the day of image aquisition
NIS (dir)	Same as msi, file name is sanyydd.hdf
XGRS (dir)	Same as msi, file name is saxyyddd.hdf
MAG (dir)	Same as msi, file name is samyyddd.hdf
NLR (dir)	Same as msi, file name is salyyddd.hdf
AIU (dir)	Same as msi, file name is saayyddd.hdf
CTP (dir)	Same as msi, file name is sacyyddd.hdf

of data. For example, for the Multi-Spectral Imager (MSI), an EDR is a single image. The processing of these EDRs takes place at 4-h intervals. This interval can be changed, but the process needs to be completed within a telemetry update interval.

6.1. SCIENCE DATA ARCHIVE

The Science Data Archive is the source data for all products provided to science teams and the data archived to the PDS. The Science Data Archive consists of multiple types of data. The largest data set is the instrument EDR's and the spacecraft housekeeping. These data are stored HDF files. HDF was developed by the National Center for Supercomputer Applications (NCSA) for use in storing scientific data. HDF was selected for the SA because the format and the associated libraries are platform independent. HDF is a self-documenting scientific data format with data type, names, and dimensionality built into the file. This provides the basis for automated access by the SDC and the science teams.

The HDF files are organized by instrument and then time as shown in Table IV.

The HDF file contains the telemetry data reformatted as EDRs. The fields are broken out to machine recognized objects (e.g., byte or integer). All information from the telemetry packet files is preserved in the transfer to the HDF files. The data are not stored in engineering units but converted upon access. A description of all of the fields in the HDF would be quite voluminous. Table A6 of Appendix A gives the layout for the MSI instrument. This information is from the prototype software development and will change with subsequent software builds. A current complete list of variables will be on-line at our WWW site (http://sd-www.jhuapl.edu/NEAR/). The XGRS and MAG teams are directly using the HDF files and are notified of changes to the HDF format.

6.2. CATALOG

The purpose of the catalog is to enhance access to the Science Data Archive. The catalog is a database created from criteria discussed in the next section. Using the catalog, the science teams will be able to rapidly carry out complex queries of the database to determine what measurements satisfy the criteria and to produce products based on those criteria. As discussed previously, the Science Data Archive is entirely on-line and can be scanned to resolve a query not covered by the catalog. The catalog will store state, position, pointing, and ancillary data information that will make the queries easier and faster. We can, during the course of the mission, add items to the catalog that would make subsequent analysis easier. The catalog will also be used to support automation of the data product generation process. Limited WWW access to the catalog will be provided by the product server interface.

The Science Data Archive has a traditional organization that is optimized for time series access. For most types of analysis, this is the appropriate organization. The NEAR SDC has added an RDBMS (an Oracle Relational Database Management System) to the tools available to the science teams. This tool will allow them to access sets of their data without having to download and read all possibly related data. The RDBMS will be populated during normal processing and added to later, as a result of analysis.

The database will include asteroidal locations, spacecraft locations, mission phases, instrument phases, instrument conditions, and pointing. The asteroidal location is the location of the instrument look direction projected to the asteroid surface. It will cross index images and instrument sequences to plates on the asteroid. A query will be able to retrieve indexed data related to a particular plate. To determine spacecraft location, the database will also include a table that has 10-min position, range, and pointing information. A query will be able to disclose times when the spacecraft was at a certain range from the asteroid and use these times to access the science archive. Information about where and in what direction the spacecraft was pointing when the data were recorded will be gathered. A query can access data when angles between the Sun and the surface of the asteroid were at a particular value.

A table in the database will relate mission phase with time. A scientist can query for information gathered during a certain mission phase, obtain the related times, and use these times to access the SA.

The Instrument Phase table relates instrument times of interest to time. It is anticipated that the table will result from analysis by the science teams. For example, the Magnetometer Science Team would determine the time the spacecraft was within the sphere of influence of the asteroid. They would update the table with this information. They could then use the table for subsequent analysis looking at data only during times when the spacecraft was within the sphere of influence of the asteroid.

The database will also include information about the condition of the instruments. Voltages and temperatures, for example, can be recorded at the time the instrument samples were taken. A database query can then access information based on when the instrument was within a certain operating temperature.

Not every instrument sample will be indexed in the database. In certain cases such as NIS (Near-Infrared Spectrograph), groups of instrument samples rather than individual samples will be indexed, and in the case of NLR (NEAR Laser Rangefinder), none of the samples are indexed. The catalog is designed to provide access to the SA (Science Archive) data. A preliminary list of the tables that will be present in the database is found in Table V, and a detailed look at one of the tables is found in Table A7 of Appendix A. Since the database is a query tool, not an analysis tool, the absolute fidelity of the database is not necessary. Information based on spacecraft and asteroid ephemeris will not be updated every time a new ephemerides is derived. Users of the information are cautioned to access a range of information and use the SA with the most recent SPICE kernels to validate the information to obtain the best possible values.

7. Product Server Process

We have described data that go into the SDC and how data are internally stored in the SDC. The next section discusses the format and distribution of data to the science teams and to the science community.

The NEAR SDC is somewhat unique because it will be using the facilities of the Internet to distribute data to the science teams. Data is stored on a server system within JHU/APL. This server is connected to the Applied Physics Laboratory Backbone Network (APLNIS) using 10BaseT technology. The backbone is a 100-mega bit per second (MBS) Fiber Distribution Data Interface (FDDI) network (see Figure 4). This network is connected to the external world using dual T1 lines to Alternet. Alternet is in turn connected to the NFS backbone via T3 lines. It is assumed that science team members have access to the Internet, and therefore the SDC does not provide that infrastructure.

The SDC was designed to generate products for science teams that reflect the data that were obtained that day and to update summary products based on the most recent data. The daily products will remain online for a short period of time (7 days), allowing the science teams to download them.

Science teams have specified what products they would like the SDC to provide. The SDC is committed to making those products available shortly after the receipt of data. For example, the MSI/NIS team along with the SDC and the Navigation team decided to use the imager data in the form of a FITS file. This file is a flat binary file with an ASCII header containing ancillary information that is of interest to both the MSI/NIS science team and the Navigation team.

Table V
Science data catalog tables

Table	Description
IMAGER	Contains search data for imager data
IMAGERlrp	Contains a list of all of the low resolution plates that are associated with each of the image
IMAGERhrp	Contains a list of all of the high resolution plates that are associated with each of the images
IMAGERtoi	Contains times of interest to the Imager science team
NIS	Contains search data for the Near Infrared Spectrograph
NISlrp	Contains a list of all of the low resolution plates that are associated with each spectrograph
NIShrp	Contains a list of all of the high resolution plates that are associated with each spectrograph
NIStoi	Contains times of interest to the Near Infrared Spectrograph science team
XR	Contains search data for the X-ray spectrometer
Xrtoi	Contains times of interest to the X-ray spectrometer science team
GR	Contains search data for the gamma-ray spectrometer
Grtoi	Contains time of interest to the gamma-ray spectrometer team
MAGtoi	Contains times of interest to the magnetometer team
LIDARtoi	Contains the search data for the laser altimeter
Position	Contains the position of the spacecraft in the asteroid body fixed coordinate system
Mission phase	Contains the start and stop times for different mission phases
Science sequence	Contains time of interest when science sequences occurred
Science sequence codes	Contains the science sequence codes
LRP	Contains the low resolution plate model
HRP	Contains the high resolution plate model
Asteroid spectral properties	Contains pointers to asteroid spectral properties datasets
DPID	Contains pointers to science data

Science teams are required to provide the SDC calibration algorithms for their instruments so that the SDC can provide calibrated data to other science teams and the science community. This will be the method for the science teams to share their knowledge with other teams and contribute to mission combined data analysis goals.

The SDC will also provide an interface that will allow the science team to create products based on criteria in the database. In the simplest case, they could ask for all of the images taken during a specific time period. The current prototype user interface for this design is a WWW-type access and anonymous file transfer protocol (FTP) (nearsdc-ftp.jhuapl). The uniform resource locator (URL) for the NEAR project is http://sd-www.jhuapl.edu/NEAR/. Full development will include a database WWW-access and will be available by encounter. The current home page is based on the Netscape 2.0 browser for access and will continue to follow browser and server development until final configuration prior to encounter.

The prototype user interface provides complete access to the SDC for the science teams and restricted access for the general public. Some science teams have requested that the data products created by the SDC be restricted to the science team for a fixed period of time. Currently that time is 7 days. After that time, the data will be available to the general public.

8. Hardware Configuration

The hardware configuration (Figure 4) for the NEAR SDC is being purchased in two phases (see Table VI). Phase 1 has been completed and the second phase will occur 6 months prior to encounter.

The main processing will be done on UNIX workstations from Hewlett Packard (HP). The HP workstations were chosen because JHU/APLÆs current infrastructure includes HP workstations that were used for development and initial processing prior to the availability of NEAR equipment. The HP workstation which hosts the Telemetry Server and Archive, is the machine 'morgan'. The HP workstation that supports Science Data Processing is the machine 'loring'.

The main user interface and display systems are on IBM PC clones from Gateway 2000 and X Window System terminals from Network Computing Devices. The PCs were chosen because the systems hosted instrument GSE software that JHU/APL continues to use. The NEAR SDC hardware configuration supports the concept of data reduction on UNIX workstations and data displays on IBM PC clones, taking advantage of the current strengths of both architectures. Prior to rendezvous, JHU/APL will be purchasing additional hardware to support the rest of the mission.

Since the Science Teams are from various organizations the SDC equipment is available remotely or when the teams are at JHU/APL. The Laboratory is providing an Instrument GSE and an X-Terminal to each group of MSI/NIS, XGRS, and NLR-MAG for their use in the team's office while they are resident at JHU/APL.

9. Development and Operations Plan

Because the NEAR mission was planned with a long cruise phase, with no science activities, the SDC development was scheduled to occur between launch and encounter. Early funding allowed development of the TSP, which was ready at launch. This prototype system will be the working SDC software system for the early cruise period. The Build 1 SDC system will be available by the encounter with the asteroid Mathilde in June of 1997. The Build 1 development will have all elements of the final system including TSP, SA and PS software sections and will

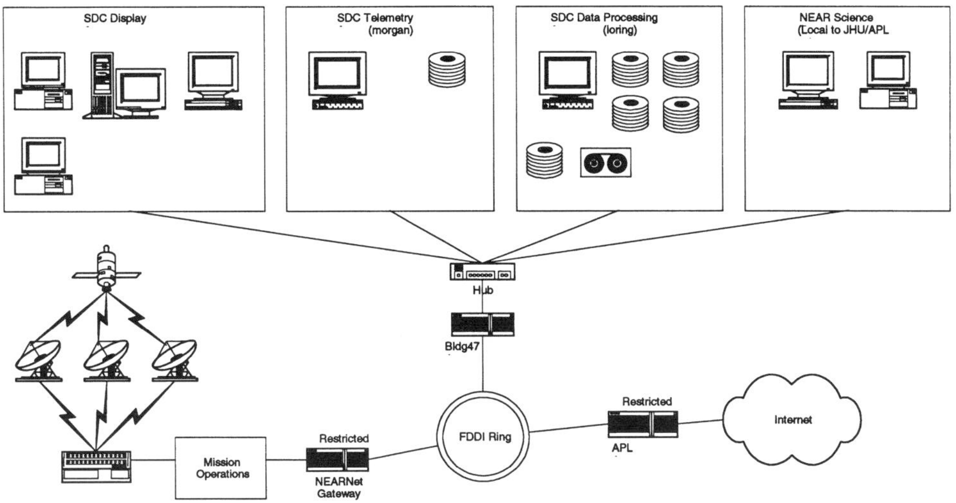

Figure 4. SDC hardware/network configuration.

include database access in the PS. Build 2 will be completed and tested prior to the encounter with EROS. After a review of the Build 1 system, changes and new products requested by the science teams will be included at that time.

The SDC will need minimal operations staff during encounter. A software development staff will be available to make limited revisions and to create new products as directed by the science teams. The software will also be used to support incorporation of science team analysis products into the SA.

During the encounter period, the SDC will prepare sample sets of science archive data in PDS format for science team review. Operation of the SDC is planned for a year after the end of the encounter. At that time, the SA will be prepared for transfer of the long term archive to the PDS.

10. Data Management

The SDC concentrates on the delivery of spacecraft and ancillary data to the science teams and the general science community. In addition the SDC has the responsibility to archive mission data to the PDS at the end of the mission.

The SDC is making spacecraft and ancillary data available to the science teams in formats developed in cooperation with the science teams. New products are being developed to address new requirements. Processes are being installed to notify the science teams when new products are available. These data are available on a daily basis. The science teams are given a short period of time ($\sim$7 days) to examine the data when they are available for general use. Since additional requirements are anticipated at rendezvous SDC will provide staff to support new requests. Software to facilitate this development is currently being designed.

Table VI
SDC hardware configuration listing

Original purchase

Processors
  1. loring – HP 9000/J200 dual processor, 384 MB
      24 bit color, 21″ display
      4 – 4 × 9 GB disk arrays, 1–4 × 4 GB disk array
      2 – 2 GB system disks, 1 – CDROM drive
      1 – DAT drive, 1 – DLT 4000 stacker
  2. morgan – HP 9000/715–64, 96 MB
      8 bit color, 17″ display
      4 – 4 × 9 GB disk arrays, 2–4 GB disks
      1 – 1 GB system disk, 1 – DAT drive
  3. nearsdc-ntas 1 – Gateway 2000 P5–1250, 32 MB
      17″ display
      1–2 GB system drive, DAT drive
  4. nearsdc-gse1 – Gateway 2000 P5–75, 16 MB
      17″ display
      1 – 1 GB system disk

Secondary purchase

Processors
  1. Data distribution processor
      Distribute data to the science community
      Binary compatible with science data processor
      Serve as backup to the science data processor
  2. Image processor
      Provide specialized image processing for project scientists
Peripherals
  1. Additional disk capacity
  2. CDROM writer for PDS archival

The SDC has the responsibility to make 'calibrated' data available to non-primary science team members and the general science community. There are two competing goals in the project. The first goal is to ensure that the science data are available to the larger community as quickly as possible. The second goal is to make sure that the data are correct. The following compromises were made. The science teams will deliver algorithms to the SDC that will allow it to correct the results for known gross errors based on bench and flight calibrations. The SDC will implement these algorithms so that it can produce these corrected data. These data are intended for the use of the non-primary science team and the general science team. At the end of the mission, the science teams must deliver best and final algorithms for the calibration of their data to the PDS so that the algorithms can be archived with the data. The format of the delivery has not yet been specified.

The science teams are developing science results that will be archived to the PDS. At the end of the mission the science teams are responsible for converting their results to PDS format. The SDC will convert the spacecraft data and the science results and produce a PDS archive.

11. Additional SDC Services

The SDC will provide a number of additional services to the NEAR project. An archive of documentation will be maintained for science team access, a mailing list and a mail archive system will be available, and a set of tools to display telemetry data to aid in data transmission and processing error detection is being developed. The SDC will also aid the mission planning process by maintaining a mosaic of instrument coverage on EROS and by constructing SPICE kernels from modeled data.

12. Summary

This paper documents the current configuration and design of the NEAR SDC. The NEAR SDC is being built during the cruise phase of the mission. A prototype of the system was available at launch and a Build 1 design was implemented by the time of the Mathilde fly-by. A second Build will be in place prior to rendezvous. Changes to the design details will occur subsequent to the publication of this paper but the general system will remain the same. Software development personnel will be available during the rendezvous period to provide support for revised science team requests.

Acronyms

AIU	Attitude Interface Unit
AMPTE	Active Magnestospheric Particle Tracer Explorer
BGO	Bismuth Germanate
C&DH	Command and Data Handling
CCSDS	Consultative Committee for Data Systems
CTP	Command and Telemetry Processor
DPU	Data Processing Unit
DSN	Deep Space Network
EDR	Experimental Data Record
FC	Flight Computer
FDDI	Fiber Distribution Data Interface
FITS	Flexible Image Transport System

FTP	File Transfer Protocol
GB	Giga-Byte
GSE	Ground Support Equipment
HDF	Hierarchical Data Format
JHU/APL	The Johns Hopkins University Applied Physics Laboratory
JPL	Jet Propulsion Laboratory
MAG	Magnetometer
MBS	Mega Bit Per Second
NCSA	National Center for Super Computer Applications
NEAR	Near Earth Asteroid Rendezvous
NFS	Network File System
NIS	Near-Infrared Spectrograph
NLR	NEAR Laser Rangefinder
PDS	Planetary Data System
PSP	Product Server Processing
RDBMS	Relational Database Management System
SA	Science Archive
SAP	Science Archive Processing
SDC	Science Data Center
SPICE	Spacecraft, Planet, Instrument, 'C-matrix', Events
TSP	Telemetry Server Processing
URL	Uniform Resource Locator
UT	Universal Time
WWW	World Wide Web
XGRS	X-ray/Gamma Ray Spectrometer

Appendix A. Tables

Table A1. EPOCH data quality header

Bit field	Length	Value	Description
Ground receipt time	32		Seconds portion of GRT
	32		Microseconds portion of GRT
Frame type	2	0, 1, 2, 3	0 – packet, 1 – transfer frame, 2 - nascom block, 3 – fill
Frame quality	1	0, 1	0 – good, 1 – questionable
Quality mode	1	0, 1	0 – ignore questionable frame, 1 – process questionable frame
CRC status	1	0, 1	0 – NASCOM block CRC passed, 1 – NASCOM block CRC error
Reed Solomon status	1	0, 1	0 – RS correction succeeded, 1 – RS correction failed
Source	2	0, 1, 2	0 – GSE, 1 – DSN station, 2 – DSB station 2
Spare	10		
Frame sync state	2	0, 1, 2, 3	0 – search, 1 – check, 2 – flywheel, 3 – locked
Bit slips	4	0, 9, 10, 11, 13, 14, 15	0 – no slip, 9 – 1 bit late, 10 – 2 bits late, 11 – 3 bits late, 13 – 1 bit early, 14 – 2 bits early, 15 – 3 bits early
New time	1		not used for NEAR
Parity error	1		not used for NEAR
Data polarity	1	0, 1	0 – normal, 1 – frame complemented
Spare			
Frame sync pattern errors	4	xxxx	Number of errors detected in frame sync pattern
Frame size	32		Size of frame in bytes
Total	128		

Table A2
Transfer frame primary header

Bit field	Length	Value	Description
Frame version number	2	00	Version 1, defined by CCSDS standard
Spacecraft ID	10	xx xxx xxx	Assigned to NEAR
Virtual channel ID	3	001	VC1: non-packetized recorder play back data
		011	VC3: packetized real-time data
		000	VC0: packetized recorder data
		001	VC2: non-packetized image data
Operational control field	1	1	Indicates that a Command Link Control Word (CLCW) is present
Master channel frame count	8	xxxx xxxx	Sequential count of all transmitted frames, modulo 256
Virtual channel frame count	8	xxxx xxxx	Sequential count of all transmitted frames for this virtual channel, modulo 256
Secondary header flag	1	1	Indicates presence of secondary header
Sync flag	1	0	for VC0 or VC3 (packetized)
		1	for VC1 or VC2 (non-packetized)
Packet order flag	1	0	Reserved for future use by CCSDS
Segment length ID	2	11	Segmentation not used in NEAR
First header pointer	11	000-0000-0000	For NEAR the location of the first octet of the first packet primary header is always 0
Total	48		

Table A3
Transfer frame secondary header

Bit field	Length	Value	Description
Secondary header version	2	00	Version 1, defined by CCSDS standard
Secondary header length	6	001111	Length in bytes
Frame counter	8		VC0: TF's to SSR this second
			VC1: VC1 frames since VC3
			(0–31 if interleaved, rolls over if not)
			VC2: imager frame counter (0–183)
			VC3: No. of VC3 frames this second
			(0 if interleaved, 0–2 if VC3 only)
LVS flags	1	0, 1	LVS 1
	1	0, 1	LVS 2
Command reject flag	1	0, 1	1 if a command has been rejected from any source
Mission elapsed time	29		MET time of frame creation
Fist autonomy rule triggered	8		Bin number for first autonomy rule triggered since last telltale clear
Last autonomy rule triggered	8		Bin number for most recently triggered autonomy rule
Parameter No. 1	8		HSKP byte in part 1 of last triggered autonomy rule
Parameter No. 2	8		HSKP byte in part 2 of last triggered autonomy rule
Spacecraft configuration flags			
C& DH bus controller	1	0, 1	C/TP 1,2 is bus controller
AIU1	1	0, 1	Active line asserted, active line not asserted
AIU2	1	0, 1	Active line asserted, active line not asserted
Flight computer 1	1	0, 1	Not powered, powered
Flight computer 2	1	0, 1	Not powered, powered
Solid state recorder 1	1	0, 1	Not powered, powered
Solid state recorder 2	1	0, 1	Not powered, powered
Battery charger selected	1	0, 1	Not powered, powered
Bus voltage regulator 1	1	0, 1	Not powered, powered
Bus voltage regulator 2	1	0, 1	Not powered, powered
Auto, rule execution counter	6		Counter of triggered autonomy rules
C/TP reset status flags			
LRT	1		Last resort timer caused reset
C&D watchdog	1		C&DH watchdog timer caused reset
EDAC	1		EDAC double bit error caused reset
C&DH bus controller	1		Internal bus error caused reset
Spare	1		Unused
Power up	1		Power-up caused last C/TP reset
RF watchdog status	1		RF watchdog times out
Uplink TF reject flag	1		1 if uplink transfer frame rejected due to bad command
Spare	1		Unused
All status flags			
AIU sun safe	1	0, 1	Not asserted, asserted
AIU sun safe freeze	1	0, 1	Not asserted, asserted
MET of first auto, rule triggered	21		Lower 21 bits of MET at time of first triggered autonomy rule
Total	128		

Table A4
Packet primary header

Bit field	Length	Value	Description
Version number	3	000	Source packet
Type indicator	1	0	Telemetry packet
Secondary header flag	1	1	Secondary header present
Application ID	11	xxx xxxx xxxx	0 – C/TP No. 1
			1 – C/TP No. 2
			2 – AIU No. 1
			3 – AIU No. 2
			4 – MSI DPU
			5 – XGRS DPU
			6 – NIS/MAG DPU
			7 – NLR
			12 – Flight computer No. 1
			13 – Flight computer No. 2
Grouping flags	2	xx	0 – Intermediate packet
			1 – First packet
			2 – Last packet
			4 – Not part of a group
Source sequence count	14		Modulo 16384 counter for this APID
Packet length	16		Length of date + secondary header
Total	48		

Table A5
Packet secondary header

Bit field	Length	Description
Time	32	MET of when packet was sampled
Subprocess ID	16	

Table A6
MSI HDF description

HDF type	HDF name	Type	Description
SDS	FULL_IMAGE	unsigned short	'n' full images for the day [244 × 537]
SDS	SUMMARY_IMAGE	unsigned short	'k' summary images for the day, images are [22 × 26]
Vdata	Full_Image_Headers		on record per full image
	ACQTIME	unsigned long	Image acquisition time from the science housekeeping
	UTC	unsigned long	UTC of ACQTIME computed using SPICE [2]
	DQI	unsigned long	Data Quality Indicator
	CAMERA_PLUS_28V_CURRENT	unsigned char	Camera electronics DC/DC +28V input current
	DPU_DC_DC_TEMPERATURE	unsigned char	DPU DC/DC temperature
	CAMERA-TEMPERATURE	unsigned char	Camera electronics DC/DC temperature
	CCD_TEMPERATURE	unsigned char	CCD temperature
	RECORD_4_TELEM_CONFIG	unsigned char	Record type 4 telemetry configuration
	RECORD_5_TELEM_CONFIG	unsigned char	Record type 5 telemetry configuration
	REPEATING_SEQUENCES	unsigned char	Repeating sequences enable/disable
	FILTER WHEEL_POSITION	unsigned char	Commanded filter wheel position from SHK
	FILTER_WHEEL_MOTOR	unsigned char	Filter wheel motor power level
	FILTER_WHEEL_LED_PWR	unsigned char	Filter wheel LED power
	FILTER_WHEEL_LED_THR	unsigned char	Filter wheel LED threshold
	FILTER_WHEEL_FIDUCIALS	unsigned char	Filter wheel fiducials
	ACTIVE_CTP	unsigned char	Active CTP
	MAX_PIXEL_VALUE	unsigned short	Maximum pixel value
	EXPOSURE_TIME_MODE	unsigned char	Exposure time mode
	EXPOSURE_TIME_ALGORITHM	unsigned short	Exposure time algorithm parameter
	AIU_RT_DATA_CONTROL	unsigned short	AIU RT → RT data control
	IMAGER_CURRENT_LMT	unsigned short	Imager electronics current limit
	FILTER_WHEEL_STEP_TM	unsigned char	Filter wheel motor step time
	FILTER_WHEEL_STEPS_CMD	unsigned short	Filter wheel command number of steps
	EXPOSURE_TIME_MS	unsigned short	Exposure time parameter
	MOST_RECENT_IMAGE	unsigned char	Most recent image sequence executed
	IMG_NUM_IN_SEQ	unsigned char	Image number in sequence from SHK
	SC_COMPRESS_ALG	unsigned char	Compression algorithm used
	GS_COMPRESS_ALG	unsigned char	Ground decompression algorithm used
	IMAGE_NUM_DAY	unsigned short	Image number in day (computed)
	TRANSMIT_CHANNEL	unsigned char	VC0VC3/VC2
	IMAGE_XE	unsigned char	Full/Summary
	COMPRESSION_FLAG	unsigned char	Compressed/Not compressed
	NUMBER_OF_PACKETS	unsigned short	Number of packets or transfer frames
Vdata	Full_Image_Sequence_Headers		one record per full image
	SEQ_SCHEDULED_TIME	unsigned long	MET at which sequence was to begin
	IMAGE_NUMBER_SEQ	unsigned char	Image number in sequence
	SEQUENCE_ID	unsigned char	Sequence ID

Table A6
Continued

HDF type	HDF name	Type	Description
	NUMBER_OF_IMAGES	unsigned char	Number of images in sequence
	EXPOSURE_TIME_MODE	unsigned char	Exposure time mode
	DELTA_TIME_TO_IMAGE	unsigned char	Delta time between images
	REPEAT_LAG	unsigned char	Repeat/no repeat
	COMPRESSION_CONTRO	unsigned short	Compression control parameters
	IMAGE_N_EXPOSURE_TIME	unsigned short	Exposure time parameter [8]
	IMAGE_N_FILTER_WHEEL_POS	unsigned char	Filter wheel position [8]
	IMAGE_NUM_IN_DAY	unsigned short	Image number in day
Vdata	Summary_linage_Headers		one record per summary image
	ACQ_TIME	unsigned long	Image acquisition time from the science housekeeping
	UTC	unsigned long	UTC of ACQ_TIME computed using SPICE
	CAMERA_PLUS_28V_CURRENT	unsigned char	Camera electronics DC/DC +28V input current
	DPU_DC_DC_TEMPERATURE	unsigned char	DPU DC/DC temperature
	CAMERA_TEMPERATURE	unsigned char	Camera electronics DC/DC temperature
	CCD_TEMPERATURE	unsigned char	CCD temperature
	RECORD_4_TELEM_CONFIG	unsigned char	Record type 4 telemetry configuration
	RECORD_5_TELEM_CONFIG	unsigned char	Record type 5 telemetry configuration
	REPEATING_SEQUENCES	unsigned char	Repeating sequences enable/disable
	FILTER_WHEEL_POSITION	unsigned char	Commanded filter wheel position from
	FILTER_WHEEL_MOTOR	unsigned char	Filter wheel motor power level
	FILTER_WHEEL_LED_PWR	unsigned char	Filter wheel LED power
	FILTER_WHEEL_LED_THR	unsigned char	Filter wheel LED threshold
	FILTER_WHEEL_FIDUCIALS	unsigned char	Filter wheel fiducials
	ACTIVE_CTP	unsigned char	Active CTP
	MAX_PIXEL_VALUE	unsigned short	Maximum pixel value
	EXPOSURE_TIME_MODE	unsigned char	Exposure time mode
	EXPOSURE_TIME_ALGORITHM	unsigned short	Exposure time algorithm parameter
	AIU_RT_DATA_CONTROL	unsigned short	AIU RT → RT data control
	IMAGER_CURRENT_LMT	unsigned short	Imager electronics current limit
	FILTER_WHEEL_STEP_TM	unsigned char	Filter wheel motor step time
	FILTER_WHEEL_STEPS_CMD	unsigned short	Filter wheel command number of steps
	EXPOSURE_TIME_MS	unsigned short	Exposure time parameter
	MOST_RECENT_IMAGE	unsigned char	Most recent image sequence executed
	SEQ_ID	unsigned char	Sequence ID
	IMG_NUM_IN_SEQ	unsigned char	Image number in sequence from
	SC_COMPRESS_ALG	unsigned char	Compression algorithm used
	GS_COMPRESS_ALG	unsigned char	Ground decompression algorithm used
	IMAGE_NUM_DAY	unsigned short	Image number in day (computed)
	TRANSMIT_CHANNEL	unsigned char	VCSVC3/VC2
	IMAGE_TYPE	unsigned char	Full/Summary
	COMPRESSION_FLAG	unsigned char	Compressed/Not compressed
	NUMBER_OF_PACKETS	unsigned short	Number of packets or transfer frames

Table A6
Continued

HDF type	HDF name	Type	Description
Vdata	Summary_ Image_ Sequence_ Headers		one record per full image
	SEQ SCHEDULED_ TIME	unsigned long	MET at which sequence was to begin
	IMAGE_ NUMBER_ SEQ	unsigned char	Image number in sequence
	SEQUENCE_ ID	unsigned char	Sequence ID
	NUMBER_ OF_ IMAGES	unsigned char	Number of images in sequence
	EXPOSURE_ TIME_ MODE	unsigned char	Exposure time mode
	DELTA_ TIME_ TO_ IMAGE	unsigned char	Delta time between images
	REPEAT_ FLAG	unsigned char	Repeat/no repeat
	COMPRESSION_ CONTRO	unsigned short	Compression control parameters
	IMAGE_ N_ EXPOSURE_ TIME	unsigned short	Exposure time parameter [8]
	IMAGE_ N_ FILTER_ WHEEL_ POS	unsigned char	Filter wheel position [8]
	IMAGE_ NUM_ IN_ DAY	unsigned short	Image number in day

Table A7
Example catalog table contents (IMAGER table, item 1 in Table V)

Item	Size (byte)	Type	Description
Image ID	4	int	A unique number per image
DPID	4	int	A data product identifier
DOI	4	int	An indicator of the data quality
SSID	4	int	A pointer to a science sequence
Image No.	4	int	Image number within day, this is used to label image products
UTC	8	int	UTC time at the time image was taken
MINpos	3 × 4	float	The minimum x, y, and z position values in the image based on the current plate model
MAXpos	3 × 4	float	The maximum x, y, and z position values in the image based on the current plate model
Filter No.	1	char	Filter number for this image
Exposure time	4	float	Exposure time in milliseconds
Temperature CCD	4	float	Temperature of the CCD in degrees Celsius
Temperature Tele	4	float	Temperature of the telescope body in degrees Celsius
Incidence Angle	4	float	The angle between the surface normal and the sun in degrees
Emission Angle	4	float	The angle between the surface normal and the imager pointing vector in degrees
Range	4	float	Range from spacecraft to asteroid surface in kilometers using ephemeris data
Phase Angle	4	float	The angle between the sun and the imager pointing vector as measured from the surface in degrees

References

1. NEAR Command and Data Handling Software Requirements Specification: 1994, JHU/APL 7352–9066.
2. NEAR Instrument DPU Common Software Requirements Specification: 1995, JHU/APL 7358–9001.
3. NEAR Multi-Spectral Imager (MSI) DPU Software Requirements Specification: 1995, JHU/APL 7356–9001.
4. NEAR X-ray/Gamma Ray Spectrometer (XGRS) DPU Software Requirements Specification: 1994, JHU/APL 7358–9002.
5. NEAR Laser Rangefinder Flight Software Requirements Specification: 1994, JHU/APL 7352–9069, Version 3.1.
6. NEAR Infrared Spectrometer/Magnetometer (NIS/MAG) DPU Software Requirements Specification: 1995, JHU/APL 7357–9011 [KJH1].